STUDENT SOLUTIONS MANUAL

JAY R. SCHAFFER
University of Northern Colorado

ELEMENTARY STATISTICS: PICTURING THE WORLD

FIFTH EDITION

Ron Larson
The Pennsylvania State University
The Behrend College

Betsy Farber
Bucks County Community College

Prentice Hall
is an imprint of

Reproduced by Pearson Prentice Hall from electronic files supplied by the author.

Publishing as Prentice Hall, 75 Arlington Street, Boston, MA 02116.

ISBN-13: 978-0-321-69373-0
ISBN-10: 0-321-69373-6

5 6 EBM 15 14 13

Prentice Hall
is an imprint of

www.pearsonhighered.com

CONTENTS

Introduction to Statistics

1.1 AN OVERVIEW OF STATISTICS

1.1 Try It Yourself Solutions

1a. The population consists of the prices per gallon of regular gasoline at all gasoline stations in the United States. The sample consists of the prices per gallon of regular gasoline at the 900 surveyed stations.

b. The data set consists of the 900 prices.

2a. Because the numerical measure of $2,655,395,194 is based on the entire collection of player's salaries, it is from a population.

b. Because the numerical measure is a characteristic of a population, it is a parameter.

3a. Descriptive statistics involve the statement "76% of women and 60% of men had a physical examination within the previous year."

b. An inference drawn from the study is that a higher percentage of women had a physical examination within the previous year.

1.1 EXERCISE SOLUTIONS

1. A sample is a subset of a population.

3. A parameter is a numerical description of a population characteristic. A statistic is a numerical description of a sample characteristic.

5. False. A statistic is a numerical measure that describes a sample characteristic.

7. True

9. False. A population is the collection of *all* outcomes, responses, measurements, or counts that are of interest.

11. The data set is a population because it is a collection of the heights of all the players on a school's basketball team.

13. The data set is a sample because the collection of the 500 spectators is a subset within the population of the stadium's 42,000 spectators.

15. Sample, because the collection of the 20 patients is a subset within the population

17. Population, because it is a collection of all golfers' scores in the tournament

19. Population, because it is a collection of all the U.S. presidents' political parties

21. Population: Party of registered voters in Warren County

Sample: Party of Warren County voters responding to online survey

23. Population: Ages of adults in the United States who own cellular phones

Sample: Ages of adults in the United States who own Samsung cellular phones

25. Population: Collection of all adults in the United States

Sample: Collection of 1000 adults surveyed

27. Population: Collection of all adults in the U.S.

Sample: Collection of 1442 adults surveyed

29. Population: Collection of all registered voters

Sample: Collection of 800 registered voters surveyed

31. Population: Collection of all women in the U.S.

Sample: Collection of the 546 U.S. women surveyed

33. Population: Collection of all Fortune magazine's top 100 companies to work for

Sample: Collection of the 85 companies who responded to the questionnaire

35. Statistic. The value $68,000 is a numerical description of a sample of annual salaries.

37. Parameter. The 62 surviving passengers out of 97 total passengers is a numerical description of all of the passengers of the Hindenburg that survived.

39. Statistic. 8% is a numerical description of a sample of computer users.

41. Statistic. 44% is a numerical description of a sample of all people.

43. The statement "56% are the primary investors in their household" is an application of descriptive statistics.

An inference drawn from the sample is that an association exists between U.S. women and being the primary investor in their household.

45. Answers will vary.

47. (a) An inference drawn from the sample is that senior citizens who live in Florida have better memory than senior citizens who do not live in Florida.

(b) It implies that if you live in Florida, you will have better memory.

49. Answers will vary.

1.2 DATA CLASSIFICATION

1.2 Try It Yourself Solutions

1a. One data set contains names of cities and the other contains city populations.

b. City: Nonnumerical
Population: Numerical

c. City: Qualitative
Population: Quantitative

2a. (1) The final standings represent a ranking of basketball teams.

(2) The collection of phone numbers represents labels. No mathematical computations can be made.

b. (1) Ordinal, because the data can be put in order

(2) Nominal, because you cannot make calculations on the data

3a. (1) The data set is the collection of body temperatures.

(2) The data set is the collection of heart rates.

b. (1) Interval, because the data can be ordered and meaningful differences can be calculated, but it does not make sense writing a ratio using the temperatures

(2) Ratio, because the data can be ordered, can be written as a ratio, you can calculate meaningful differences, and the data set contains an inherent zero

1.2 EXERCISE SOLUTIONS

1. Nominal and ordinal

3. False. Data at the ordinal level can be qualitative or quantitative.

5. False. More types of calculations can be performed with data at the interval level than with data at the nominal level.

7. Qualitative, because telephone numbers are merely labels

9. Quantitative, because the body temperatures of patients is a numerical measure.

11. Quantitative, because the lengths of songs on an MP3 player are numerical measures

13. Qualitative, because the player numbers are merely labels

15. Quantitative, because weights of infants are a numerical measure

17. Qualitative, because the poll results are merely responses

19. Qualitative. Ordinal. Data can be arranged in order, but differences between data entries make no sense.

21. Qualitative. Nominal. No mathematical computations can be made and data are categorized using names.

23. Qualitative. Ordinal. The data can be arranged in order, but differences between data entries are not meaningful.

25. Ordinal

27. Nominal

29. (a) Interval (b) Nominal (c) Ratio (d) Ordinal

31. An inherent zero is a zero that implies "none." Answers will vary.

1.3 DATA COLLECTION AND EXPERIMENTAL DESIGN

1.3 Try It Yourself Solutions

1a. (1) Focus: Effect of exercise on relieving depression

(2) Focus: Success of graduates

b. (1) Population: Collection of all people with depression

(2) Population: Collection of all university graduates at this large university

c. (1) Experiment

(2) Survey

2a. There is no way to tell why people quit smoking. They could have quit smoking either from the gum or from watching the DVD.

b. Two experiments could be done; one using the gum and the other using the DVD.

3a. Example: start with the first digits 92630782 …

b. 92 | 63 | 07 | 82 | 40 | 19 | 26

c. 63, 7, 40, 19, 26

4a. (1) The sample was selected by only using the students in a randomly chosen class. Cluster sampling

(2) The sample was selected by numbering each student in the school, randomly choosing a starting number, and selecting students at regular intervals from the starting number. Systematic sampling

b. (1) The sample may be biased because some classes may be more familiar with stem cell research than other classes and have stronger opinions.

(2) The sample may be biased if there is any regularly occurring pattern in the data.

1.3 EXERCISE SOLUTIONS

1. In an experiment, a treatment is applied to part of a population and responses are observed. In an observational study, a researcher measures characteristics of interest of part of a population but does not change existing conditions.

3. In a random sample, every member of the population has an equal chance of being selected. In a simple random sample, every possible sample of the same size has an equal chance of being selected.

5. True

7. False. Using stratified sampling guarantees that members of each group within a population will be sampled.

9. False. To select a systematic sample, a population is ordered in some way and then members of the population are selected at regular intervals.

11. Use a census because all the patients are accessible and the number of patients is not too large.

13. In this study, you want to measure the effect of a treatment (using a fat substitute) on the human digestive system. So, you would want to perform an experiment.

15. Because it is impractical to create this situation, you would want to use a simulation.

17. (a) The experimental units are the 30–35 year old females being given the treatment. One treatment is used.

(b) A problem with the design is that there may be some bias on the part of the researchers if he or she knows which patients were given the real drug. A way to eliminate this problem would be to make the study into a double-blind experiment.

(c) The study would be a double-blind study if the researcher did not know which patients received the real drug or the placebo.

19. Each U.S. telephone number has an equal chance of being dialed and all samples of 1400 phone numbers have an equal chance of being selected, so this is a simple random sample. Telephone sampling only samples those individuals who have telephones, are available, and are willing to respond, so this is a possible source of bias.

21. Because the students were chosen due to their convenience of location (leaving the library), this is a convenience sample. Bias may enter into the sample because the students sampled may not be representative of the population of students. For example, there may be an association between time spent at the library and drinking habits.

23. Simple random sampling is used because each customer has an equal chance of being contacted, and all samples of 580 customers have an equal chance of being selected.

25. Because a sample is taken from each one-acre subplot (stratum), this is a stratified sample.

27. Answers will vary.

29. Answers will vary.

31. Census, because it is relatively easy to obtain the ages of the 115 residents

33. Question is biased because it already suggests that eating whole-grain foods is good for you. The question might be rewritten as "How does eating whole-grain foods affect your health?"

35. Question is unbiased because it does not imply how much exercise is good or bad.

37. The households sampled represent various locations, ethnic groups, and income brackets. Each of these variables is considered a stratum.

39. Observational studies may be referred to as natural experiments because they involve observing naturally occurring events that are not influenced by the study.

41. (a) Advantage: Usually results in a savings in the survey cost

(b) Disadvantage: There tends to be a lower response rate and this can introduce a bias into the sample.
Sampling Technique: Convenience sampling

43. If blinding is not used, then the placebo effect is more likely to occur.

45. Both a randomized block design and a stratified sample split their members into groups based on similar characteristics.

CHAPTER 1 REVIEW EXERCISE SOLUTIONS

1. Population: Collection of all U.S. adults

Sample: Collection of the 1000 U.S. adults that were sampled

3. Population: Collection of all credit cards

Sample: Collection of 39 credit cards that were sampled

5. The team payroll is a parameter since it is a numerical description of a population (entire baseball team) characteristic.

7. Since "10 students" is describing a characteristic of a population of math majors, it is a parameter.

9. The average APR of 12.83% charged by credit cards is representative of the descriptive branch of statistics. An inference drawn from the sample is that all credit cards charge an APR of 12.83%.

11. Quantitative, because monthly salaries are numerical measurements

13. Quantitative, because ages are numerical measurements

15. Quantitative, because revenues are numerical measures

17. Interval. It makes no sense saying that 100 degrees is twice as hot as 50 degrees.

19. Nominal. The data are qualitative and cannot be arranged in a meaningful order.

21. Because CEOs keep accurate records of charitable donations, you could take a census.

23. Perform an experiment because you want to measure the effect of training from animal shelters on inmates.

25. The subjects could be split into male and female and then be randomly assigned to each of the five treatment groups.

27. Answers will vary.

29. Because random telephone numbers were generated and called, this is a simple random sample.

31. Because each community is considered a cluster and every pregnant woman in a selected community is surveyed, this is a cluster sample.

33. Because grade levels are considered strata and 25 students are sampled from each stratum, this is a stratified sample.

35. Telephone sampling only samples individuals who have telephones, are available, and are willing to respond.

37. The selected communities may not be representative of the entire area.

CHAPTER 1 QUIZ SOLUTIONS

1. Population: Collection of all men

   Sample: Collection of 20,000 men in study

2. (a) Statistic. 19% is a characteristic of a sample of Internet users.

(b) Parameter. 90% is a characteristic of the entire Board of Trustees (population).

(c) Statistic. 55% is a characteristic of a sample of chief financial officers and senior comptrollers.

3. (a) Qualitative, since debit card pin numbers are merely labels.

(b) Quantitative, since a final score is a numerical measure.

4. (a) Ordinal. Badge numbers may be ordered numerically according to seniority of service, but no mathematical computations can be made.

(b) Ratio. It makes sense to say that the horsepower of one car was twice as many as another car.

(c) Ordinal, because data can be arranged in order but the differences between data entries make no sense

(d) Interval, because meaningful differences between entries can be calculated, but a zero entry is not an inherent zero.

5. (a) In this study, you want to measure the effect of a treatment (low dietary intake of vitamin C and iron) on lead levels in adults. You want to perform an experiment.

(b) Because it would be difficult to survey every individual within 500 miles of your home, sampling should be used.

6. Randomized Block Design

7. (a) Because people were chosen due to their convenience of location (on the campground), this is a convenience sample.

(b) Because every tenth part is selected from an assembly line, this is a systematic sample.

(c) Stratified sample because the population is first stratified and then a sample is collected from each stratum

8. Convenience

Descriptive Statistics

2.1 FREQUENCY DISTRIBUTIONS AND THEIR GRAPHS

2.1 Try It Yourself Solutions

1a. The number of classes is 8.

b. Min = 35, Max = 89, Class width = $\dfrac{\text{Range}}{\text{Number of classes}} = \dfrac{89-35}{8} = 6.75 \Rightarrow 7$

c.

Lower limit	Upper limit
35	41
42	48
49	55
56	62
63	69
70	76
77	83
84	90

d. See part (e).

e.

Class	Frequency, f
35-41	2
42-48	5
49-55	7
56-62	7
63-69	10
70-76	5
77-83	8
84-90	6

2a. See part (b).

b.

Class	Frequency, f	Midpoint	Relative frequency	Cumulative frequency
35-41	2	38	0.04	2
42-48	5	45	0.10	7
49-55	7	52	0.14	14
56-62	7	59	0.14	21
63-69	10	66	0.20	31
70-76	5	73	0.10	36
77-83	8	80	0.16	44
84-90	6	87	0.12	50
	$\sum f = 50$		$\sum \frac{f}{n} = 1$	

c. 72% of the 50 richest people are older than 55.
4% of the 50 richest people are younger than 42.
The most common age bracket for the 50 richest people is 63-69.

3a.

Class Boundaries
34.5-41.5
41.5-48.5
48.5-55.5
55.5-62.5
62.5-69.5
69.5-76.5
76.5-83.5
83.5-90.5

b. Use class midpoints for the horizontal scale and frequency for the vertical scale. (Class boundaries can also be used for the horizontal scale.)

c.

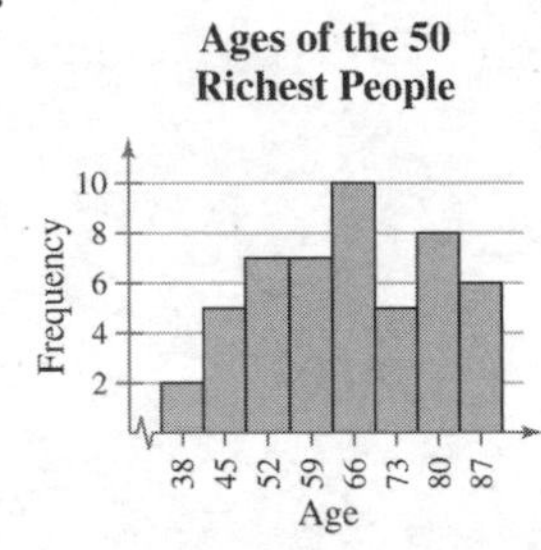

d. 72% of the 50 richest people are older than 55.
4% of the 50 richest people are younger than 42.
The most common age bracket for the 50 richest people is 63-69.

4a. Use class midpoints for the horizontal scale and frequency for the vertical scale. (Class boundaries can also be used for the horizontal scale.)

b. See part (c).

c.

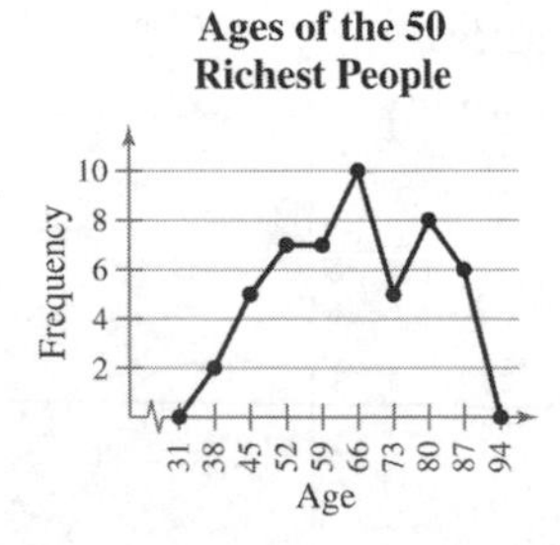

d. The frequency of ages increases up to 66 and then decreases.

5abc.

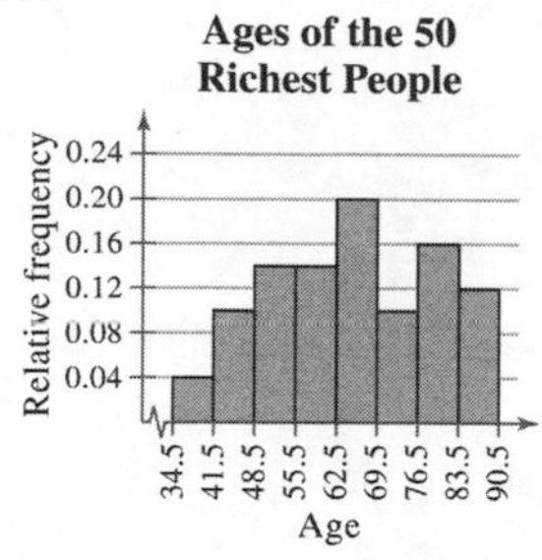

6a. Use upper class boundaries for the horizontal scale and cumulative frequency for the vertical scale.

b. See part (c).

c.

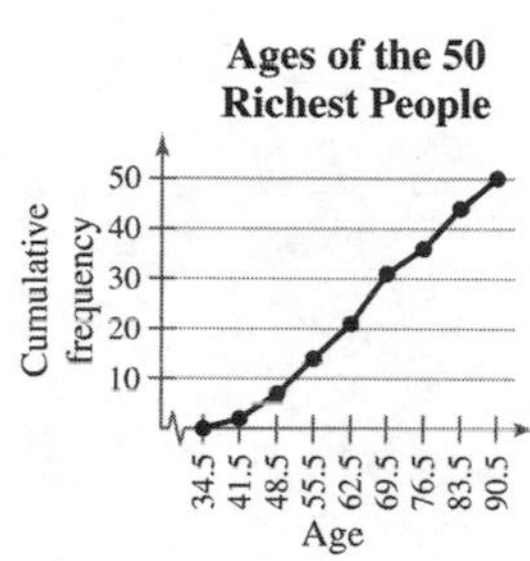

d. Approximately 40 of the 50 richest people are 80 years or younger.

e. Answers will vary.

7ab.

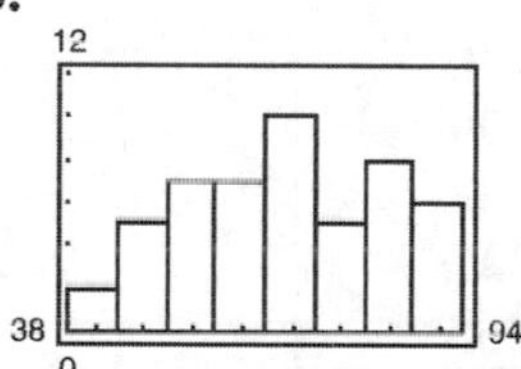

2.1 EXERCISE SOLUTIONS

1. Organizing the data into a frequency distribution may make patterns within the data more evident. Sometimes it is easier to identify patterns of a data set by looking at a graph of the frequency distribution.

3. Class limits determine which numbers can belong to that class.
   Class boundaries are the numbers that separate classes without forming gaps between them.

5. The sum of the relative frequencies must be 1 or 100% because it is the sum of all portions or percentages of the data.

7. False. Class width is the difference between the lower (or upper limits) of consecutive classes.

9. False. An ogive is a graph that displays cumulative frequencies.

11. Class width $= \dfrac{\text{Range}}{\text{Number of classes}} = \dfrac{64-9}{7} \approx 7.9 \Rightarrow 8$

Lower class limits: 9, 17, 25, 33, 41, 49, 57

Upper class limits: 16, 24, 32, 40, 48, 56, 64

13. Class width $= \dfrac{\text{Range}}{\text{Number of classes}} = \dfrac{135-17}{8} = 14.75 \Rightarrow 15$

Lower class limits: 17, 32, 47, 62, 77, 92, 107, 122

Upper class limits: 31, 46, 61, 76, 91, 106, 121, 136

15a. Class width = 31 – 20 =11

b. and c.

Class	Frequency, f	Midpoint	Class boundaries
20-30	19	25	19.5-30.5
31-41	43	36	30.5-41.5
42-52	68	47	41.5-52.5
53-63	69	58	52.5-63.5
64-74	74	69	63.5-74.5
75-85	68	80	74.5-85.5
86-96	24	91	85.5-96.5

17.

Class	Frequency, f	Midpoint	Relative frequency	Cumulative frequency
20-30	19	25	0.05	19
31-41	43	36	0.12	62
42-52	68	47	0.19	130
53-63	69	58	0.19	199
64-74	74	69	0.20	273
75-85	68	80	0.19	341
86-96	24	91	0.07	365
	$\sum f = 365$		$\sum \frac{f}{n} \approx 1$	

19a. Number of classes = 7

b. Least frequency ≈ 10

c. Greatest frequency ≈ 300

d. Class width = 10

21a. 50

b. 22.5-23.5 pounds

23a. 42

b. 29.5 pounds

c. 35

d. 2

25a. Class with greatest relative frequency: 8-9 inches
Class with least relative frequency: 17-18 inches

b. Greatest relative frequency ≈ 0.195
Least relative frequency ≈ 0.005

c. Approximately 0.015

27. Class with greatest frequency: 29.5-32.5
Classes with least frequency: 11.5-14.5 and 38.5-41.5

29. Class width $= \dfrac{\text{Range}}{\text{Number of classes}} = \dfrac{39-0}{5} = 7.8 \Rightarrow 8$

Class	Frequency, f	Midpoint	Relative frequency	Cumulative frequency
0-7	8	3.5	0.32	8
8-15	8	11.5	0.32	16
16-23	3	19.5	0.12	19
24-31	3	27.5	0.12	22
32-39	3	35.5	0.12	25
	$\sum f = 25$		$\sum \frac{f}{n} = 1$	

Classes with greatest frequency: 0-7, 8-15
Classes with least frequency: 16-23, 24-31, 32-39

31. Class width $= \dfrac{\text{Range}}{\text{Number of classes}} = \dfrac{7119-1000}{6} \approx 1019.83 \Rightarrow 1020$

Class	Frequency, f	Midpoint	Relative frequency	Cumulative frequency
1000-2019	12	1509.5	0.5455	12
2020-3039	3	2529.5	0.1364	15
3040-4059	2	3549.5	0.0909	17
4060-5079	3	4569.5	0.1364	20
5080-6099	1	5589.5	0.0455	21
6100-7119	1	6609.5	0.0455	22
	$\sum f = 22$		$\sum \frac{f}{n} \approx 1$	

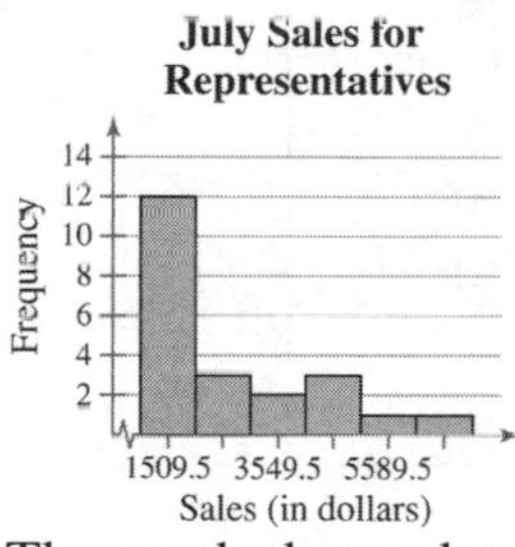

The graph shows that most of the sales representatives at the company sold between \$1000 and \$2019. (Answers will vary.)

33. Class width $= \dfrac{\text{Range}}{\text{Number of classes}} = \dfrac{514-291}{8} = 27.875 \Rightarrow 28$

Class	Frequency, f	Midpoint	Relative frequency	Cumulative frequency
291-318	5	304.5	0.1667	5
319-346	4	332.5	0.1333	9
347-374	3	360.5	0.1000	12
375-402	5	388.5	0.1667	17
403-430	6	416.5	0.2000	23
431-458	4	444.5	0.1333	27
459-486	1	472.5	0.0333	28
487-514	2	500.5	0.0667	30
	$\sum f = 30$		$\sum \frac{f}{n} = 1$	

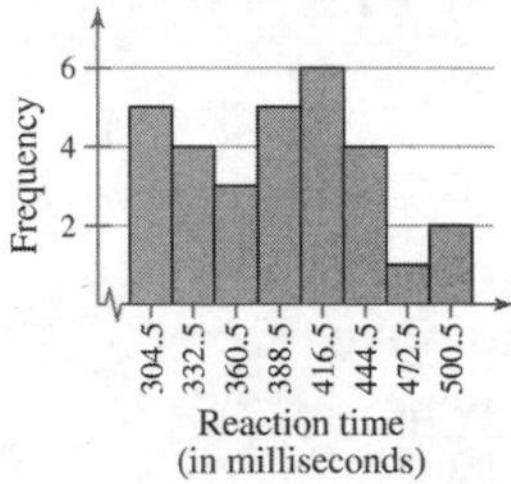

The graph shows that the most frequent reaction times were between 403 and 430 milliseconds. (Answers will vary.)

35. Class width $= \dfrac{\text{Range}}{\text{Number of classes}} = \dfrac{55-24}{5} = 6.2 \Rightarrow 7$

Class	Frequency, f	Midpoint	Relative frequency	Cumulative frequency
24-30	9	27	0.30	9
31-37	8	34	0.27	17
38-44	10	41	0.33	27
45-51	2	48	0.07	29
52-58	1	55	0.03	30
	$\sum f = 30$		$\sum \frac{f}{n} = 1$	

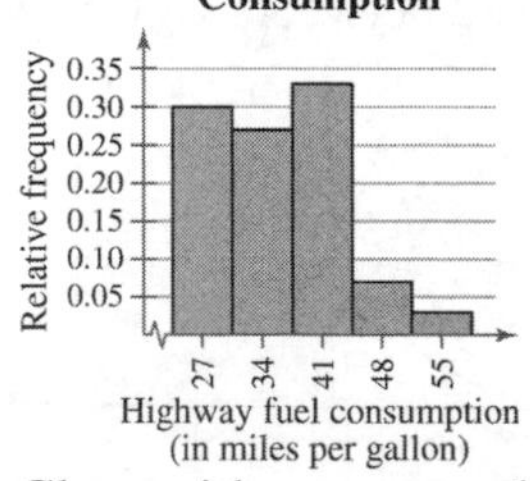

Class with greatest relative frequency: 38-44
Class with least relative frequency: 52-58

37. Class width $= \dfrac{\text{Range}}{\text{Number of classes}} = \dfrac{462-138}{5} = 64.8 \Rightarrow 65$

Class	Frequency, f	Midpoint	Relative frequency	Cumulative frequency
138-202	12	170	0.46	12
203-267	6	235	0.23	18
268-332	4	300	0.15	22
333-397	1	365	0.04	23
398-462	3	430	0.12	26
	$\sum f = 26$		$\sum \frac{f}{n} = 1$	

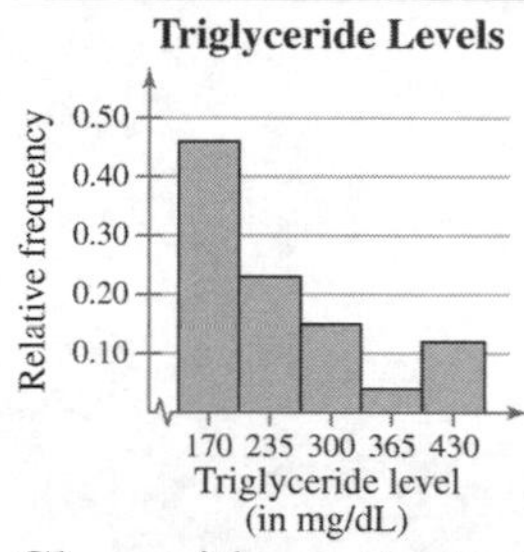

Class with greatest relative frequency: 138-202
Class with least relative frequency: 333-397

39. Class width $= \dfrac{\text{Range}}{\text{Number of classes}} = \dfrac{73-52}{6} = 3.5 \Rightarrow 4$

Class	Frequency, f	Relative frequency	Cumulative frequency
52-55	3	0.125	3
56-59	3	0.125	6
60-63	9	0.375	15
64-67	4	0.167	19
68-71	4	0.167	23
72-75	1	0.042	24
	$\sum f = 24$	$\sum \frac{f}{n} \approx 1$	

Location of the greatest increase in frequency: 60-63

41. Class width $= \dfrac{\text{Range}}{\text{Number of classes}} = \dfrac{98-47}{5} = 10.2 \Rightarrow 11$

Class	Frequency, f	Midpoint	Relative frequency	Cumulative frequency
47-57	1	52	0.05	1
58-68	1	63	0.05	2
69-79	5	74	0.25	7
80-90	8	85	0.40	15
91-101	5	96	0.25	20
	$\sum f = 20$		$\sum \frac{f}{N} = 1$	

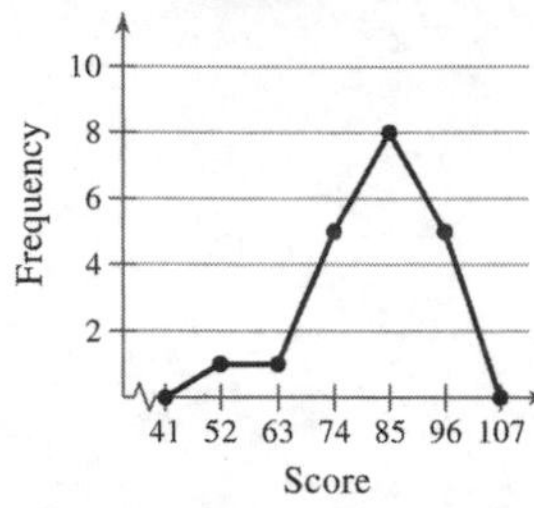

The graph shows that the most frequent exam scores were between 80 and 90. (Answers will vary.)

43a. Class width $= \dfrac{\text{Range}}{\text{Number of classes}} = \dfrac{120-65}{6} \approx 9.2 \Rightarrow 10$

Class	Frequency, f	Midpoint	Relative frequency	Cumulative frequency
65-74	4	69.5	0.17	4
75-84	7	79.5	0.29	11
85-94	4	89.5	0.17	15
95-104	5	99.5	0.21	20
105-114	3	109.5	0.13	23
115-124	1	119.5	0.04	24
	$\sum f = 24$		$\sum \frac{f}{N} \approx 1$	

b.

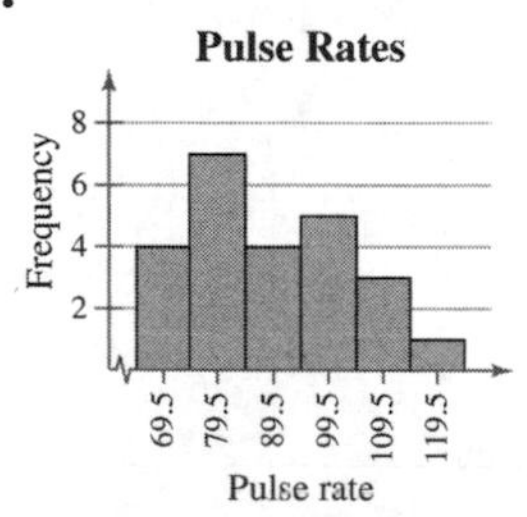

c.

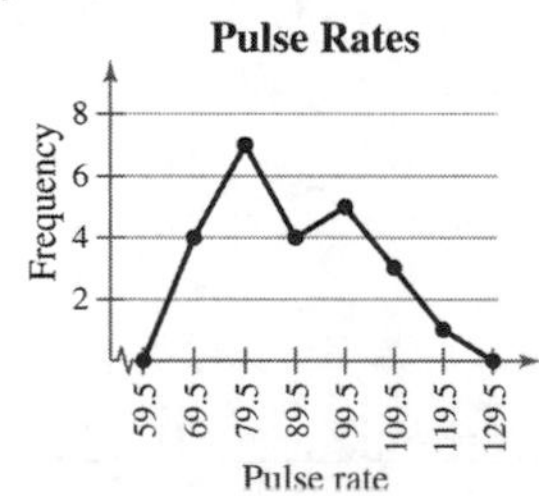

d.

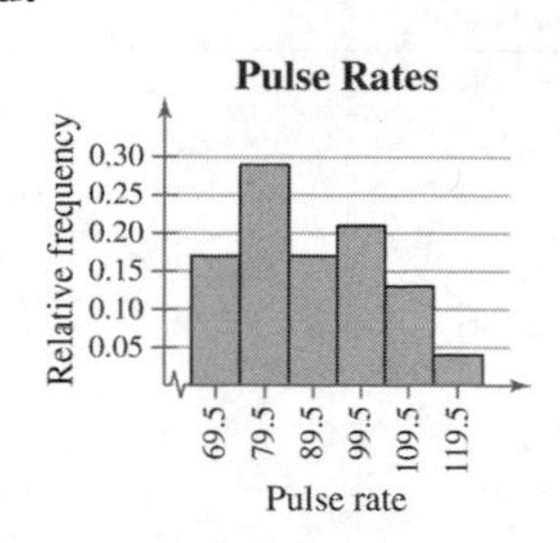

e.

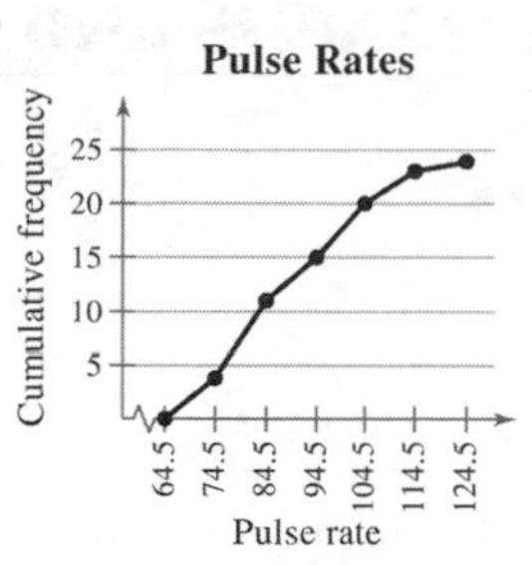

45.

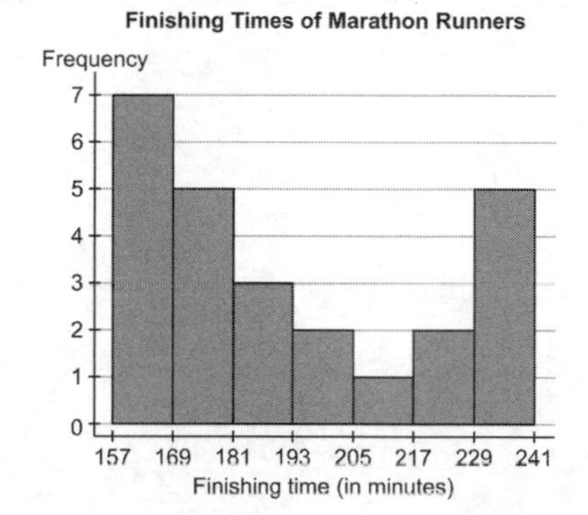

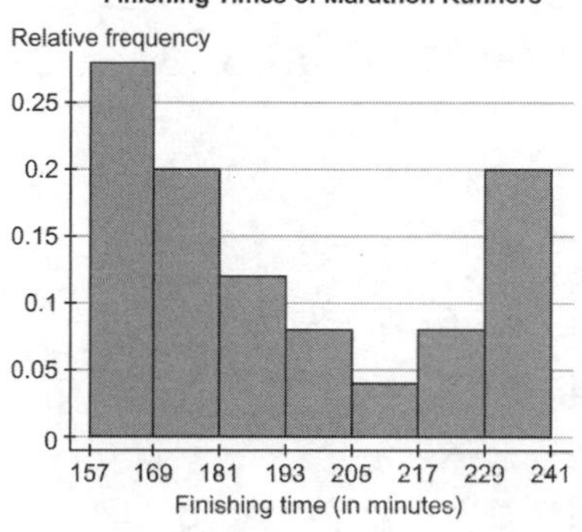

47a. Class width $= \dfrac{\text{Range}}{\text{Number of classes}} = \dfrac{104-61}{8} = 5.375 \Rightarrow 6$

Class	Frequency, f	Midpoint	Relative frequency
61-66	1	63.5	0.0333
67-72	3	69.5	0.1000
73-78	6	75.5	0.2000
79-84	10	81.5	0.3333
85-90	5	87.5	0.1667
91-96	2	93.5	0.0667
97-102	2	99.5	0.0667
103-108	1	105.5	0.0333
	$\sum f = 30$		$\sum \frac{f}{N} = 1$

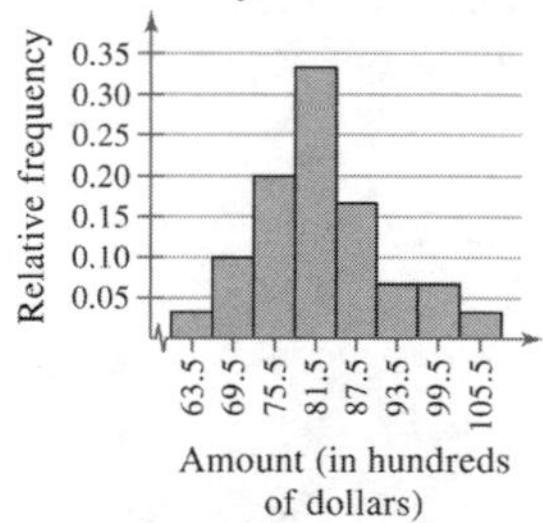

b. 16.7%, because the sum of the relative frequencies for the last three classes is 0.167.

c. \$9600, because the sum of the relative frequencies for the last two classes is 0.10.

2.2 MORE GRAPHS AND DISPLAYS

2.2 Try It Yourself Solutions

1a.

```
3|
4|
5|
6|
7|
8|
```

b. Key 3|6 = 36

```
3|6 5
4|9 7 6 4 3 2
5|9 8 7 6 4 4 3 3 1 1
6|9 9 8 7 6 6 5 5 4 3 1 1 0
7|8 8 7 6 3 3 3 2
8|9 9 7 6 6 5 3 3 2 1 0
```

c. Key 3|5 = 35

```
3|5 6
4|2 3 4 6 7 9
5|1 1 3 3 4 4 6 7 8 9
6|0 1 1 3 4 5 5 6 6 7 8 9 9
7|2 3 3 3 6 7 8 8
8|0 1 2 3 3 5 6 6 7 9 9
```

d. More than 50% of the 50 richest people are older than 60. (Answers will vary.)

2a, b. Key 3|5 = 35

```
3|
3|5 6
4|2 3 4
4|6 7 9
5|1 1 3 3 4 4
5|6 7 8 9
6|0 1 1 3 4
6|5 5 6 6 7 8 9 9
7|2 3 3 3
7|6 7 8 8
8|0 1 2 3 3
8|5 6 6 7 9 9
```

c. Most of the 50 richest people are older than 60. (Answers will vary.)

3a. Use the age for the horizontal axis.

b.

Ages of the 50 Richest People

30 40 50 60 70 80 90

Age

c. A large percentage of the ages are over 60. (Answers will vary.)

4a.

Type of Degree	f	Relative Frequency	Angle
Associate's	455	0.23	82.8°
Bachelor's	1052	0.54	194.4°
Master's	325	0.17	61.2°
First Professional	71	0.04	14.4°
Doctoral	38	0.02	7.2°
	$\sum f = 50$	$\sum \frac{f}{N} = 1$	$\sum \ = 360°$

b.

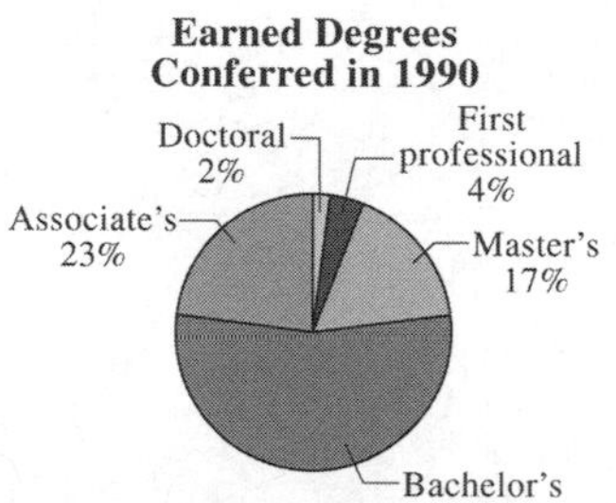

c. From 1990 to 2007, as percentages of total degrees conferred, associate's degrees increased by 1%, bachelor's degrees decreased by 3%, master's degrees increased by 3%, first professional degrees decreased by 1%, and doctoral degrees remained unchanged.

5a.

Cause	Frequency, f
Auto Dealers	14,668
Auto Repair	9,728
Home Furnishing	7,792
Computer Sales	5,733
Dry Cleaning	4,649

b.

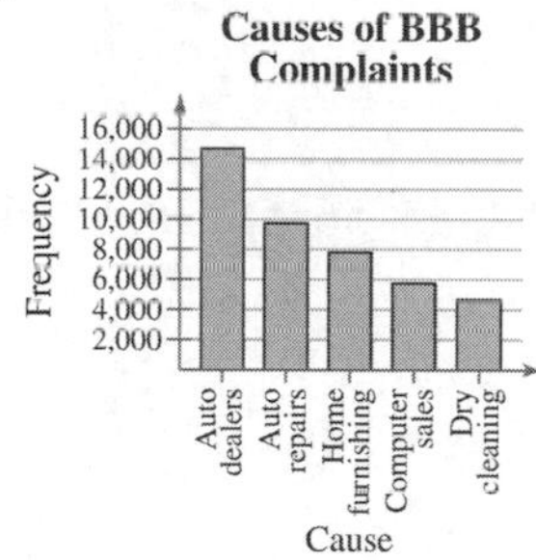

c. It appears that the auto industry (dealers and repair shops) account for the largest portion of complaints filed at the BBB. (Answers will very.)

6a, b.

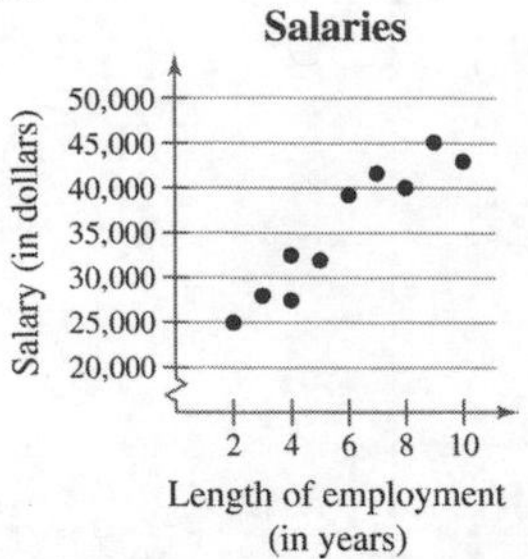

c. It appears that the longer an employee is with the company, the larger the employee's salary will be.

7a, b.

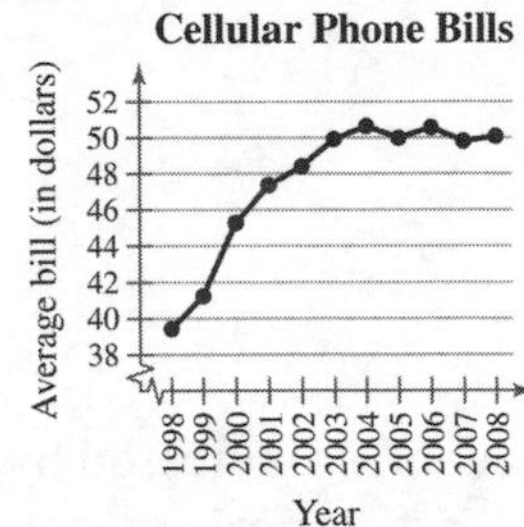

c. The average bill increased from 1998 to 2004, then it hovered around $50.00 from 2004 to 2008.

2.2 EXERCISE SOLUTIONS

1. Quantitative: stem-and-leaf plot, dot plot, histogram, time series chart, scatter plot.
Qualitative: pie chart, Pareto chart

3. Both the stem-and-leaf plot and the dot plot allow you to see how data are distributed, determine specific data entries, and identify unusual data values.

5. b **7.** a

9. 27, 32, 41, 43, 43, 44, 47, 47, 48, 50, 51, 51, 52, 53, 53, 53, 54, 54, 54, 54, 55, 56, 56, 58, 59, 68, 68, 68, 73, 78, 78, 85
Max: 85 Min: 27

11. 13, 13, 14, 14, 14, 15, 15, 15, 15, 15, 16, 17, 17, 18, 19
Max: 19 Min: 13

13. Sample answer: Users spend the most amount of time on MySpace and the least amount of time on Twitter. Answers will vary.

15. Answers will vary. Sample answer: Tailgaters irk drivers the most, while too cautious drivers irk drivers the least.

17. Key: $6|7 = 67$

6 | 7 8
7 | 3 5 5 6 9
8 | 0 0 2 3 5 5 7 7 8
9 | 0 1 1 1 2 4 5 5

It appears that most grades for the biology midterm were in the 80s or 90s. (Answers will vary.)

19. Key: $4|3 = 4.3$

4 | 3 9
5 | 1 8 8 8 9
6 | 4 8 9 9 9
7 | 0 0 2 2 2 5
8 | 0 1

It appears that most ice had a thickness of 5.8 centimeters to 7.2 centimeters. (Answers will vary.)

21.

Systolic Blood Pressures

100 110 120 130 140 150 160 170 180 190 200

Systolic blood pressure (in mmHg)

It appears that systolic blood pressure tends to be between 120 and 150 millimeters of mercury. (Answers will vary.)

23.

Category	Frequency, f	Relative Frequency	Angle
United States	15	0.375	135°
Italy	4	0.100	36°
Ethiopia	1	0.025	9°
South Africa	2	0.050	18°
Tanzania	1	0.025	9°
Kenya	8	0.200	72°
Mexico	4	0.100	36°
Morocco	1	0.025	9°
Great Britain	1	0.025	9°
Brazil	2	0.050	18°
New Zealand	1	0.025	9°
	$\sum f = 40$	$\sum \frac{f}{N} = 1$	$\sum = 360°$

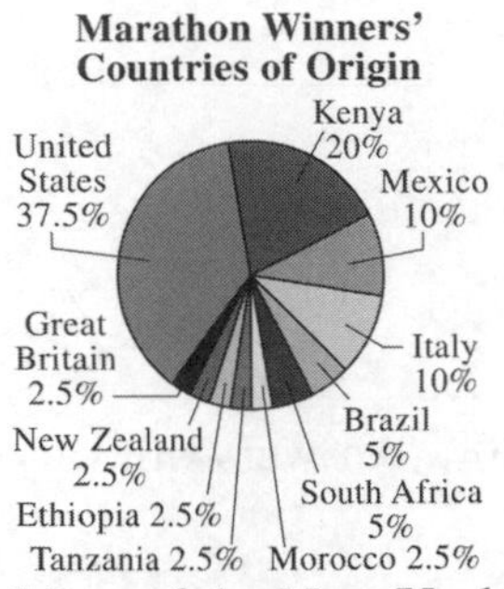

Most of the New York City Marathon winners are from the United States and Kenya. (Answers will vary.)

25.

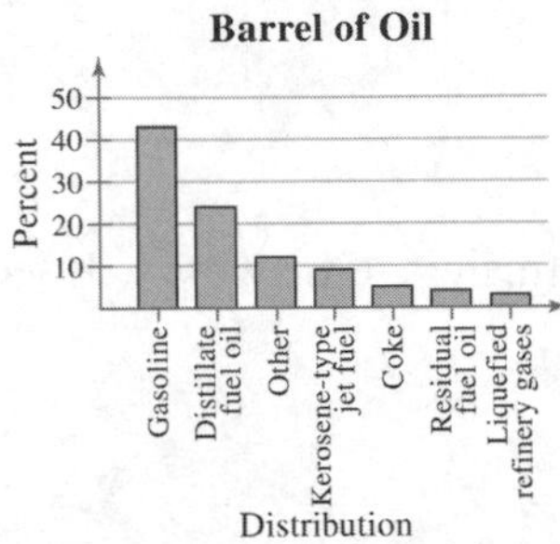

It appears that the largest portion of a 42-gallon barrel of crude oil is used for making gasoline. (Answers will vary.)

27.

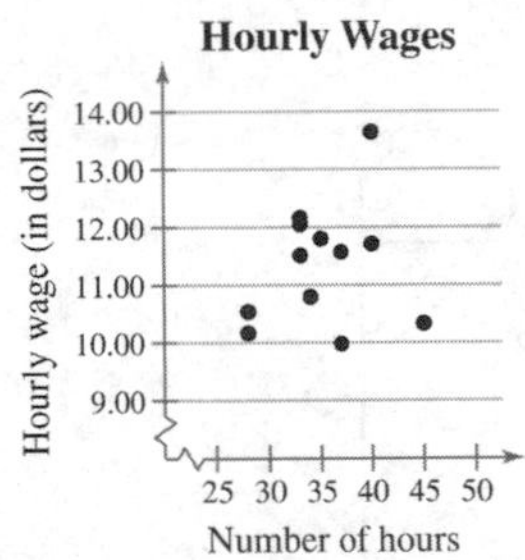

It appears that there is no relation between wages and hours worked. (Answers will vary.)

29.

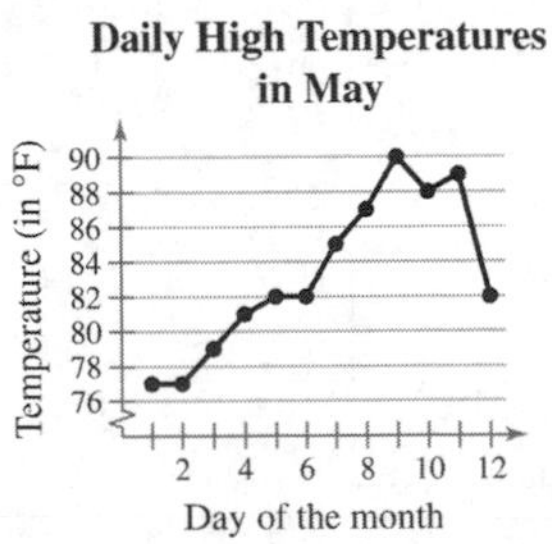

It appears that it was hottest from May 7 to May 11. (Answers will vary.)

31. Variable: Scores

Key: 5|5 = 5.5

Stem	Leaves
5	5
6	2
6	8
7	0 1
7	5 6
8	0 2 3
8	5 6 7 8 8 9
9	0 3 3
9	5 5 8 9
10	0

It appears that most scores on the final exam in economics were in the 80's and 90's. (Answers will vary.)

33a.

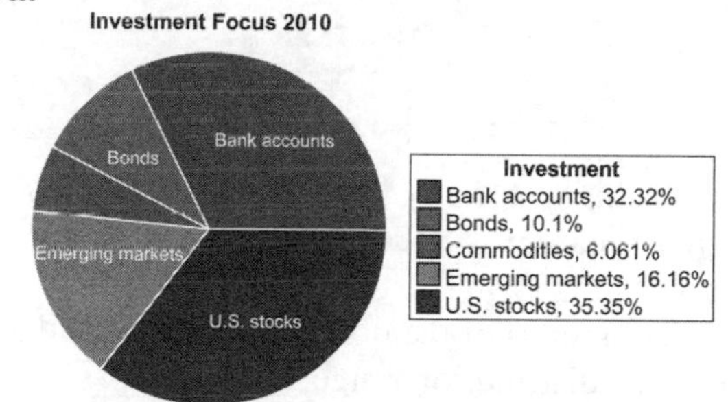

It appears that a large portion of adults said that the type of the investment that they would focus on in 2010 was U.S. stocks or bank accounts. (Answers will vary.)

b.

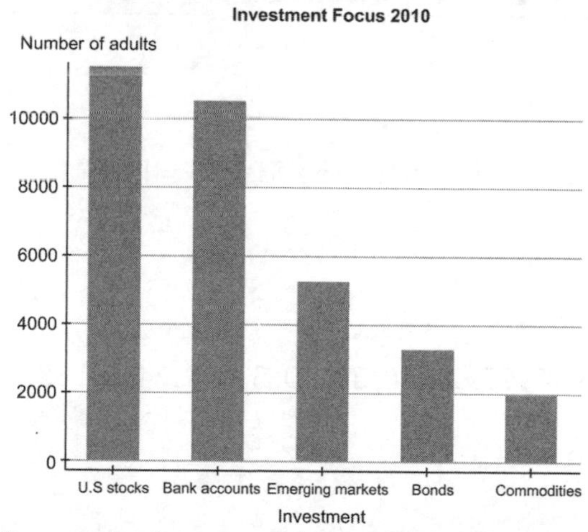

It appears that most adults said that the type of investment that they would focus on in 2010 was U.S. stocks or bank accounts. (Answers will vary.)

35a. The graph is misleading because the large gap from 0 to 90 makes it appear that the sales for the 3rd quarter are disproportionately larger than the other quarters. (Answers will vary.)

b.

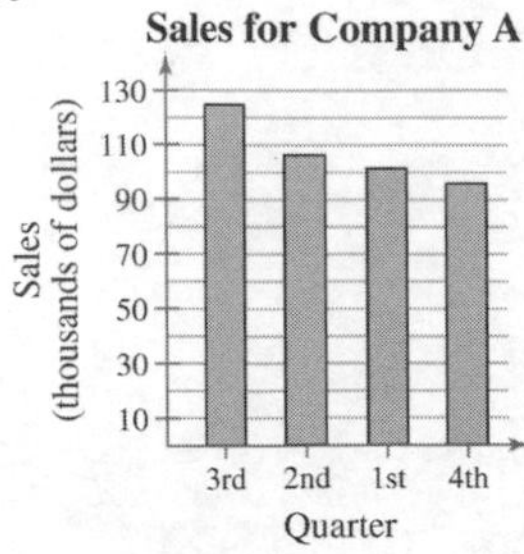

37a. The graph is misleading because the angle makes it appear as though the 3rd quarter had a larger percent of sales than the others, when the 1st and 3rd quarters have the same percent.

b.

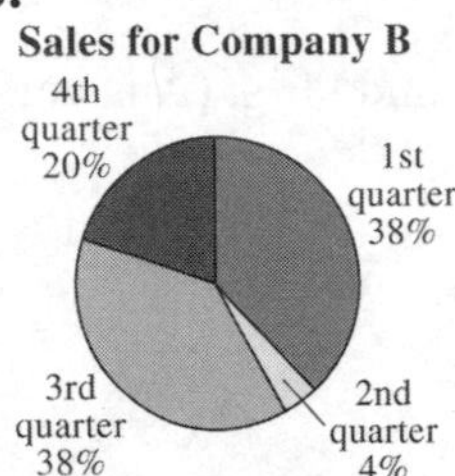

39a. At Law Firm A, the lowest salary was $90,000 and the highest salary was $203,000. At Law Firm B, the lowest salary was $90,000 and the highest salary was $190,000.

b. There are 30 lawyers at Law Firm A and 32 lawyers at Law Firm B.

c. At Law Firm A, the salaries tend to be clustered at the far ends of the distribution range and at Law Firm B, the salaries tend to fall in the middle of the distribution range.

2.3 MEASURES OF CENTRAL TENDENCY

2.3 Try It Yourself Solutions

1a. $\sum x = 74 + 78 + 81 + 87 + 81 + 80 + 77 + 80 + 85 + 78 + 80 + 83 + 75 + 81 + 73 = 1193$

b. $\bar{x} = \dfrac{\sum x}{n} = \dfrac{1193}{15} \approx 79.5$

c. The mean height of the player is about 79.5 inches.

2a. 18, 18, 19, 19, 19, 20, 21, 21, 21,21, 23, 24, 24, 26, 27, 27, 29, 30, 30, 30, 33, 33, 34, 35, 38

b. median = 24

c. The median age of the sample of fans at the concert is 24.

3a. 25, 60, 80, 97, 100, 130, 140, 200, 220, 250

b. median $= \dfrac{100 + 130}{2} = 115$

c. The median price of the sample of digital photo frames is $115.

4a. 324, 385, 450, 450, 462, 475, 540, 540, 564, 618, 624, 638, 670, 670, 670, 705, 720, 723, 750, 750, 825, 830, 912, 975, 980, 980, 1100, 1260, 1420, 1650

b. The price that occurs with the greatest frequency is \$670 per square foot.

c. The mode of the prices for the sample of South Beach, FL condominiums is \$670 per square foot.

5a. "Yes" occurs with the greatest frequency (510).

b. The mode of the responses to the survey is "Yes." In this sample, there were more people who thought public cell phone conversations were rude than people who did not or had no opinion.

6a. $\bar{x} = \frac{\sum x}{n} = \frac{410}{19} \approx 21.6$

median = 21

mode = 20

b. The mean in Example 6 ($\bar{x} \approx 23.8$) was heavily influenced by the entry 65. Neither the median nor the mode was affected as much by the entry 65.

7a, b.

Source	Score, x	Weight, w	$x \cdot w$
Test mean	86	0.50	43.0
Midterm	96	0.15	14.4
Final exam	98	0.20	19.6
Computer lab	98	0.10	9.8
Homework	100	0.05	5.0
		$\sum w = 1$	$\sum (x \cdot w) = 91.8$

c. $\bar{x} = \frac{\sum (x \cdot w)}{\sum w} = \frac{91.8}{1} = 91.8$

d. The weighted mean for the course is 91.8. So, you did get an A.

8a, b, c.

Class	Midpoint, x	Frequency, f	$x \cdot f$
35-41	38	2	76
42-48	45	5	225
49-55	52	7	364
56-62	59	7	413
63-69	66	10	660
70-76	73	5	365
77-83	80	8	640
84-90	87	6	552
		$N = 50$	$\sum (x \cdot f) = 3265$

d. $\mu = \frac{\sum (x \cdot f)}{N} = \frac{3265}{50} = 65.3$

The mean age of the 50 richest people is 65.3

2.3 EXERCISE SOLUTIONS

1. True

3. True

5. 1, 2, 2, 2, 3 (Answers will vary.)

7. 2, 5, 7, 9, 35 (Answers will vary.)

9. Skewed right because the "tail" of the distribution extends to the right.

11. Uniform because the bars are approximately the same height.

13. (11), because the distribution values range from 1 to 12 and has (approximately) equal frequencies.

15. (12), because the distribution has a maximum value of 90 and is skewed left due to a few students scoring much lower than the majority of the students.

17. $\bar{x} = \frac{\sum x}{n} = \frac{64}{13} \approx 4.9$

2 2 3 4 4 4 (5) 5 6 6 7 8 8

median = 5

mode = 4 (occurs 3 times)

19. $\bar{x} = \frac{\sum x}{n} = \frac{76.8}{7} = 11.0$

9.7 10.3 10.7 (11.0) 11.7 11.7 11.7

median = 11.0

mode = 11.7 (occurs 3 times)

21. $\bar{x} = \frac{\sum x}{n} = \frac{686.8}{32} = 21.46$

10.9 12.3 15.2 15.3 16.1 16.4 16.6 17.5 18.1 18.4 19.1 19.7 20.4 20.4 20.6
21.8 22.1 22.5 22.6 22.7 23.0 23.4 24.2 24.4 25.1 26.0 26.7 26.8 28.4 28.8 29.4 31.9

median = $\frac{21.8 + 22.1}{2}$ =21.95

mode = 20.4 (occurs 2 times)

23. $\bar{x}$ = not possible (nominal data)
median = not possible (nominal data)
mode = "Eyeglasses"
The mean and median cannot be found because the data are at the nominal level of measurement.

25. $\bar{x} = \frac{\sum x}{n} = \frac{1194.4}{7} \approx 170.63$

155.7 158.1 162.2 (169.3) 180 181.8 187.3

median = 169.3

mode = none
The mode cannot be found because no data points are repeated.

27. $\bar{x} = \frac{\sum x}{n} = \frac{1687}{10} = 168.7$

125 125 132 140 155 170 175 210 225 230

median $= \frac{155 + 170}{2} = 162.5$

mode = 125 (occurs 2 times)

The mode does not represent the center of the data because 125 is the smallest number in the data set.

29. $\bar{x} = \frac{\sum x}{n} = \frac{197.5}{14} \approx 14.11$

1.5 2.5 2.5 5 10.5 11 13 15.5 16.5 17.5 20 26.5 27 28.5

median $= \frac{13 + 15.5}{2} = 14.25$

mode = 2.5 (occurs 2 times)

The mode does not represent the center of the data set because 2.5 is much smaller than most of the data in the set.

31. $\bar{x} = \frac{\sum x}{n} = \frac{835}{28} = 29.82$

6 7 12 15 18 19 20 24 24 24 25 28 29 32 32 33 35 35 35 36 38 39 40 41 42 47 48 51

median $= \frac{32 + 32}{2} = 32$

mode = 24, 35 (occurs 3 times each)

33. $\bar{x} = \frac{\sum x}{n} = \frac{292}{15} \approx 19.47$

5 8 10 15 15 15 17 (20) 21 22 22 25 28 32 37

median = 20

mode = 15 (occurs 3 times)

35. The data are skewed right.

A = mode, because it is the data entry that occurred most often.

B = median, because the median is to the left of the mean in a skewed right distribution.

C = mean, because the mean is to the right of the median in a skewed right distribution.

37. Mode, because the data are at the nominal level of measurement.

39. Mean, because there are no outliers.

41.

Source	Score, x	Weight, w	$x \cdot w$
Homework	85	0.05	4.25
Quiz	80	0.35	28
Project	100	0.20	20
Speech	90	0.15	13.5
Final exam	93	0.25	23.25
		$\sum w = 1$	$\sum(x \cdot w) = 89$

$$\bar{x} = \frac{\sum(x \cdot w)}{\sum w} = \frac{89}{1} = 89$$

43.

Balance, x	Days, w	$x \cdot w$
\$523	24	12,552
\$2415	2	4830
\$250	4	1000
	$\sum w = 30$	$\sum(x \cdot w) = 18{,}382$

$$\bar{x} = \frac{\sum(x \cdot w)}{\sum w} = \frac{18{,}382}{30} \approx \$612.73$$

45.

Grade	Points, x	Credits, w	$x \cdot w$
B	3	3	9
B	3	3	9
A	4	4	16
D	1	2	2
C	2	3	6
		$\sum w = 15$	$\sum(x \cdot w) = 42$

$$\bar{x} = \frac{\sum(x \cdot w)}{\sum w} = \frac{42}{15} = 2.8$$

47.

Source	Score, x	Weight, w	$x \cdot w$
Homework	85	0.05	4.25
Quiz	80	0.35	28
Project	100	0.20	20
Speech	90	0.15	13.5
Final exam	85	0.25	21.25
		$\sum w = 1$	$\sum(x \cdot w) = 87$

$$\bar{x} = \frac{\sum(x \cdot w)}{\sum w} = \frac{87}{1} = 87$$

49.

Class	Midpoint, x	Frequency, f	$x \cdot f$
29-33	31	11	341
34-38	36	12	432
39-43	41	2	82
44-48	46	5	230
		$n = 30$	$\sum(x \cdot f) = 1085$

$$\bar{x} = \frac{\sum(x \cdot f)}{n} = \frac{1085}{30} \approx 36.2 \text{ miles per gallon}$$

51.

Class	Midpoint, x	Frequency, f	$x \cdot f$
0-9	4.5	55	247.5
10-19	14.5	70	1015
20-29	24.5	35	857.5
30-39	34.5	56	1932
40-49	44.5	74	3293
50-59	54.5	42	2289
60-69	64.5	38	2451
70-79	74.5	17	1266.5
80-89	84.5	10	845
		$n = 397$	$\sum(x \cdot f) = 14{,}196.5$

$$\bar{x} = \frac{\sum(x \cdot f)}{n} = \frac{14{,}196.5}{397} \approx 35.8 \text{ years old}$$

53. Class width $= \dfrac{\text{Range}}{\text{Number of classes}} = \dfrac{297 - 127}{5} = 34 \Rightarrow 35$

Class	Midpoint	Frequency, f
127-161	144	9
162-196	179	8
197-231	214	3
232-266	249	3
267-301	284	1
		$\sum f = 24$

Shape: Positively skewed

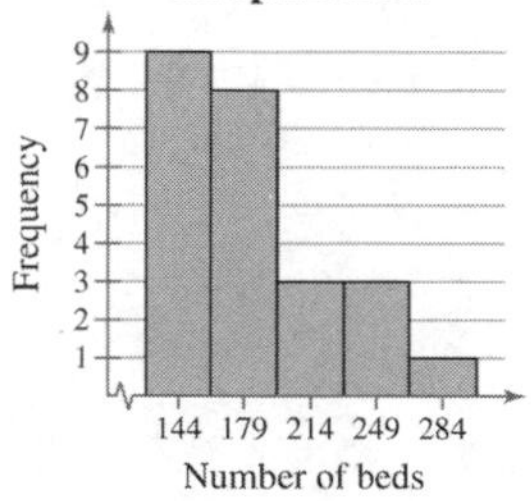

55. Class width = $\frac{\text{Range}}{\text{Number of classes}} = \frac{76-62}{5} = 2.8 \Rightarrow 3$

Class	Midpoint	Frequency, f
62-64	63	3
65-67	66	7
68-70	69	9
71-73	72	8
74-76	75	3
		$\sum f = 30$

Shape: Symmetric

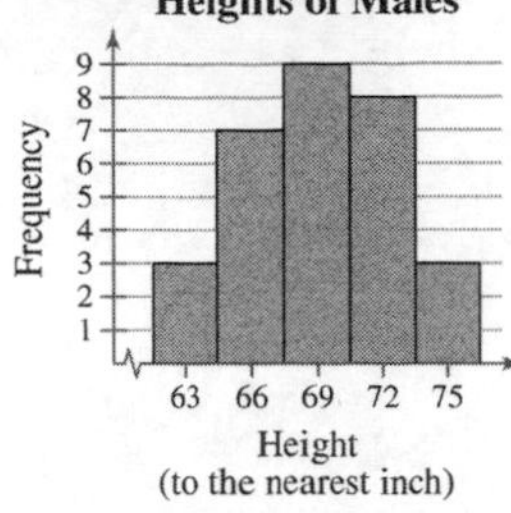

57a. $\bar{x} = \frac{\sum x}{n} = \frac{36.03}{6} = 6.005$

5.59 5.99 6 6.02 6.03 6.40

median $= \frac{6 + 6.02}{2} = 6.01$

b. $\bar{x} = \frac{\sum x}{n} = \frac{35.67}{6} = 5.945$

5.59 5.99 6 6.02 6.03 6.04

median $= \frac{6 + 6.02}{2} = 6.01$

c. The mean was affected more.

59. Summary Statistics:

Column	n	Mean	Median	Min	Max
Amount (in dollars)	11	112.11364	105.25	79	151.5

61a. $\bar{x} = \frac{\sum x}{n} = \frac{3222}{9} = 358$

147 177 336 360 (375) 393 408 504 522

median = 375

b. $\bar{x} = \frac{\sum x}{n} = \frac{9666}{9} = 1074$

441 531 1008 1080 (1125) 1179 1224 1512 1566

median = 1125

c. The mean and median in part (b) are three times the mean and median in part (a).

d. If you multiply the mean and median from part (b) by 3, you will get the mean and median of the data set in inches.

63. Car A: Midrange $= \dfrac{34+28}{2} = 31$

Car B: Midrange $= \dfrac{31+29}{2} = 30$

Car C: Midrange $= \dfrac{32+28}{2} = 30$

Car A because it has the highest midrange of the three.

65a. Order the data values.

11 13 22 28 36 36 36 37 37 37 38 41 43 44 46
47 51 51 51 53 61 62 63 64 72 72 74 76 85 90

Delete the lowest 10%, smallest 3 observations (11, 13, 22).
Delete the highest 10%, largest 3 observations (76, 85, 90).
Find the 10% trimmed mean using the remaining 24 observations.

10% trimmed mean ≈ 49.2

b. $\bar{x} \approx 49.2$

median = 46.5

mode = 36, 37, 51

midrange $= \dfrac{90+11}{2} = 50.5$

c. Using a trimmed mean eliminates potential outliers that may affect the mean of all the observations.

2.4 MEASURES OF VARIATION

2.4 Try It Yourself Solutions

1a. Min = 23, or \$23,000 and Max = 58, or \$58,000

b. Range = Max – Min = 58 – 23 = 35, or \$35,000

c. The range of the starting salaries for Corporation B is 35, or \$35,000. This is much larger than the range for Corporation A.

2a. $\mu = \dfrac{\sum x}{N} = \dfrac{415}{10} = 41.5$, or \$41,500

b.

Salary, x (1000s of dollars)	**Deviation, $x - \mu$ (100s of dollars)**
23	–18.5
29	–12.5
32	–9.5
40	–1.5
41	–0.5
41	–0.5
49	7.5
50	8.5
52	10.5
58	16.5
$\sum x = 415$	$\sum(x-\mu) = 0$

3ab. $\mu = 41.5$, or $41,500

Salary, x	**$x - \mu$**	**$(x - \mu)^2$**
23	–18.5	342.25
29	–12.5	156.25
32	–9.5	90.25
40	–1.5	2.25
41	–0.5	0.25
41	–0.5	0.25
49	7.5	56.25
50	8.5	72.25
52	10.5	110.25
58	16.5	272.25
$\sum x = 415$	$\sum(x-\mu) = 0$	$\sum(x-\mu)^2 = 1102.5$

c. $\sigma^2 = \frac{\Sigma(x-\mu)^2}{N} = \frac{1102.5}{10} \approx 110.3$

d. $\sigma = \sqrt{\sigma^2} = \sqrt{\frac{1102.5}{10}} = 10.5$, or $10,500

e. The population variance is about 110.3 and the population standard deviation is 10.5, or $10,500.

4a. From 3ab, $SS_x = \Sigma(x - \bar{x})^2 = 1102.5.$

b. $s^2 = \frac{\Sigma(x-\bar{x})^2}{n-1} = \frac{1102.5}{9} = 122.5$

c. $s = \sqrt{s^2} = \sqrt{122.5} \approx 11.1$, or $11,100

d. The sample variance is 122.5 and the sample standard deviation is 11.1, or $11,100.

5a. Enter the data in a computer or a calculator.

b. $\bar{x} = 37.89$, $s = 3.98$

6a. 7, 7, 7, 7, 7, 13, 13, 13, 13, 13

b.

Salary, x	$x - \mu$	$(x - \mu)^2$
7	–3	9
7	–3	9
7	–3	9
7	–3	9
7	–3	9
13	3	9
13	3	9
13	3	9
13	3	9
13	3	9
$\sum x = 100$	$\sum (x-\mu) = 0$	$\sum (x-\mu)^2 = 90$

$$\mu = \frac{\sum x}{N} = \frac{100}{10} = 10$$

$$\sigma = \sqrt{\frac{\sum (x-\mu)^2}{N}} = \sqrt{\frac{90}{10}} = \sqrt{9} = 3$$

7a. 66.92 – 64.3 = 2.62 = 1 standard deviation

b. 34%

c. Approximately 34% of women ages 20-29 are between 64.3 and 66.92 inches tall.

8a. 31.6 – 2(19.5) = –7.4

Because –7.4 does not make sense for an age, use 0.

b. 31.6 + 2(19.5) = 70.6

c. $1 - \frac{1}{k^2} = 1 - \frac{1}{(2)^2} = 1 - \frac{1}{4} = 0.75$

At least 75% of the data lie within 2 standard deviations of the mean. At least 75% of the population of Alaska is between 0 and 70.6 years old.

9a.

x	f	xf
0	10	0
1	19	19
2	7	14
3	7	21
4	5	20
5	1	5
6	1	6
	$n = 50$	$\sum xf = 85$

b. $\bar{x} = \frac{\sum xf}{n} = \frac{85}{50} = 1.7$

c.

$x-\bar{x}$	$(x-\bar{x})^2$	$(x-\bar{x})^2 f$
−1.7	2.89	28.90
−0.7	0.49	9.31
0.3	0.09	0.63
1.3	1.69	11.83
2.3	5.29	26.45
3.3	10.89	10.89
4.3	18.49	18.49
		$\sum(x-\bar{x})^2 f = 106.5$

d. $s=\sqrt{\frac{\Sigma(x-\bar{x})^2 f}{n-1}}=\sqrt{\frac{106.5}{49}}\approx 1.5$

10a.

Class	x	f	xf
1-99	49.5	380	18,810
100-199	149.5	230	34,385
200-299	249.5	210	52,395
300-399	349.5	50	17,475
400-499	449.5	60	26,970
500+	650	70	45,500
		$n =$ 1000	$\sum xf = 195{,}535$

b. $\bar{x}=\frac{\Sigma xf}{n}=\frac{195{,}535}{1000}\approx 195.5$

c.

$x-\bar{x}$	$(x-\bar{x})^2$	$(x-\bar{x})^2 f$
−146	21,316	8,100,080
−46	2116	486,680
54	2916	612,360
154	23,716	1,185,800
254	64,516	3,870,960
454.5	206,570.25	14,459,917.5
		$\sum(x-\bar{x})^2 f = 28{,}715{,}797.5$

d. $s=\sqrt{\frac{\Sigma(x-\bar{x})^2 f}{n-1}}=\sqrt{\frac{28{,}715{,}797.5}{999}}\approx 169.5$

2.4 EXERCISE SOLUTIONS

1. The range is the difference between the maximum and minimum values of a data set. The advantage of the range is that it is easy to calculate. The disadvantage is that it uses only two entries from the data set.

3. The units of variance are squared. Its units are meaningless. (Example: dollars^2)

5. $\{9, 9, 9, 9, 9, 9, 9\}$

$n = 7$

$$\bar{x} = \frac{\sum x}{n} = \frac{63}{7} = 9$$

x	$x - \bar{x}$	$(x - \bar{x})^2$
9	0	0
9	0	0
9	0	0
9	0	0
9	0	0
9	0	0
9	0	0
$\sum x = 63$	$\sum(x - \bar{x}) = 0$	$\sum(x - \bar{x})^2 = 0$

$$s = \sqrt{\frac{\sum(x - \bar{x})^2}{n - 1}} = \sqrt{\frac{0}{6}} = 0$$

7. When calculating the population standard deviation, you divide the sum of the squared deviations by N, then take the square root of that value. When calculating the sample standard deviation, you divide the sum of the squared deviations by $n - 1$, then take the square root of that value.

9. Similarity: Both estimate proportions of the data contained within k standard deviations of the mean.
Difference: The Empirical Rule assumes the distribution is bell-shaped. Chebychev's Theorem makes no such assumption.

11. Range = Max – Min = 12 – 5 = 7

$$\mu = \frac{\sum x}{N} = \frac{90}{10} = 9$$

x	$x - \mu$	$(x - \mu)^2$
9	0	0
5	–4	16
9	0	0
10	1	1
11	2	4
12	3	9
7	–2	4
7	–2	4
8	–1	1
12	3	9
$\sum x = 90$	$\sum(x - \mu) = 0$	$\sum(x - \mu)^2 = 48$

$$\sigma^2 = \frac{\sum(x - \mu)^2}{N} = \frac{48}{10} = 4.8, \qquad \sigma = \sqrt{\frac{\sum(x - \mu)^2}{N}} = \sqrt{4.8} \approx 2.2$$

13. Range = Max – Min = 19 – 4 = 15

$$\bar{x} = \frac{\sum x}{n} = \frac{108}{9} = 12$$

x	$x - \bar{x}$	$(x - \bar{x})^2$
4	–8	64
15	3	9
9	–3	9
12	0	0
16	4	16
8	–4	16
11	–1	1
19	7	49
14	2	4
$\sum x = 108$	$\sum (x - \bar{x}) = 0$	$\sum (x - \bar{x})^2 = 168$

$$s^2 = \frac{\sum (x - \bar{x})^2}{n - 1} = \frac{168}{9 - 1} = 21$$

$$s = \sqrt{\frac{\sum (x - \bar{x})^2}{n - 1}} = \sqrt{21} \approx 4.6$$

15. Range = Max – Min = 96 – 23 = 73

17. Range = Max – Min = 98 – 74 = 24

19a. Range = Max – Min = 38.5 – 20.7 = 17.8
b. Range = Max – Min = 60.5 – 20.7 = 39.8

21. Graph (a) has a standard deviation of 24 and graph (b) has a standard deviation of 16 because graph (a) has more variability.

23. Company B. An offer of $33,000 is two standard deviations from the mean of Company A's starting salaries, which makes it unlikely. The same offer is within one standard deviation of the mean of Company B's starting salaries, which makes the offer likely.

25a. Dallas:

$$\bar{x} = \frac{\sum x}{n} = \frac{398.5}{9} \approx 44.28$$

38.7 39.9 40.5 41.6 (44.7) 45.8 47.8 49.5 50.0

median = 44.7

Range = Max – Min = 50.0 – 38.7 = 11.3

x	$x-\bar{x}$	$(x-\bar{x})^2$
38.7	–5.58	31.1364
39.9	–4.38	19.1844
40.5	–3.78	14.2884
41.6	–2.68	7.1824
44.7	0.42	0.1764
45.8	1.52	2.3104
47.8	3.52	12.3904
49.5	5.22	27.2484
50.0	5.72	32.7184
		$\sum(x-\bar{x})^2 = 146.6356$

$$s^2 = \frac{\sum(x-\bar{x})^2}{n-1} = \frac{146.6356}{8} \approx 18.33$$

$$s = \sqrt{\frac{\sum(x-\bar{x})^2}{n-1}} = \sqrt{\frac{146.6356}{8}} \approx 4.28$$

New York City:

$$\bar{x} = \frac{\sum x}{n} = \frac{458.2}{9} \approx 50.91$$

41.5 42.3 45.6 47.2 (50.6) 55.1 57.6 59.0 59.3

median = 50.6

Range = Max – Min = 59.3 – 41.5 = 17.8

x	$x-\bar{x}$	$(x-\bar{x})^2$
41.5	–9.41	88.5481
42.3	–8.61	74.1321
45.6	–5.31	28.1961
47.2	–3.71	13.7641
50.6	–0.31	0.0961
55.1	4.19	17.5561
57.6	6.69	44.7561
59.0	8.09	65.4481
59.3	8.39	70.3921
		$\sum(x-\bar{x})^2 = 402.8889$

$$s^2 = \frac{\sum(x-\bar{x})^2}{n-1} = \frac{402.8889}{9-1} \approx 50.36 \qquad s = \sqrt{\frac{\sum(x-\bar{x})^2}{n-1}} = \sqrt{\frac{402.8889}{8}} \approx 7.10$$

b. It appears from the data that the annual salaries in New York City are more variable than the annual salaries in Dallas. The annual salaries in Dallas have a lower mean and a lower median than the annual salaries in New York City.

27a. Male:

$$\bar{x} = \frac{\sum x}{n} = \frac{13{,}144}{8} = 1643$$

1033 1380 1520 1645 1714 1750 1982 2120

$$\text{median} = \frac{1645 + 1714}{2} = 1679.5$$

Range = Max – Min = 2120 – 1033 = 1087

x	$x - \bar{x}$	$(x - \bar{x})^2$
1033	–610	372,100
1380	–263	69,169
1520	–123	15,129
1645	2	4
1714	71	5041
1750	107	11,449
1982	339	114,921
2120	477	227,529
		$\sum(x - \bar{x})^2 = 815{,}342$

$$s^2 = \frac{\sum(x-\bar{x})^2}{n-1} = \frac{815{,}342}{8-1} \approx 116{,}477.4$$

$$s = \sqrt{\frac{\sum(x-\bar{x})^2}{n-1}} = \sqrt{\frac{815{,}342}{7}} \approx 341.3$$

Female:

$$\bar{x} = \frac{\sum x}{n} = \frac{13{,}673}{8} \approx 1709.1$$

1263 1497 1507 1588 1785 1871 1952 2210

$$\text{median} = \frac{1588 + 1785}{2} = 1686.5$$

Range = Max – Min = 2210 – 1263 = 947

x	$x - \bar{x}$	$(x - \bar{x})^2$
1263	–446.1	199,005.21
1497	–212.1	44,986.41
1507	–202.1	40,844.41
1588	–121.1	14,665.21
1785	75.9	5760.81
1871	161.9	26,211.61
1952	242.9	59,000.41
2210	500.9	250,900.81
		$\sum(x - \bar{x})^2 = 641{,}374.88$

$$s^2 = \frac{\sum(x-\bar{x})^2}{n-1} = \frac{641{,}374.88}{8-1} \approx 91{,}625.0$$

$$s = \sqrt{\frac{\sum(x-\bar{x})^2}{n-1}} = \sqrt{\frac{641{,}374.88}{7}} = 302.7$$

b. It appears from the data that the SAT scores for males are more variable than the SAT scores for females. The SAT scores for males have a lower mean and median than the SAT scores for females.

29a. Greatest sample standard deviation: (ii)
Data set (ii) has more entries that are farther away from the mean.
Least sample standard deviation: (iii)
Data set (iii) has more entries that are close to the mean.

b. The three data sets have the same mean but have different standard deviations.

31a. Greatest sample standard deviation: (ii)
Data set (ii) has more entries that are farther away from the mean.
Least sample standard deviation: (iii)
Data set (iii) has more entries that are close to the mean.

b. The three data sets have the same mean, median, and mode, but have different standard deviations.

33. $(1300,\ 1700) \rightarrow (1500-1(200),\ 1500+1(200)) \rightarrow (\bar{x}-s,\ \bar{x}+s)$
68% of the farms have values between \$1300 and \$1700 per acre.

35a. $n = 75$
68%(75) = (0.68)(75) ≈ 51 farms have values between \$1300 and \$1700 per acre.

b. $n = 25$
68%(25) = (0.68)(25) ≈ 17 farms have values between \$1300 and \$1700 per acre.

37. $\bar{x} = 1500 \qquad s = 200$
{\$950, \$1000, \$2000, \$2180} are outliers. They are more than 2 standard deviations from the mean (1100, 1900). \$2180 is very unusual because it is more than 3 standard deviations from the mean.

39. $(\bar{x}-2s,\ \bar{x}+2s) \rightarrow (1.14,\ 5.5)$ are 2 standard deviations from the mean.

$1-\frac{1}{k^2} = 1-\frac{1}{(2)^2} = 1-\frac{1}{4} = 0.75 \Rightarrow$ At least 75% of the eruption times lie between 1.14 and 5.5 minutes.

If $n = 32$, at least (0.75)(32) = 24 eruptions will lie between 1.14 and 5.5 minutes.

41.

x	f	xf	$x-\bar{x}$	$(x-\bar{x})^2$	$(x-\bar{x})^2 f$
0	5	0	–2.1	4.41	22.05
1	11	11	–1.1	1.21	13.31
2	7	14	–0.1	0.01	0.07
3	10	30	0.9	0.81	8.10
4	7	28	1.9	3.61	25.27
	$n=40$	$\sum xf = 83$			$\sum(x-\bar{x})^2 f = 68.8$

$$\bar{x} = \frac{\sum xf}{n} = \frac{83}{40} \approx 2.1$$

$$s = \sqrt{\frac{\sum(x-\bar{x})^2 f}{n-1}} = \sqrt{\frac{68.8}{39}} \approx 1.3$$

43. Class width $= \dfrac{\text{Max} - \text{Min}}{5} = \dfrac{14-1}{5} = \dfrac{13}{5} = 2.6 \Rightarrow 3$

Class	Midpoint, x	f	xf
1-3	2	3	6
4-6	5	6	30
7-9	8	13	104
10-12	11	7	77
13-15	14	3	42
		$N=32$	$\sum xf = 259$

$$\mu = \frac{\sum xf}{N} = \frac{259}{32} \approx 8.1$$

$x-\mu$	$(x-\mu)^2$	$(x-\mu)^2 f$
–6.1	37.21	111.63
–3.1	9.61	57.66
–0.1	0.01	0.13
2.9	8.41	58.87
5.9	34.81	104.43
		$\sum(x-\mu)^2 f = 332.72$

$$\sigma = \sqrt{\frac{\sum(x-\mu)^2 f}{N}} = \sqrt{\frac{332.72}{32}} \approx 3.2$$

45.

Midpoint, x	f	xf
70.5	1	70.5
92.5	12	1110.0
114.5	25	2862.5
136.5	10	1365.0
158.5	2	317.0
	$n=50$	$\sum xf = 5725$

$$\bar{x} = \frac{\sum xf}{n} = \frac{5725}{50} = 114.5$$

$x - \bar{x}$	$(x - \bar{x})^2$	$(x - \bar{x})^2 f$
−44	1936	1936
−22	484	5808
0	0	0
22	484	4840
44	1936	3872
		$\sum(x-\bar{x})^2 f = 16,456$

$$s = \sqrt{\frac{\sum(x-\bar{x})^2 f}{n-1}} = \sqrt{\frac{16,456}{49}} \approx 18.33$$

47.

Class	Midpoint, x	f	xf
0-4	2.0	22.1	44.20
5-14	9.5	43.4	412.30
15-19	17.0	21.2	360.40
20-24	22.0	22.3	490.60
25-34	29.5	44.5	1312.75
35-44	39.5	41.3	1631.35
45-64	54.5	83.9	4572.55
65+	70.0	46.8	3276.00
		$n = 325.5$	$\sum xf = 12,100.15$

$$\bar{x} = \frac{\sum xf}{n} = \frac{12,100.15}{325.5} \approx 37.17$$

$x - \bar{x}$	$(x - \bar{x})^2$	$(x - \bar{x})^2 f$
−35.17	1236.9289	27,336.12869
−27.67	765.6289	33,228.29426
−20.17	406.8289	8624.77268
−15.17	230.1289	5131.87447
−7.67	58.8289	2617.88605
2.33	5.4289	224.21357
17.33	300.3289	25,197.59471
32.83	1077.8089	50,441.45642
		$\sum(x-\bar{x})^2 f = 152,802.22085$

$$s = \sqrt{\frac{\sum(x-\bar{x})^2 f}{n-1}} = \sqrt{\frac{152,802.22085}{324.5}} \approx 21.70$$

49. **Summary Statistics:**

Column	n	Mean	Variance
Amount (in dollars)	15	58.8	239.74286

Std. Dev.	Median	Range	Min	Max
15.483632	60	59	30	89

51. Heights:

$$\mu = \frac{\sum x}{N} = \frac{873}{12} = 72.75$$

x	$x-\mu$	$(x-\mu)^2$
68	−4.75	22.5625
69	−3.75	14.0625
69	−3.75	14.0625
70	−2.75	7.5625
72	−0.75	0.5625
72	−0.75	0.5625
73	0.25	0.0625
74	1.25	1.5625
74	1.25	1.5625
76	3.25	10.5625
77	4.25	18.0625
79	6.25	39.0625
		$\sum(x-\mu)^2 = 130.25$

$$\sigma = \sqrt{\frac{\sum(x-\mu)^2}{N}} = \sqrt{\frac{130.25}{12}} \approx 3.29$$

$$\text{CV}_{\text{heights}} = \frac{\sigma}{\mu}\cdot 100\% = \frac{3.29}{72.75}\cdot 100 \approx 4.5\%$$

Weights:

$$\mu = \frac{\sum x}{N} = \frac{2254}{12} = 187.83$$

x	$x-\mu$	$(x-\mu)^2$
162	−25.83	667.1889
168	−19.83	393.2289
171	−16.83	283.2489
174	−13.83	191.2689
180	−7.83	61.3089
185	−2.83	8.0089
189	1.17	1.3689
192	4.17	17.3889
197	9.17	84.0889
201	13.17	173.4489
210	22.17	491.5089
225	37.17	1381.6089
		$\sum(x-\mu)^2 = 3753.6668$

$$\sigma = \sqrt{\frac{\sum(x-\mu)^2}{N}} = \sqrt{\frac{3753.6668}{12}} \approx 17.69$$

$$CV_{\text{weights}} = \frac{\sigma}{\mu} \cdot 100\% = \frac{17.69}{187.83} \cdot 100 \approx 9.4\%$$

It appears that weight is more variable than height.

53a. $\bar{x} \approx 41.5$ $s \approx 5.3$

b. $\bar{x} \approx 43.6$ $s \approx 5.6$

c. $\bar{x} \approx 3.5$ $s \approx 0.4$

d. By multiplying each entry by a constant *k*, the new sample mean is $k \cdot \bar{x}$ and the new sample standard deviation is $k \cdot s$.

55a. Male: $\bar{x} = 1643$

x	$\lvert x-\bar{x} \rvert$
1520	123
1750	107
2120	477
1380	263
1982	339
1645	2
1033	610
1714	71
	$\sum\lvert x-\bar{x} \rvert = 1992$

$$\frac{\sum\lvert x-\bar{x} \rvert}{n} = \frac{1992}{8} = 249$$

$s = 341.3$

Female: $\bar{x} \approx 1709.1$

x	$\lvert x-\bar{x} \rvert$
1785	75.9
1507	202.1
1497	212.1
1952	242.9
2210	500.9
1871	161.9
1263	446.1
1588	121.1
	$\sum\lvert x-\bar{x} \rvert = 1963$

$$\frac{\sum\lvert x-\bar{x} \rvert}{n} = \frac{1963}{8} \approx 245.4$$

$s = 302.7$

The mean absolute deviation is less than the sample standard deviation.

b. Team A: $\bar{x} \approx 0.2993$

x	$\lvert x-\bar{x}\rvert$
0.295	0.0043
0.310	0.0107
0.325	0.0257
0.272	0.0273
0.256	0.0433
0.297	0.0023
0.320	0.0207
0.384	0.0847
0.235	0.0643
	$\sum\lvert x-\bar{x}\rvert = 0.2833$

$$\frac{\sum\lvert x-\bar{x}\rvert}{n} = \frac{0.2833}{9} \approx 0.0315$$

$s = 0.0435$

Team B: $\bar{x} \approx 0.2989$

x	$\lvert x-\bar{x}\rvert$
0.285	0.0139
0.305	0.0061
0.315	0.0161
0.270	0.0289
0.292	0.0069
0.330	0.0311
0.335	0.0361
0.268	0.0309
0.290	0.0089
	$\sum\lvert x-\bar{x}\rvert = 0.1789$

$$\frac{\sum\lvert x-\bar{x}\rvert}{n} = \frac{0.1789}{9} \approx 0.0199$$

$s = 0.0242$

The mean absolute deviation is less than the sample standard deviation.

57a. $P = \dfrac{3(\bar{x} - \text{median})}{s} = \dfrac{3(17-19)}{2.3} \approx -2.61$; skewed left

b. $P = \dfrac{3(\bar{x} - \text{median})}{s} = \dfrac{3(32-25)}{5.1} \approx 4.12$; skewed right

c. $P = \dfrac{3(\bar{x} - \text{median})}{s} = \dfrac{3(9.2-9.2)}{1.8} = 0$; symmetric

d. $P = \dfrac{3(\bar{x} - \text{median})}{s} = \dfrac{3(42-40)}{6.0} = 1$; skewed right

2.5 MEASURES OF POSITION

2.5 Try It Yourself Solutions

1a. 35, 36, 42, 43, 44, 46, 47, 49, 51, 51, 53, 53, 54, 54, 56, 57, 58, 59, 60, 61, 61, 63, 64, 65, 65, 66, 66, 67, 68, 69, 69, 72, 73, 73, 73, 76, 77, 78, 78, 80, 81, 82, 83, 83, 85, 86, 86, 87, 89, 89

b. $Q_2 = 65.5$

c. $Q_1 = 54,\ Q_3 = 78$

d. About one fourth of the 50 richest people are 54 years old or younger; one half are 65.5 years old or younger; and about three fourths of the 50 richest people are 78 years old or younger.

2a. (Enter the data)

b. $Q_1 = 17,\ Q_2 = 23\ Q_3 = 28.5$

c. One quarter of the tuition costs is $17,000 or less, one half is $23,000 or less, and three quarters is $28,500 or less.

3a. $Q_1 = 54,\ Q_3 = 78$

b. IQR $= Q_3 - Q_1 = 78 - 54 = 24$

c. The ages of the 50 richest people in the middle portion of the data set vary by at most 24 years.

4a. Min = 35, $Q_1 = 54,\ Q_2 = 65.5,\ Q_3 = 78,$ Max = 89

b, c.

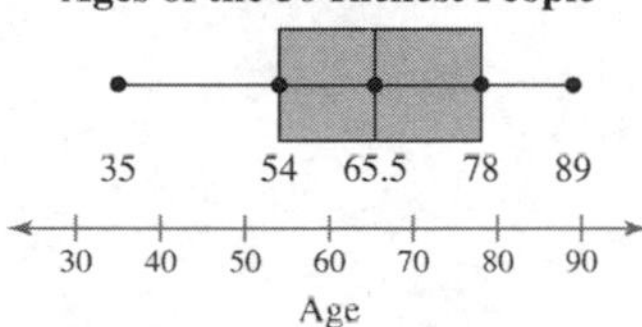

d. It appears that half of the ages are between 54 and 78.

5a. 50th percentile

b. 50% of the 50 richest people are younger than 66.

6a. $x = 60:\ z = \dfrac{x-\mu}{\sigma} = \dfrac{60-70}{8} = -1.25$

$x = 71:\ z = \dfrac{x-\mu}{\sigma} = \dfrac{71-70}{8} = 0.125$

$x = 92:\ z = \dfrac{x-\mu}{\sigma} = \dfrac{92-70}{8} = 2.75$

b. From the z-scores, the utility bill of $60 is 1.25 standard deviations below the mean, the bill of $71 is 0.125 standard deviation above the mean, and the bill of $92 is 2.75 standard deviations above the mean.

7a. Best actor: $\mu = 43.7,\ \sigma = 8.7$

Best actress: $\mu = 35.9,\ \sigma = 11.4$

b. Sean Penn: $x = 48$: $z = \dfrac{x-\mu}{\sigma} = \dfrac{48-43.7}{8.7} \approx 0.49$

Kate Winslet: $x = 33$: $z = \dfrac{x-\mu}{\sigma} = \dfrac{33-35.9}{11.4} \approx -0.25$

c. Sean Penn's age is 0.49 standard deviation above the mean of the best actors. Kate Winslet's age is -0.25 standard deviation below the mean of the best actresses. Neither actor's age is unusual.

2.5 EXERCISE SOLUTIONS

1. The soccer team scored fewer points per game than 75% of the teams in the league.

3. The student scored higher than 78% of the students who took the actuarial exam.

5. The interquartile range of a data set can be used to identify outliers because data values that are greater than $Q_3 + 1.5(\text{IQR})$ or less than $Q_1 - 1.5(\text{IQR})$ are considered outliers.

7. False. The median of a data set is a fractile, but the mean may or may not be fractile depending on the distribution of the data.

9. True

11. False. The 50th percentile is equivalent to Q_2.

13. False. A z-score of -2.5 is considered unusual.

15a. Min = 10, $Q_1 = 13$, $Q_2 = 15$, $Q_3 = 17$, Max = 20

b. IQR = $Q_3 - Q_1 = 17 - 13 = 4$

17a. Min = 900, $Q_1 = 1250$, $Q_2 = 1500$, $Q_3 = 1950$, Max = 2100

b. IQR = $Q_3 - Q_1 = 1950 - 1250 = 700$

19a. Min = -1.9, $Q_1 = -0.5$, $Q_2 = 0.1$, $Q_3 = 0.7$, Max = 2.1

b. IQR = $Q_3 - Q_1 = 0.7 - (-0.5) = 1.2$

21a.

lower half (24 26 27 28 30 32 35) | upper half (36 39 39 41 50 51 60)

24 26 27 28 30 32 35 35 36 39 39 41 50 51 60

Q_1 = 28 (↑), Q_2 = 35 (↑), Q_3 = 41 (↑)

Min = 24, $Q_1 = 28$, $Q_2 = 35$, $Q_3 = 41$, Max = 60

b.

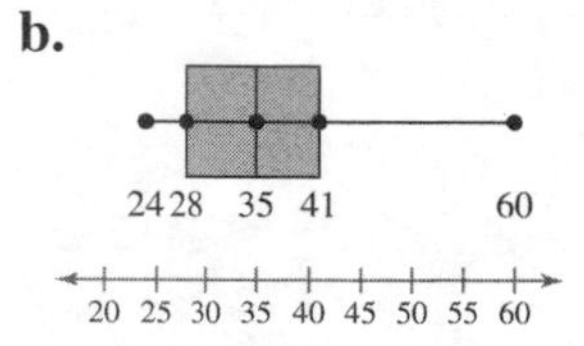

23a.

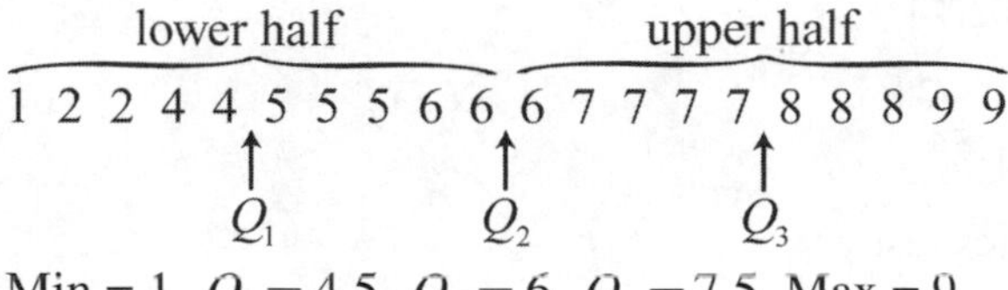

Min = 1, $Q_1 = 4.5$, $Q_2 = 6$, $Q_3 = 7.5$, Max = 9

b.

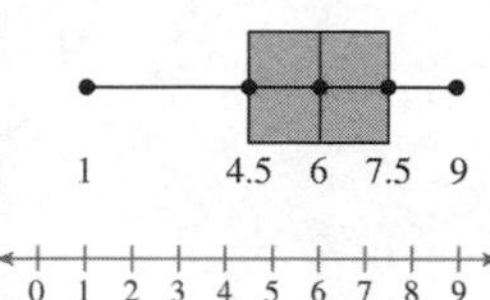

25. None. The Data are not skewed or symmetric.

27. Skewed left. Most of the data lie to the right in the box-and-whisker plot.

29. $Q_1 = \text{B}$, $Q_2 = \text{A}$, $Q_3 = \text{C}$

25% of the values are below B, 50% of the values are below A, and 75% of the values are below C.

31a. $Q_1 = 2$, $Q_2 = 4$, $Q_3 = 5$

b.

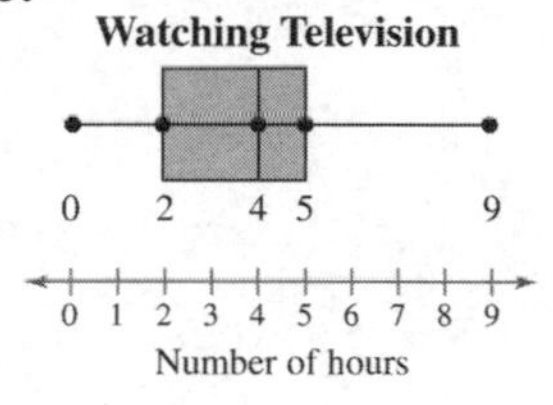

33a. $Q_1 = 3$, $Q_2 = 3.85$, $Q_3 = 5.2$

b.

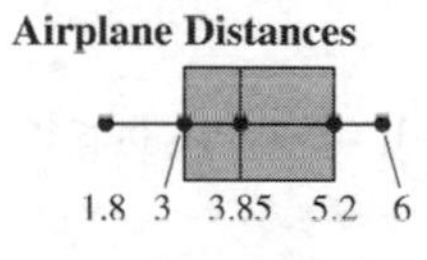

35a. 5 **b.** 50% **c.** 25%

37. A $\Rightarrow z = -1.43$

B $\Rightarrow z = 0$

C $\Rightarrow z = 2.14$

The z-score 2.14 is unusual because it is so large.

39a. Statistics: $x = 75 \Rightarrow z = \dfrac{x-\mu}{\sigma} = \dfrac{75-63}{7} \approx 1.71$

Biology: $x = 25 \Rightarrow z = \dfrac{x-\mu}{\sigma} = \dfrac{25-23}{3.9} \approx 0.51$

b. The student had a better score on the statistics test.

41a. Statistics: $x=78 \Rightarrow z=\dfrac{x-\mu}{\sigma}=\dfrac{78-63}{7} \approx 2.14$

Biology: $x=29 \Rightarrow z=\dfrac{x-\mu}{\sigma}=\dfrac{29-23}{3.9} \approx 1.54$

b. The student had a better score on the statistics test.

43a. $x=34{,}000 \Rightarrow z=\dfrac{x-\mu}{\sigma}=\dfrac{34{,}000-35{,}000}{2{,}250} \approx -0.44$

$x=37{,}000 \Rightarrow z=\dfrac{x-\mu}{\sigma}=\dfrac{37{,}000-35{,}000}{2{,}250} \approx 0.89$

$x=30{,}000 \Rightarrow z=\dfrac{x-\mu}{\sigma}=\dfrac{30{,}000-35{,}000}{2{,}250} \approx -2.22$

The tire with a life span of 30,000 miles has an unusually short life span.

b. $x=30{,}500 \Rightarrow z=\dfrac{x-\mu}{\sigma}=\dfrac{30{,}500-35{,}000}{2{,}250}=-2 \Rightarrow$ 2.5th percentile

$x=37{,}250 \Rightarrow z=\dfrac{x-\mu}{\sigma}=\dfrac{37{,}250-35{,}000}{2{,}250}=1 \Rightarrow$ 84th percentile

$x=35{,}000 \Rightarrow z=\dfrac{x-\mu}{\sigma}=\dfrac{35{,}000-35{,}000}{2{,}250}=0 \Rightarrow$ 50th percentile

45. 72 inches

60% of the heights are below 72 inches.

47. $x=74 \Rightarrow z=\dfrac{x-\mu}{\sigma}=\dfrac{74-69.9}{3.0} \approx 1.37$

$x=62 \Rightarrow z=\dfrac{x-\mu}{\sigma}=\dfrac{62-69.9}{3.0} \approx -2.63$

$x=80 \Rightarrow z=\dfrac{x-\mu}{\sigma}=\dfrac{80-69.9}{3.0} \approx 3.37$

The height of 62 inches is unusual due to a rather small z-score. The height of 80 inches is very unusual due to a rather large z-score.

49. $x=71.1 \Rightarrow z=\dfrac{x-\mu}{\sigma}=\dfrac{71.1-71.1}{3.0}=0.0$

Approximately the 50th percentile.

51a.

27 28 31 32 32 33 35 36 36 36 36 37 38 39 39 40 40 40 41 41
41 42 42 42 42 42 42 43 43 43 44 44 45 45 46 47 47 47 47 47
48 48 48 48 48 | 49 49 49 49 49 49 50 50 51 51 51 51 51 51 52
52 52 53 53 54 | 54 54 54 54 54 | 54 54 55 56 56 56 57 57 57 59
59 59 60 60 60 | 61 61 61 62 62 | 63 63 63 63 64 | 65 67 68 74 82

Q_1 Q_2 Q_3

Min = 27, $Q_1=42$, $Q_2=49$, $Q_3=56$, Max = 82

b.

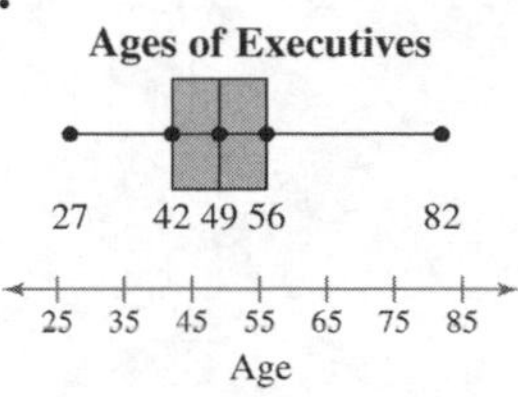

c. Half of the executives are between 42 and 56 years old.

d. About 49 years old because half of the executives are older and half are younger.

e. The age groups 20-29, 70-79, and 80-89 would all be considered unusual because they are more than two standard deviations from the mean.

53.

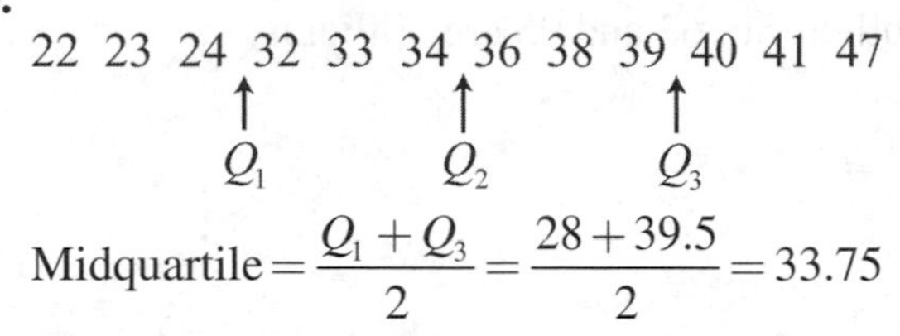

$$\text{Midquartile} = \frac{Q_1 + Q_3}{2} = \frac{28 + 39.5}{2} = 33.75$$

55.

13.4 15.2 15.6 16.7 17.2 18.7 19.7 19.8 19.8 20.8 21.4 22.9 28.7 30.1 31.9

(Q_1 = 16.7, Q_2 = 19.8, Q_3 = 22.9)

$$\text{Midquartile} = \frac{Q_1 + Q_3}{2} = \frac{16.7 + 22.9}{2} = 19.8$$

57.

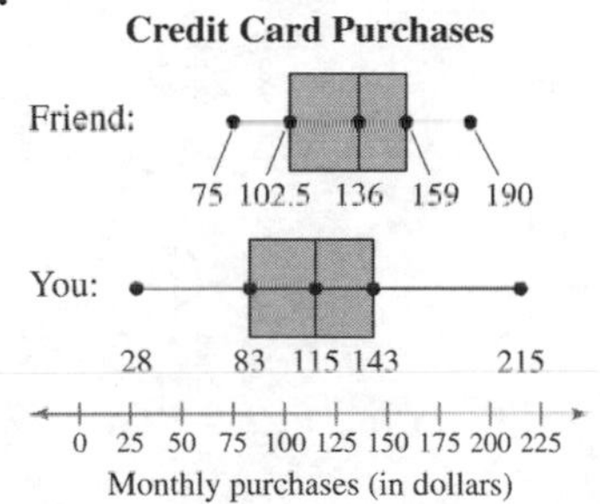

Your distribution is symmetric and your friend's distribution is uniform.

59.
$$\text{Percentile} = \frac{\text{Number of data values less than } x}{\text{Total number of data values}} \cdot 100$$
$$= \frac{30}{73} \cdot 100 \approx 41\text{st percentile}$$

61a.

lower half / upper half

62 72 72 74 75 75 75 76 78 80 80 95

Q_1 (between 72 and 74), Q_2 (between 75 and 75), Q_3 (between 78 and 80)

$Q_1 = 73,\ Q_2 = 75,\ Q_3 = 79$

$\text{IQR} = Q_3 - Q_1 = 79 - 73 = 6$

$1.5 \times \text{IQR} = 9$

$Q_1 - (1.5 \times \text{IQR}) = 73 - 9 = 64$

$Q_3 + (1.5 \times \text{IQR}) = 79 + 9 = 88$

Any values less than 64 or greater than 88 is an outlier. So, 62 and 95 are outliers.

b.

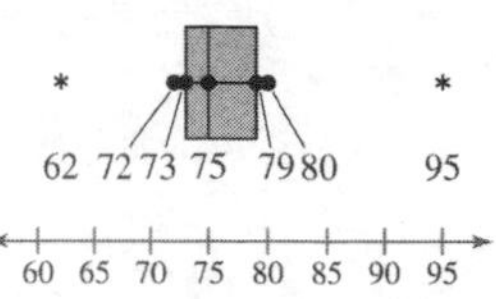

63a. **Summary statistics:**

Column	Min	Q_1	Median	Q_3	Max
Weight (in pounds)	165	230	262.5	294	395

b.

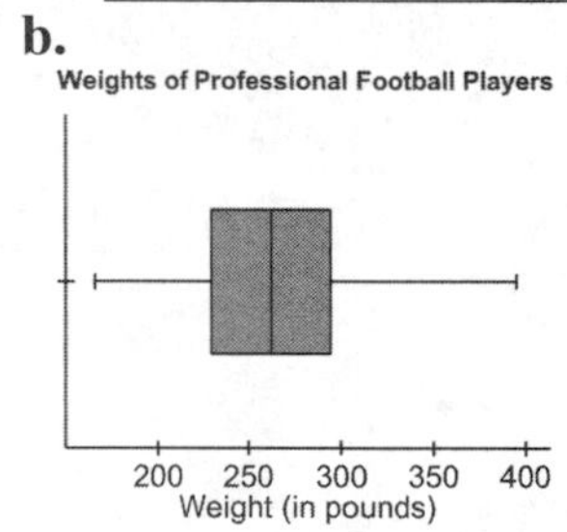

c.

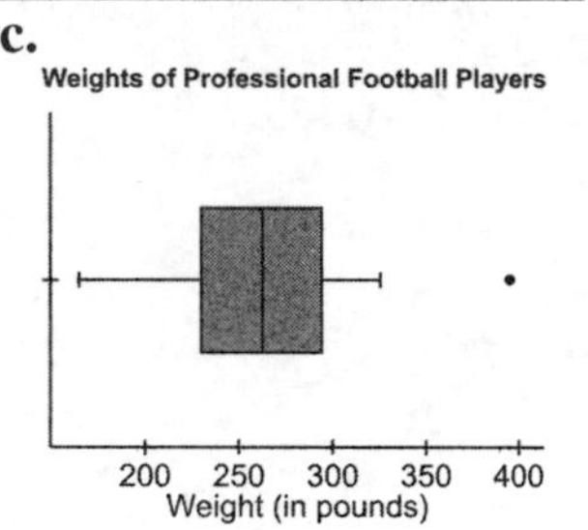

CHAPTER 2 REVIEW EXERCISE SOLUTIONS

1. $\text{Class width} = \dfrac{\text{Max} - \text{Min}}{\text{Number of classes}} = \dfrac{30 - 8}{5} = 4.4 \Rightarrow 5$

Class	Midpoint, x	Boundaries	Frequency, f	Relative frequency	Cumulative frequency
8-12	10	7.5-12.5	2	0.10	2
13-17	15	12.5-17.5	10	0.50	12
18-22	20	17.5-22.5	5	0.25	17
23-27	25	22.5-27.5	1	0.05	18
28-32	30	27.5-32.5	2	0.10	20
			$\sum f = 20$	$\sum \frac{f}{n} = 1$	

3. $\text{Class width} = \dfrac{\text{Max} - \text{Min}}{\text{Number of classes}} = \dfrac{12.10 - 11.86}{7} \approx 0.03 \Rightarrow 0.04$

Class	Midpoint	Frequency, f	Relative frequency
11.86-11.89	11.875	3	0.125
11.90-11.93	11.915	5	0.208
11.94-11.97	11.955	8	0.333
11.98-12.01	11.995	7	0.292
12.02-12.05	12.035	0	0
12.06-12.09	12.075	0	0
12.10-12.13	12.115	1	0.042
		$\sum f = 24$	$\sum \frac{f}{n} = 1$

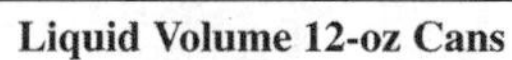

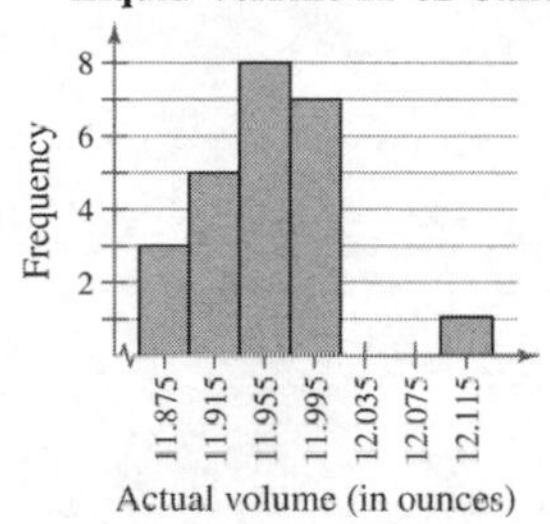

5.

Class	Midpoint	Frequency, f	Cumulative frequency
79-93	86	9	9
94-108	101	12	21
109-123	116	5	26
124-138	131	3	29
139-153	146	2	31
154-168	161	1	32
		$\sum f = 32$	

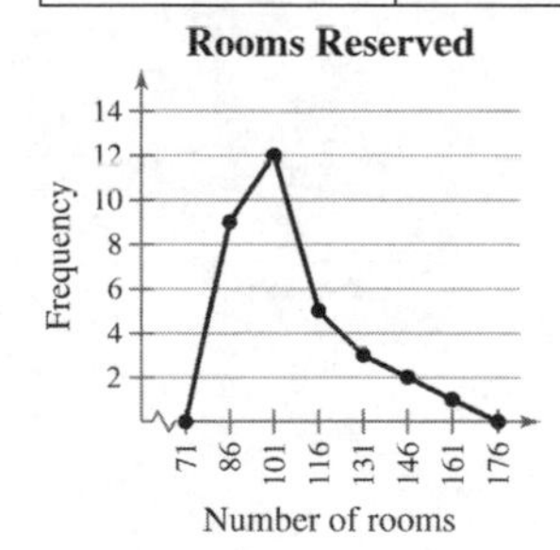

7.

		Key: 1|0 = 10
1	0 0	
2	0 0 2 5 5	
3	0 3 4 5 5 8	
4	1 2 4 4 7 8	
5	2 3 3 7 9	
6	1 1 5	
7	1 5	
8	9	

9.

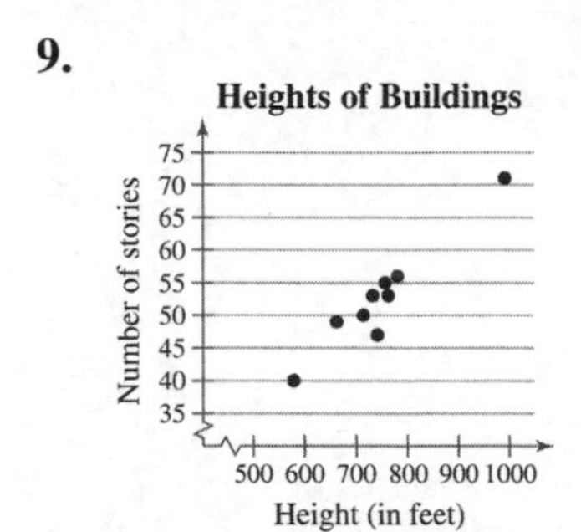

The number of stories appears to increase with height.

11.

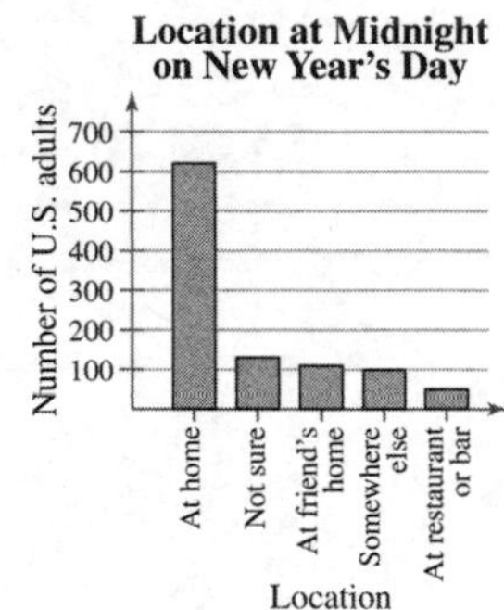

13. $\bar{x} = \dfrac{\sum x}{n} = \dfrac{291.5}{10} = 29.15$

25.0 26.0 27.0 27.5 29.5 29.5 30.5 31.5 32.0 33.0

median = 29.5

Mode = 29.5 (occurs 2 times)

15.

Midpoint, x	**Frequency, f**	$x \cdot f$
10	2	20
15	10	15
20	5	100
25	1	25
30	2	60
	$n = 20$	$\sum(x \cdot f) = 355$

$$\bar{x} = \frac{\Sigma(x \cdot f)}{n} = \frac{355}{20} \approx 17.8$$

17.

Source	**Score, x**	**Weight, w**	$x \cdot w$
Test 1	78	0.15	11.7
Test 2	72	0.15	10.8
Test 3	86	0.15	12.9
Test 4	91	0.15	13.65
Test 5	87	0.15	13.05
Test 6	80	0.25	20
		$\sum w = 1$	$\sum(x \cdot w) = 82.1$

$$\bar{x} = \frac{\Sigma(x \cdot w)}{\Sigma w} = \frac{82.1}{1} = 82.1$$

19. Skewed

21. Skewed left

23. Median, because the mean is to the left of the median in a skewed left distribution.

25. Range = Max – Min = 8.26 – 5.46 = \$2.80

27. $\mu = \dfrac{\sum x}{N} = \dfrac{96}{14} \approx 6.9$

x	$x - \mu$	$(x - \mu)^2$
4	−2.9	8.41
2	−4.9	24.01
9	2.1	4.41
12	5.1	26.01
15	8.1	65.61
3	−3.9	15.21
6	−0.9	0.81
8	1.1	1.21
1	−5.9	34.81
4	−2.9	8.41
14	7.1	50.41
12	5.1	26.01
3	−3.9	15.21
3	−3.9	15.21
$\sum x = 96$	$\sum (x - \mu) \approx 0$	$\sum (x - \mu)^2 = 295.74$

$$\sigma = \sqrt{\frac{\sum (x - \mu)^2}{N}} = \sqrt{\frac{295.74}{14}} \approx 4.6$$

29. $\bar{x} = \dfrac{\sum x}{n} = \dfrac{36,801}{15} = 2453.4$

x	$x - \bar{x}$	$(x - \bar{x})^2$
2445	−8.4	70.56
2940	486.6	236,779.56
2399	−54.4	2959.36
1960	−493.4	243,443.56
2421	−32.4	1049.76
2940	486.6	236,779.56
2657	203.6	41,452.96
2153	−300.4	90,240.16
2430	−23.4	547.56
2278	−175.4	30,765.16
1947	−506.4	256,440.96
2383	−70.4	4956.16
2710	256.6	65,843.56
2761	307.6	94,617.76
2377	−76.4	5836.96
$\sum x = 36,801$	$\sum (x - \bar{x}) = 0$	$\sum (x - \bar{x})^2 = 1,311,783.6$

$$s = \sqrt{\frac{\sum (x - \bar{x})^2}{n - 1}} = \sqrt{\frac{1,311,783.6}{14}} \approx 306.1$$

31. 99.7% of the distribution lies within 3 standard deviations of the mean.

$$\mu - 3\sigma = 49 - (3)(2.50) = 41.5$$

$$\mu + 3\sigma = 49 + (3)(2.50) = 56.5$$

99.7% of the distribution lies between \$41.50 and \$56.50.

33. $(\bar{x} - 2s,\ \bar{x} + 2s) \rightarrow (20,\ 52)$ are 2 standard deviations from the mean.

$$1 - \frac{1}{k^2} = 1 - \frac{1}{(2)^2} = 1 - \frac{1}{4} = 0.75$$

At least (40)(0.75) = 30 customers have a mean sale between \$20 and \$52.

35.

x	f	xf	$x - \bar{x}$	$(x - \bar{x})^2$	$(x - \bar{x})^2 f$
0	1	0	−2.5	6.25	6.25
1	8	8	−1.5	2.25	18.00
2	13	26	−0.5	0.25	3.25
3	10	30	0.5	0.25	2.50
4	5	20	1.5	2.25	11.25
5	3	15	2.5	6.25	18.75
	$n = 40$	$\sum xf = 99$			$\sum (x - \bar{x})^2 f = 60$

$$\bar{x} = \frac{\sum xf}{n} = \frac{99}{40} \approx 2.5$$

$$s = \sqrt{\frac{\sum (x - \bar{x})^2 f}{n - 1}} = \sqrt{\frac{60}{39}} \approx 1.2$$

37.

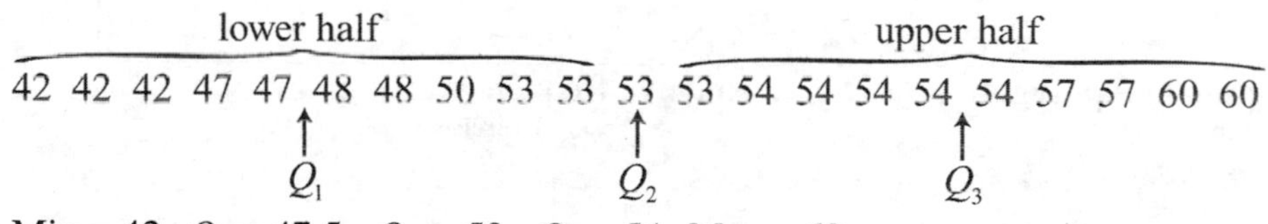

Min = 42, $Q_1 = 47.5$, $Q_2 = 53$, $Q_3 = 54$, Max = 60

39.

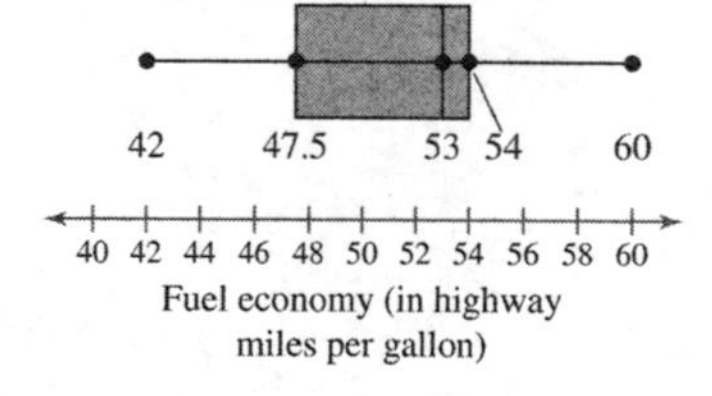

41.

$$\overbrace{25.0\ 26.0\ 27.0\ 27.5\ 29.5}^{\text{lower half}}\ \overbrace{29.5\ 30.5\ 31.5\ 32.0\ 33.0}^{\text{upper half}}$$

$Q_1 = 27.0$, $Q_2 = 29.5$, $Q_3 = 31.5$

$\text{IQR} = Q_3 - Q_1 = 31.5 - 27.0 = 4.5$

43. The 65th percentile means that 65% had a test grade of 75 or less. So, 35% scored higher than 75.

45. $z = \dfrac{16{,}500 - 11{,}830}{2370} = 1.97$

47. $z = \dfrac{18{,}000 - 11{,}830}{2370} = 2.60$

CHAPTER 2 QUIZ SOLUTIONS

1a. Class width $= \dfrac{\text{Max} - \text{Min}}{\text{Number of classes}} = \dfrac{157 - 101}{5} = 11.2 \Rightarrow 12$

Class	Midpoint	Class boundaries	Frequency, f	Relative frequency	Cumulative frequency
101-112	106.5	100.5-112.5	3	0.12	3
113-124	118.5	112.5-124.5	11	0.44	14
125-136	130.5	124.5-136.5	7	0.28	21
137-148	142.5	136.5-148.5	2	0.08	23
149-160	154.5	148.5-160.5	2	0.08	25
			$\sum f = 25$	$\sum \frac{f}{n} = 1$	

b. Frequency histogram and polygon

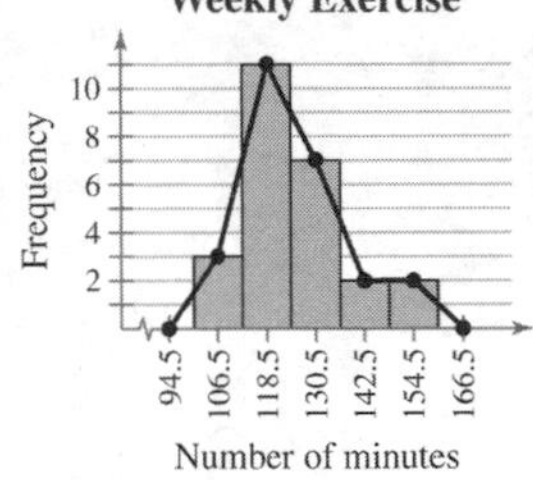

c. Relative frequency histogram

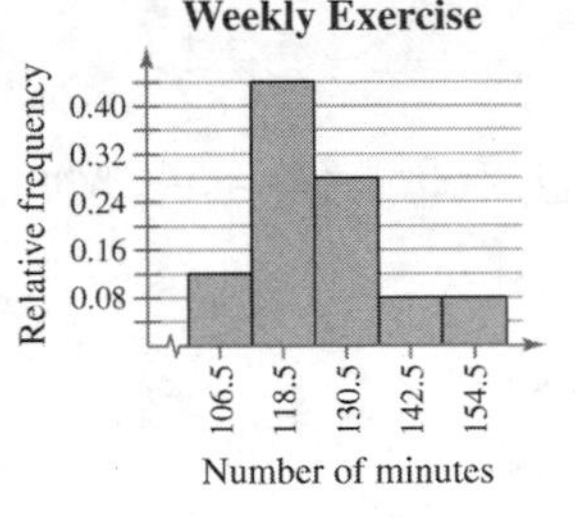

d. Skewed

e.

10	1 8
11	1 4 6 7 8 9 9
12	0 0 3 3 4 7 7 8
13	1 1 2 5 9 9
14	
15	0 7

Key: $10|8 = 108$

f.

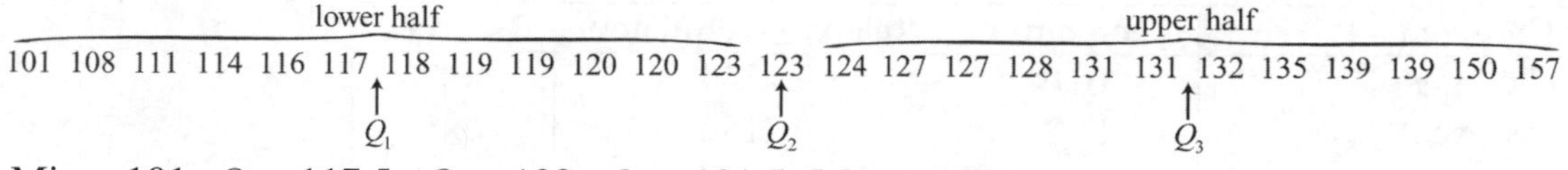

Min = 101, $Q_1 = 117.5$, $Q_2 = 123$, $Q_3 = 131.5$, Max = 157

Weekly Exercise

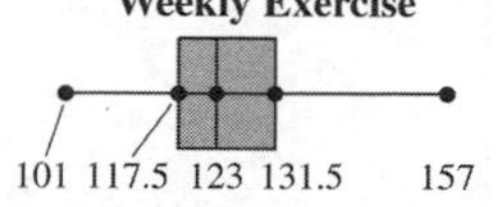

100 110 120 130 140 150 160

Number of minutes

g.

Weekly Exercise

Cumulative frequency

25 20 15 10 5

100.5 112.5 124.5 136.5 148.5 160.5

Number of minutes

2.

Midpoint, x	Frequency, f	xf	$x-\bar{x}$	$(x-\bar{x})^2$	$(x-\bar{x})^2 f$
106.5	3	319.5	–18.7	349.69	1049.07
118.5	11	1303.5	–6.7	44.89	493.79
130.5	7	913.5	5.3	28.09	196.63
142.5	2	285.0	17.3	299.29	598.58
154.5	2	309.0	29.3	858.49	1716.98
	$n = 25$	$\sum xf = 3130.5$			$\sum(x-\bar{x})^2 f = 4055.05$

$$\bar{x} = \frac{\sum xf}{n} = \frac{3130.5}{25} \approx 125.2$$

$$s = \sqrt{\frac{\sum(x-\bar{x})^2 f}{n-1}} = \sqrt{\frac{4055.05}{24}} \approx 13.0$$

3a.

Category	Frequency	Relative frequency	Degrees
Clothing	10.6	0.1330	48°
Footwear	17.2	0.2158	78°
Equipment	24.9	0.3124	112°
Rec. Transport	27.0	0.3388	122°
	$n = 79.7$	$\sum \frac{f}{n} = 1$	

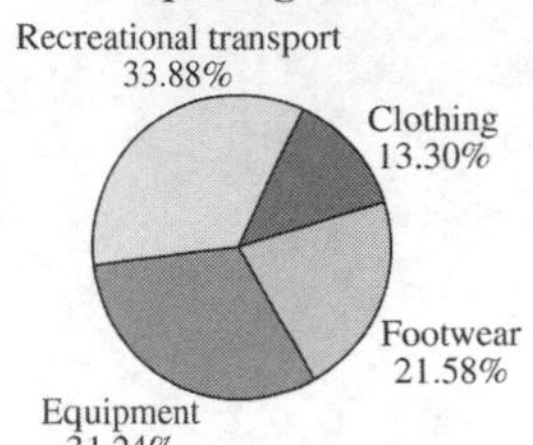

b.

4a. $\bar{x} = \frac{\sum x}{n} = \frac{6013}{8} \approx 751.6$

444 446 667 774 795 908 960 1019

$\text{median} = \frac{774 + 795}{2} = 784.5$

mode = none

The mean best describes a typical salary because there are no outliers.

b. Range = Max – Min = 1019 – 444 = 575

x	$x - \bar{x}$	$(x - \bar{x})^2$
774	22.4	501.76
446	–305.6	93,391.36
1019	267.4	71,502.76
795	43.4	1883.56
908	156.4	24,460.96
667	–84.6	7157.16
444	–307.6	94,617.76
960	208.4	43,430.56
		$\sum(x - \bar{x})^2 = 336{,}945.88$

$s^2 = \frac{\sum(x - \bar{x})^2}{n - 1} = \frac{336{,}945.88}{7} \approx 48{,}135.1$ $\qquad s = \sqrt{\frac{\sum(x - \bar{x})^2}{n - 1}} = \sqrt{\frac{336{,}945.88}{7}} \approx 219.4$

5. $\bar{x}-2s=155{,}000-2\cdot 15{,}000=\$125{,}000$

$\bar{x}+2s=155{,}000+2\cdot 15{,}000=\$185{,}000$

95% of the new home prices fall between \$125,000 and \$185,000.

6a. $x=200{,}000:\ z=\dfrac{x-\mu}{\sigma}=\dfrac{200{,}000-155{,}000}{15{,}000}=3.0\Rightarrow$ unusual price

b. $x=55{,}000:\ z=\dfrac{x-\mu}{\sigma}=\dfrac{55{,}000-155{,}000}{15{,}000}=-6.67\Rightarrow$ very unusual price

c. $x=175{,}000:\ z=\dfrac{x-\mu}{\sigma}=\dfrac{175{,}000-155{,}000}{15{,}000}\approx 1.33\Rightarrow$ not unusual

d. $x=122{,}000:\ z=\dfrac{x-\mu}{\sigma}=\dfrac{122{,}000-155{,}000}{15{,}000}=-2.2\Rightarrow$ unusual price

7a.

59 62 64 65 65 70 70 74 75 75 75 78 79 80 83 84 85 86 86 87 87 87 88 91 92 93 95 95 97 103

(arrows: 74 ↑ Q_1; between 83 and 84 ↑ Q_2; 88 ↑ Q_3)

Min = 59, $Q_1=74$, $Q_2=83.5$, $Q_3=88$, Max = 103

b. IQR = $Q_3-Q_1=88-74=14$

c.

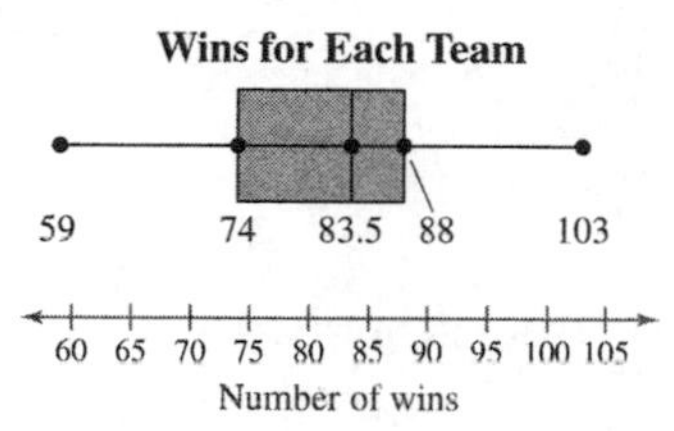

CUMULATIVE REVIEW FOR CHAPTERS 1 AND 2

1. Systematic sampling. A bias may enter this study if the machine makes a consistent error.

2. Simple Random Sampling. A bias of this type of study is that the researchers did not include people without telephones.

3.

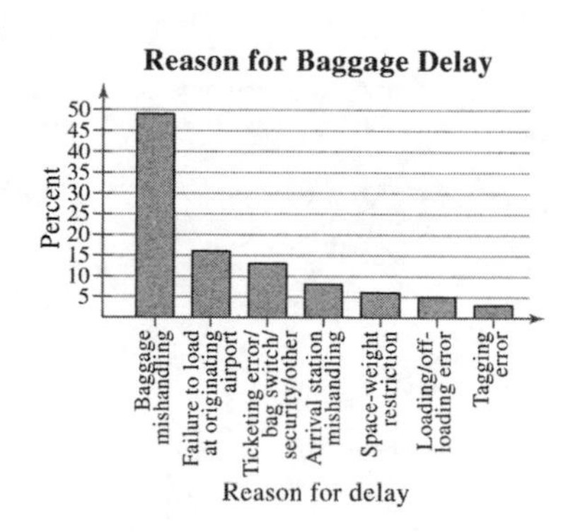

4. \$2,996,106 is a parameter because it is describing the average salary of all Major League Baseball players.

5. 19% is a statistic because it is describing a proportion within a sample of 100 adults.

6a. $\bar{x} = 83{,}500,\ s = \$1500$

$(80{,}500\ 86{,}500) = 83{,}500 \pm 2(1500) \Rightarrow$ 2 standard deviations away form the mean.

Approximately 95% of the electrical engineers will have salaries between (80,500 86,500).

b. 40(0.95) = 38

c. $x = \$90{,}500:\ z = \dfrac{x-\mu}{\sigma} = \dfrac{90{,}500 - 83{,}500}{1500} \approx 4.67$

$x = \$79{,}750:\ z = \dfrac{x-\mu}{\sigma} = \dfrac{79{,}750 - 83{,}500}{1500} = -2.5$

$x = \$82{,}600:\ z = \dfrac{x-\mu}{\sigma} = \dfrac{82{,}600 - 83{,}500}{1500} = -0.6$

The salaries of \$90,500 and \$79,750 are unusual.

7. Population: Collection of the career interests of all college and university students
Sample: Collection of the career interests of the 195 college and university students whose career counselors were surveyed

8. Population: Collection of the life spans of all people
Sample: Collection of the life spans of the 232,606 people in the study

9. Census, because there are only 100 members in the Senate.

10. Experiment, because we want to study the effects of removing recess from schools.

11. Quantitative: The data are at the ratio level.

12. Qualitative: The data are at the nominal level.

13.

0 0 0 0 0 0 0 0 1 1 1 1 2 (Q_1) 2 4 4 4 4 5 6 6 7 8 10
11 14 (Q_2) 14 17 19 21 23 23 23 24 27 30 34 39 (Q_3) 40 46 53 54 56 63 69 71 81 105 105 136

Min = 0, $Q_1 = 2$, $Q_2 = 12.5$, $Q_3 = 39$, Max = 136

b.

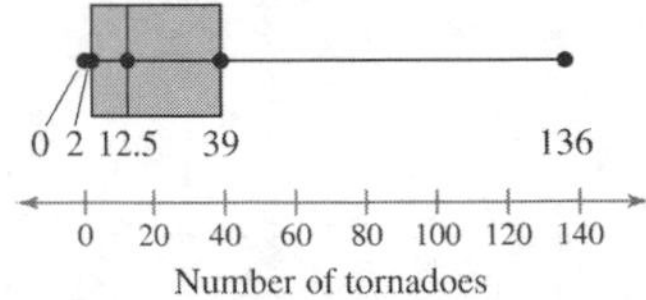

c. The distribution of the number of tornadoes is skewed right.

14.

Source	Score, x	Weight, w	$x \cdot w$
Test 1	85	0.15	12.75
Test 2	92	0.15	13.80
Test 3	84	0.15	12.60
Test 4	89	0.15	13.35
Test 5	91	0.40	36.40
		$\sum w = 1$	$\sum (x \cdot w) = 88.9$

$$\bar{x} = \frac{\sum (x \cdot w)}{\sum w} = \frac{88.9}{1} = 88.9$$

15a. $\bar{x} = \dfrac{49.4}{9} \approx 5.49$

3.4 3.9 4.2 4.6 (5.4) 6.5 6.8 7.1 7.5

median = 5.4

mode = none

Both the mean and median accurately describe a typical American alligator tail length. (Answers will vary.)

b. Range – Max – Min – 7.5 – 3.4 = 4.1

x	$x - \bar{x}$	$(x - \bar{x})^2$
3.4	–2.09	4.3681
3.9	–1.59	2.5281
4.2	–1.29	1.6641
4.6	–0.89	0.7921
5.4	–0.09	0.0081
6.5	1.01	1.0201
6.8	1.31	1.7161
7.1	1.61	2.5921
7.5	2.01	4.0401
		$\sum (x - \bar{x})^2 = 18.7289$

$$s^2 = \frac{\sum (x - \bar{x})^2}{n-1} = \frac{18.7289}{8} \approx 2.34$$

$$s = \sqrt{\frac{\sum (x - \bar{x})^2}{n-1}} = \sqrt{\frac{18.7289}{8}} \approx 1.53$$

The maximum difference in alligator tail lengths is 4.1 feet and the standard deviation of tail lengths is about 1.53 feet.

16a. The number of deaths due to heart disease for women will continue to decrease.

b. The study was only conducted over the past 5 years and deaths may not decrease in the next year.

17. Class width $= \dfrac{\text{Max} - \text{Min}}{\text{Number of classes}} = \dfrac{65-0}{8} = 8.125 \Rightarrow 9$

Class limits	Midpoint	Class boundaries	Frequency	Relative frequency	Cumulative frequency
0-8	4	–0.5-8.5	8	0.27	8
9-17	13	8.5-17.5	5	0.17	13
18-26	22	17.5-26.5	7	0.23	20
27-35	31	26.5-35.5	3	0.10	23
36-44	40	35.5-44.5	4	0.13	27
45-53	49	44.5-53.5	1	0.03	28
54-62	58	53.5-62.5	0	0.00	28
63-71	67	62.5-71.5	2	0.07	30
			$\sum f = 30$	$\sum \frac{f}{n} = 1$	

18. The distribution is skewed right.

19.

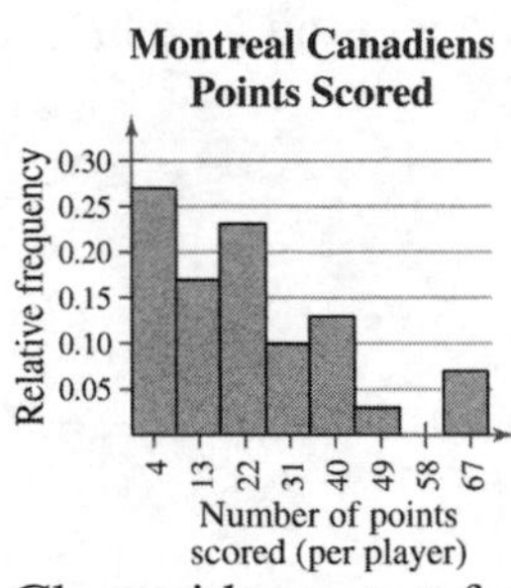

Class with greatest frequency: 0-8
Class with least frequency: 54-62

Probability

3.1 BASIC CONCEPTS OF PROBABILITY AND COUNTING

3.1 Try It Yourself Solutions

1ab. (1)

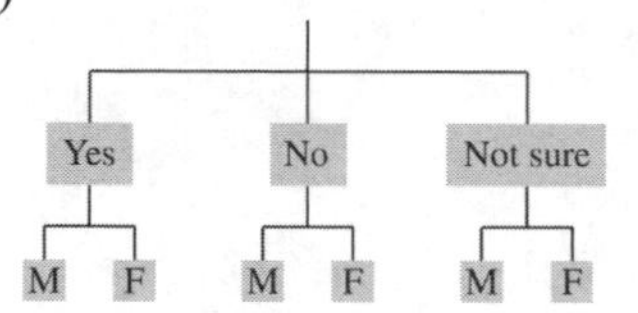

(2)

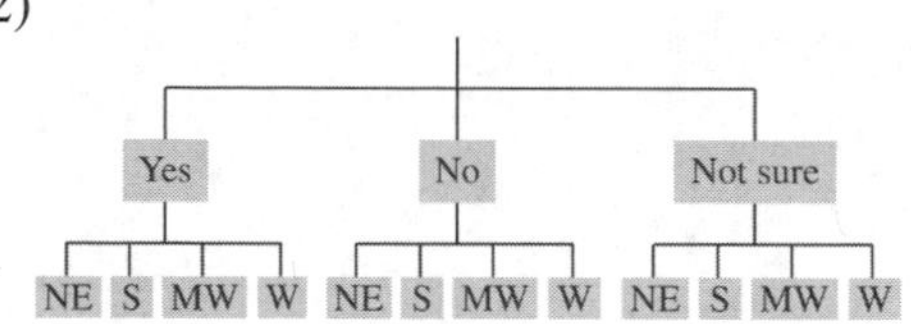

c. (1) 6 outcomes (2) 12 outcomes

d. (1) Let Y = Yes, N = No, NS = Not sure, M = Male and F = Female.
Sample space = $\{YM, YF, NM, NF, NSM, NSF\}$
(2) Let Y = Yes, N = No, NS = Not sure, NE = Northeast, S = South, MW = Midwest, and W = West
Sample space = $\{YNE, YS, YMW, YW, NNE, NS, NMW, NW, NSNE, NSS, NSMW, NSW\}$

2a. (1) Event C has six outcomes: choosing the ages 18, 19, 20, 21, 22, and 23.
(2) Event D has one outcome: choosing the age 20.

b. (1) The event is not a simple event because it consists of more than a single outcome.
(2) The event is a simple event because it consists of a single outcome.

3a. Manufacturer: 4
Size: 2
Color: 5

b. (4)(2)(5) = 40 ways

c. Tree Diagram for Car Selections

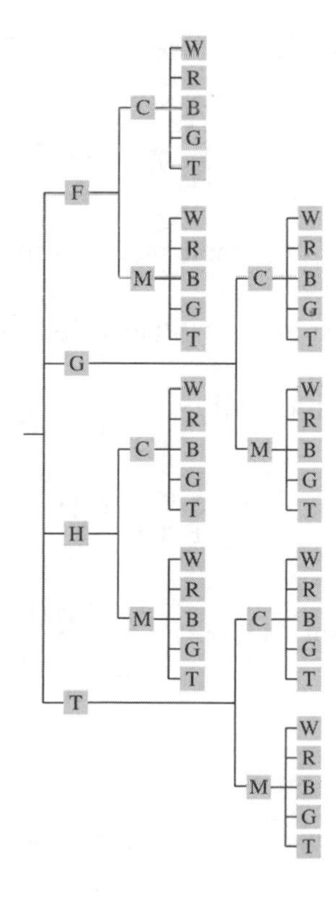

4a. (1) Each letter is an event (26 choices for each).
(2) Each letter is an event (26, 25, 24, 23, 22, and 21 choices).
(3) Each letter is an event (22, 26, 26, 26, 26, and 26 choices).

b. (1) $26 \cdot 26 \cdot 26 \cdot 26 \cdot 26 \cdot 26 = 308{,}915{,}776$
(2) $26 \cdot 25 \cdot 24 \cdot 23 \cdot 22 \cdot 21 = 165{,}765{,}600$
(3) $22 \cdot 26 \cdot 26 \cdot 26 \cdot 26 \cdot 26 = 261{,}390{,}272$

5a. (1) 52 (2) 52 (3) 52

b. (1) 1 (2) 13 (3) 52

c. (1) $P(\text{9 of clubs}) = \dfrac{1}{52} \approx 0.019$

(2) $P(\text{heart}) = \dfrac{13}{52} = 0.25$

(3) $P(\text{diamond, heart, club, or spade}) = \dfrac{52}{52} = 1$

6a. The event is "the next claim processed is fraudulent." The frequency is 4.

b. Total Frequency = 100

c. $P(\text{fraudulent claim}) = \dfrac{4}{100} = 0.04$

7a. Frequency = 54

b. Total of the Frequencies = 1000

c. $P(\text{age 15 to 24}) = \dfrac{54}{1000} = 0.054$

8a. The event is "salmon successfully passing through a dam on the Columbia River."

b. The probability is estimated from the results of an experiment.

c. Empirical probability

9a. $P(\text{age 45 to 54}) = \dfrac{180}{1000} = 0.18$

b. $P(\text{age is not 45 to 54}) = 1 - \dfrac{180}{1000} = \dfrac{820}{1000} = 0.82$

c. $\dfrac{820}{1000}$ or 0.82

10a. There are 5 outcomes in the event: {T1, T2, T3, T4, T5}.

b. $P(\text{tail and less than } 6) = \dfrac{5}{16} \approx 0.313$

11a. $10 \cdot 10 \cdot 10 \cdot 10 \cdot 10 \cdot 10 \cdot 10 = 10{,}000{,}000$

b. $\dfrac{1}{10{,}000{,}000}$

3.1 EXERCISE SOLUTIONS

1. An outcome is the result of a single trial in a probability experiment, whereas an event is a set of one or more outcomes.

3. It is impossible to have more than a 100% chance of rain.

5. The law of large numbers states that as an experiment is repeated over and over, the probabilities found in the experiment will approach the actual probabilities of the event. Examples will vary.

7. False. If you roll a six-sided die six times, the probability of rolling an even number at least once is approximately 0.984.

9. False. A probability of less than 0.05 indicates an unusual event.

11. b　　　**13.** c

15. {A, B, C, D, E, F, G, H, I, J, K, L, M, N, O, P, Q, R, S, T, U, V, W, X, Y, Z}; 26

17. {A♥, K♥, Q♥, J♥, 10♥, 9♥, 8♥, 7♥, 6♥, 5♥, 4♥, 3♥, 2♥,
A♦, K♦, Q♦, J♦, 10♦, 9♦, 8♦, 7♦, 6♦, 5♦, 4♦, 3♦, 2♦,
A♠, K♠, Q♠, J♠, 10♠, 9♠, 8♠, 7♠, 6♠, 5♠, 4♠, 3♠, 2♠,
A♣, K♣, Q♣, J♣, 10♣, 9♣, 8♣, 7♣, 6♣, 5♣, 4♣, 3♣, 2♣}; 52

19.

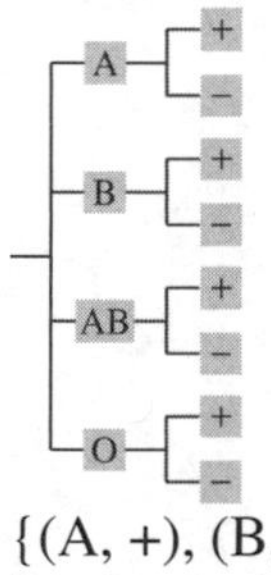

A + –
B + –
AB + –
O + –

{(A, +), (B, +), (AB, +), (O, +), (A, –), (B, –), (AB, –), (O, –)}, where (A, +) represents positive Rh-factor with blood type A and (A, –) represents negative Rh-factor with blood type A; 8.

21. 1 outcome; simple event because it is an event that consists of a single outcome.

23. Ace = {Ace of hearts, Ace of spades, Ace of clubs, Ace of diamonds}; 4 outcomes
Not a simple event because it is an event that consists of more than a single outcome.

25. $(12)(17) = 204$

27. $(9)(10)(10)(5) = 4500$

29. $P(A) = \frac{1}{12} \approx 0.083$

31. $P(C) = \frac{8}{12} = \frac{2}{3} \approx 0.667$

33. $P(E) = \frac{5}{12} \approx 0.417$

35. Empirical probability because company records were used to calculate the frequency of a washing machine breaking down.

37. $P(\text{less than } 1000) = \frac{999}{6296} \approx 0.159$

39. $P(\text{number divisible by } 1000) = \frac{6}{6296} \approx 0.000953$

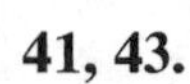

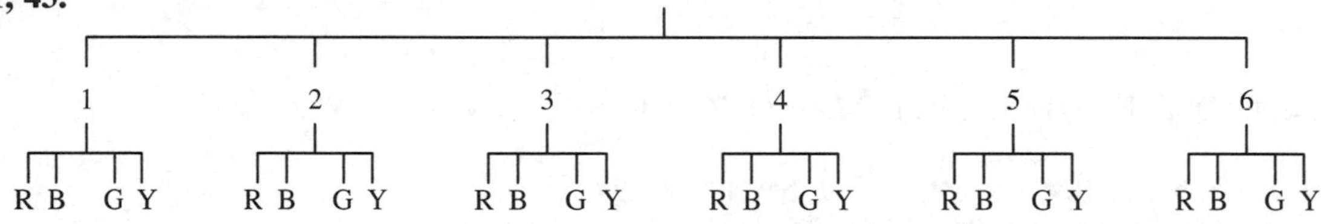

41. $P(A) = \frac{1}{24} \approx 0.042$; unusual

43. $P(C) = \frac{5}{24} \approx 0.208$; not unusual

45. (a) $10 \cdot 10 \cdot 10 = 1000$ (b) $\frac{1}{1000} = 0.001$ (c) $1 - \frac{1}{1000} = \frac{999}{1000} = 0.999$

47. {(SSS), (SSR), (SRS), (SRR), (RSS), (RSR), (RRS), (RRR)}

49. {(SSR), (SRS), (RSS)}

51. Let S = sunny day and R = rainy day.

(a)

SSSS
SSSR
SSRS
SSRR
SRSS
SRSR
SRRS
SRRR
RSSS
RSSR
RSRS
RSRR
RRSS
RRSR
RRRS
RRRR

(b) {(SSSS), (SSSR), (SSRS), (SSRR), (SRSS), (SRSR), (SRRS), (SRRR), (RSSS), (RSSR), (RSRS), (RSRR), (RRSS), (RRSR), (RRRS), (RRRR)}

(c) {(SSSR), (SSRS), (SRSS), (RSSS)}

53. $P(\text{voted in 2009 Gubernatorial election}) = \dfrac{1,982,432}{4,964,024} \approx 0.399$

55. $P(\text{between 18 and 20}) = \dfrac{5.8}{146.2} \approx 0.040$

57. $P(\text{not between 21 and 24}) = 1 - \dfrac{9.3}{146.2} \approx 1 - 0.064 \approx 0.936$

59. $P(\text{doctorate}) = \dfrac{3}{91} \approx 0.033$

61. $P(\text{master's}) = \dfrac{25}{91} \approx 0.275$

63. Yes; the event in Exercise 55 can be considered unusual because its probability is 0.05 or less.

65. (a) $P(\text{pink}) = \dfrac{2}{4} = 0.5$ (b) $P(\text{red}) = \dfrac{1}{4} = 0.25$ (c) $P(\text{white}) = \dfrac{1}{4} = 0.25$

67. $P(\text{service industry}) = \dfrac{115,498}{145,363} \approx 0.795$

69. $P(\text{not in service industry}) = 1 - P(\text{service industry}) \approx 1 - 0.795 = 0.205$

71. (a) $P(\text{at least } 51) = \frac{27}{120} = 0.225$

(b) $P(\text{between 20 and 30 inclusive}) = \frac{16}{120} \approx 0.133$

(c) $P(\text{more than } 69) = \frac{2}{120} \approx 0.017$; This event is unusual because its probability is 0.05 or less.

73. The probability of randomly choosing a tea drinker who does not have a college degree

75. (a)

Sum	Outcomes	*P*(sum)	Probability
2	(1, 1)	1/36	0.028
3	(1, 2), (2, 1)	2/36	0.056
4	(1, 3), (2, 2,), (3, 1)	3/36	0.083
5	(1, 4), (2, 3), (3, 2), (4, 1)	4/36	0.111
6	(1, 5), (2, 4), (3, 3), (4, 2), (5, 1)	5/36	0.139
7	(1, 6), (2, 5), (3, 4), (4, 3), (5, 2), (6, 1)	6/36	0.167
8	(2, 6), (3, 5), (4, 4), (5, 3), (6, 2)	5/36	0.139
9	(3, 6), (4, 5), (5, 4), (6, 3)	4/36	0.111
10	(4, 6), (5, 5), (6, 4)	3/36	0.083
11	(5, 6), (6, 5)	2/36	0.056
12	(6, 6)	1/36	0.028

(b) Answers will vary.
(c) Answers will vary.

77. The first game; the probability of winning the second game is $\frac{1}{11} \approx 0.091$, which is less than $\frac{1}{10}$.

79. 13 : 39 = 1 : 3

81. p = number of successful outcomes
q = number of unsuccessful outcomes

$$P(A) = \frac{\text{number of successful outcomes}}{\text{total number of outcomes}} = \frac{p}{p+q}$$

3.2 CONDITIONAL PROBABILITY AND THE MULTIPLICATION RULE

3.2 Try It Yourself Solutions

1a. (a) 30 and 102 (2) 11 and 50

b. $P(\text{does not have gene}) = \frac{30}{102} \approx 0.294$ (2) $P(\text{does not have gene} \mid \text{normal IQ}) = \frac{11}{50} = 0.22$

2a. (1) Yes (2) No

b. (1) Dependent (2) Independent

3a. (1) Independent (2) Dependent

b. (1) Let A = {first salmon swims successfully through the dam}
B = {second salmon swims successfully through the dam}
$P(A \text{ and } B) = P(A) \cdot P(B) = (0.85) \cdot (0.85) \approx 0.723$

(2) Let A = {selecting a heart}
B = {selecting a second heart}

$$P(A \text{ and } B) = P(A) \cdot P(B|A) = \left(\frac{13}{52}\right) \cdot \left(\frac{12}{51}\right) \approx 0.059$$

4a. (1) Find probability of the event (2) Find probability of the event
(3) Find probability of the compliment of the event

b. (1) P(3 surgeries are successful) = $(0.90) \cdot (0.90) \cdot (0.90) = 0.729$

(2) P(none are successful) = $(0.10) \cdot (0.10) \cdot (0.10) = 0.001$

(3) P(at least one rotator cuff surgery is successful) = $1 - P$(none are successful)
$= 1 - 0.001 = 0.999$

c. (1) The event cannot be considered unusual because its probability is not less than or equal to 0.05.
(2) The event can be considered unusual because its probability is less than or equal to 0.05.
(3) The event cannot be considered unusual because its probability is not less than or equal to 0.05.

5a. (1),(2) A = {is female}; B = {works in health field}

b. (1) $P(A \text{ and } B) = P(A) \cdot P(B|A) = (0.65) \cdot (0.25)$

(2) $P(A \text{ and } B') = P(A) \cdot P(B'|A) = P(A) \cdot (1 - P(B|A)) = (0.65) \cdot (0.75)$

c. (1) 0.136 (2) 0.488

3.2 EXERCISE SOLUTIONS

1. Two events are independent if the occurrence of one of the events does not affect the probability of the occurrence of the other event, whereas two events are dependent if the occurrence of one of the events does affect the probability of the occurrence of the other event.

3. The notation $P(B|A)$ means the probability of B, given A.

5. False. If two events are independent, $P(A|B) = P(A)$.

7. These events are independent because the outcome of the first card drawn does not affect the outcome of the second card drawn.

9. These events are dependent because the outcome of a father having hazel eyes affects the outcome of a daughter having hazel eyes.

11. These events are dependent because the sum of the rolls depends on which numbers were rolled first and second.

13. Events: moderate to severe sleep apnea, high blood pressure
These events are dependent because people with moderate to severe sleep apnea are more likely to have high blood pressure.

15. Events: exposure to aluminum, Alzheimer's disease
These events are independent because exposure to everyday sources of aluminum does not cause Alzheimer's disease.

17. Let A = {have mutated BRCA gene} and B = {develop breast cancer}.
So, $P(B)=\frac{1}{8}$, $P(A)=\frac{1}{600}$, and $P(B|A)=\frac{6}{10}$.

(a) $P(B|A)=\frac{6}{10}=0.6$

(b) $P(A \text{ and } B) = P(A)\cdot P(B|A)=\left(\frac{1}{600}\right)\cdot\left(\frac{6}{10}\right)=0.001$

(c) Dependent, because $P(B|A) \neq P(B)$.

19. Let A = {own a computer} and B = {summer vacation this year}.

(a) $P(B')=\frac{45}{146}\approx 0.308$

(b) $P(A)=\frac{115}{146}\approx 0.788$

(c) $P(B|A)=\frac{87}{115}\approx 0.757$

(d) $P(A \text{ and } B) = P(A)\cdot P(B|A) = \left(\frac{115}{146}\right)\left(\frac{87}{115}\right)=0.596$

(e) Dependent, because $P(B|A) \neq P(B)$.

21. Let A = {pregnant} and B = {multiple births}. So, $P(A) = 0.37$ and $P(B|A)=0.25$.

(a) $P(A \text{ and } B) = P(A)\cdot P(B|A) = (0.37)\cdot(0.25)\approx 0.093$

(b) $P(B'|A)=1-P(B|A)=1-0.25=0.75$

(c) No, this is not unusual because the probability is not less than or equal to 0.05.

23. Let A = {household in U.S. has a computer} and B = {has Internet access}.
$P(A \text{ and } B) = P(A)\cdot P(B|A) = (0.8)\cdot(0.92)\approx 0.745$

25. Let A = {first person can wiggle their ears} and B = {second person can wiggle their ears}.

(a) $P(A \text{ and } B) = P(A) \cdot P(B|A) = \left(\frac{130}{1000}\right) \cdot \left(\frac{129}{999}\right) \approx 0.017$

(b) $P(A' \text{ and } B') = P(A') \cdot P(B'|A') = \left(\frac{870}{1000}\right) \cdot \left(\frac{869}{999}\right) \approx 0.757$

(c) $P(\text{at least one can wiggle their ears}) = 1 - P(A' \text{ and } B') \approx 1 - 0.757 = 0.243$

(d) The event in part (a) is unusual because its probability is less than or equal to 0.05.

27. Let A = {have one month's income or more} and B = {male}.

(a) $P(A) = \frac{138}{287} \approx 0.481$

(b) $P(A'|B) = \frac{66}{142} \approx 0.465$

(c) $P(B'|A) = \frac{62}{138} \approx 0.449$

(d) Dependent, because $P(A') \approx 0.519 \neq 0.465 \approx P(A'|B)$

29. (a) $P(\text{all five have B+}) = (0.09) \cdot (0.09) \cdot (0.09) \cdot (0.09) \cdot (0.09) \approx 0.00000590$

(b) $P(\text{none have B+}) = (0.91) \cdot (0.91) \cdot (0.91) \cdot (0.91) \cdot (0.91) \approx 0.624$

(c) $P(\text{at least one has B+}) = 1 - P(\text{none have B+}) \approx 1 - 0.624 = 0.376$

31. (a) $P(\text{first question correct}) = 0.25$

(b) $P(\text{first two questions correct}) = (0.25) \cdot (0.25) \approx 0.063$

(c) $P(\text{all five questions correct}) = (0.25) \cdot (0.25) \cdot (0.25) \cdot (0.25) \cdot (0.25) \approx 0.000977$

(d) $P(\text{none correct}) = (0.75) \cdot (0.75) \cdot (0.75) \cdot (0.75) \cdot (0.75) \approx 0.237$

(e) $P(\text{at least one correct}) = 1 - P(\text{none correct}) \approx 1 - 0.237 = 0.763$

33. (a) $P(\text{all three products came form the third factory}) = \frac{25}{110} \cdot \frac{24}{109} \cdot \frac{23}{108} \approx 0.011$

(b) $P(\text{none of the three products came from the third factory}) = \frac{85}{110} \cdot \frac{84}{109} \cdot \frac{83}{108} \approx 0.458$

35.
$$P(A|B) = \frac{P(A) \cdot P(B|A)}{P(A) \cdot P(B|A) + P(A') \cdot P(B|A')}$$
$$= \frac{\left(\frac{2}{3}\right) \cdot \left(\frac{1}{5}\right)}{\left(\frac{2}{3}\right) \cdot \left(\frac{1}{5}\right) + \left(\frac{1}{3}\right) \cdot \left(\frac{1}{2}\right)} = \frac{\frac{2}{15}}{\frac{3}{10}} = \frac{4}{9} \approx 0.444$$

37.
$$P(A|B) = \frac{P(A) \cdot P(B|A)}{P(A) \cdot P(B|A) + P(A') \cdot P(B|A')}$$
$$= \frac{(0.25) \cdot (0.3)}{(0.25) \cdot (0.3) + (0.75) \cdot (0.5)} = \frac{0.075}{0.45} \approx 0.167$$

39. $P(A)=\dfrac{1}{200}=0.005$

$P(B|A)=0.8$

$P(B|A')=0.05$

(a) $$P(A|B)=\frac{P(A)\cdot P(B|A)}{P(A)\cdot P(B|A)+P(A')\cdot P(B|A')}$$
$$=\frac{(0.005)\cdot(0.8)}{(0.005)\cdot(0.8)+(0.995)\cdot(0.05)}=\frac{0.004}{0.05375}\approx 0.074$$

(b) $$P(A'|B')=\frac{P(A')\cdot P(B'|A')}{P(A')\cdot P(B'|A')+P(A)\cdot P(B'|A)}$$
$$=\frac{(0.995)\cdot(0.95)}{(0.995)\cdot(0.95)+(0.005)\cdot(0.2)}=\frac{0.94525}{0.94625}\approx 0.999$$

41. Let A = {flight departs on time} and B = {flight arrives on time}.

$$P(A|B)=\frac{P(A \text{ and } B)}{P(B)}=\frac{0.83}{0.87}\approx 0.954$$

3.3 THE ADDITION RULE

3.3 Try It Yourself Solutions

1a. (1) None of the statements are true.
(2) None of the statements are true.
(3) All of the statements are true.

b. (1) A and B are not mutually exclusive.
(2) A and B are not mutually exclusive.
(3) A and B are mutually exclusive.

2a. (1) Mutually exclusive (2) Not mutually exclusive

b. (1) Let A = {6} and B = {odd}.

$P(A)=\dfrac{1}{6}$ and $P(B)=\dfrac{3}{6}=\dfrac{1}{2}$

(2) Let A = {face card} and B = {heart}.

$P(A)=\dfrac{12}{52}$, $P(B)=\dfrac{13}{52}$, and $P(A \text{ and } B)=\dfrac{3}{52}$

c. (1) $P(A \text{ or } B)=P(A)+P(B)=\dfrac{1}{6}+\dfrac{1}{2}\approx 0.667$

(2) $P(A \text{ or } B)=P(A)+P(B)-P(A \text{ and } B)=\dfrac{12}{52}+\dfrac{13}{52}-\dfrac{3}{52}\approx 0.423$

3a. Let A = {sales between \$0 and \$24,999}
Let B = {sales between \$25,000 and \$49,999}.

b. A and B cannot occur at the same time. So A and B are mutually exclusive.

c. $P(A) = \frac{3}{36}$ and $P(B) = \frac{5}{36}$

d. $P(A \text{ or } B) = P(A) + P(B) = \frac{3}{36} + \frac{5}{36} \approx 0.222$

4a. (1) Let A = {type B} and B = {type AB}.
(2) Let A = {type O} and B = {Rh-positive}.

b. (1) A and B cannot occur at the same time. So, A and B are mutually exclusive.
(2) A and B can occur at the same time. So, A and B are not mutually exclusive.

c. (1) $P(A) = \frac{45}{409}$ and $P(B) = \frac{16}{409}$

(2) $P(A) = \frac{184}{409}$, $P(B) = \frac{344}{409}$, and $P(A \text{ and } B) = \frac{156}{409}$

d. (1) $P(A \text{ or } B) = P(A) + P(B) = \frac{45}{409} + \frac{16}{409} \approx 0.149$

(2) $P(A \text{ or } B) = P(A) + P(B) - P(A \text{ and } B) = \frac{184}{409} + \frac{344}{409} - \frac{156}{409} \approx 0.910$

5a. Let A = {linebacker} and B = {quarterback}.
$P(A \text{ or } B) = P(A) + P(B) = \frac{24}{256} + \frac{12}{256} \approx 0.141$

b. P(not a linebacker or quarterback) = 1 – $P(A$ or $B) \approx 1 - 0.141 = 0.859$

3.3 EXERCISE SOLUTIONS

1. $P(A$ and $B)$ = 0 because A and B cannot occur at the same time.

3. True

5. False. The probability that event A or event B will occur is
$P(A \text{ or } B) = P(A) + P(B) - P(A \text{ and } B)$

7. Not mutually exclusive because a student can be an athlete and on the Dean's list.

9. Not mutually exclusive because a public school teacher can be female and be 25 years old.

11. Mutually exclusive because a student cannot have a birthday in both months.

13. (a) Not mutually exclusive because for five weeks the events overlapped.

(b) $P(\text{OT or temp}) = P(\text{OT}) + P(\text{temp}) - P(\text{OT and temp}) = \frac{18}{52} + \frac{9}{52} - \frac{5}{52} \approx 0.423$

15. (a) Not mutually exclusive because a carton can have a puncture and a smashed corner.

(b) $P(\text{puncture or corner}) = P(\text{puncture}) + P(\text{corner}) - P(\text{puncture and corner})$
$= 0.05 + 0.08 - 0.004 = 0.126$

17. (a) $P(\text{club or 3}) = P(\text{club}) + P(3) - P(\text{club and 3})$
$= \frac{13}{52} + \frac{4}{52} - \frac{1}{52} \approx 0.308$

(b) $P(\text{red or king}) = P(\text{red}) + P(\text{king}) - P(\text{red and king})$
$= \frac{26}{52} + \frac{4}{52} - \frac{2}{52} \approx 0.538$

(c) $P(\text{9 or face card}) = P(9) + P(\text{face card}) - P(\text{9 and face card})$
$= \frac{4}{52} + \frac{12}{52} - 0 \approx 0.308$

19. (a) $P(\text{under 5}) = 0.067$

(b) $P(\text{not 65+}) = 1 - P(\text{65+}) = 1 - 0.16 = 0.84$

(c) $P(\text{between 20 and 34})$
$= P(\text{between 20 and 24 or between 25 and 34})$
$= P(\text{between 20 and 24}) + P(\text{between 25 and 34}) = 0.064 + 0.135 = 0.199$

21. (a) $P(\text{not A}) = 1 - P(\text{A}) = 1 - \frac{52}{1026} = \frac{974}{1026} \approx 0.949$

(b) $P(\text{D or F}) = P(\text{D}) \text{ or } P(\text{F}) = \frac{272}{1026} + \frac{126}{1026} = \frac{398}{1026} \approx 0.388$

23. $A = \{\text{male}\}$; $B = \{\text{nursing major}\}$

(a) $P(A \text{ or } B) = P(A) + P(B) - P(A \text{ and } B)$
$= \frac{1255}{3964} + \frac{1167}{3964} - \frac{151}{3964} = 0.573$

(b) $P(A' \text{ or } B') = P(A') + P(B') - P(A' \text{ and } B')$
$= \frac{2709}{3964} + \frac{2797}{3964} - \frac{1693}{3964} = 0.962$

(c) $P(A \text{ or } B) \approx 0.573$

(d) Not mutually exclusive. A male can be a nursing major.

25. $A = \{\text{frequently}\}$; $B = \{\text{occasionally}\}$; $C = \{\text{not at all}\}$; $D = \{\text{male}\}$

(a) $P(A \text{ or } B) = \frac{428}{2850} + \frac{886}{2850} \approx 0.461$

(b) $P(D' \text{ or } C) = P(D') + P(C) - P(D' \text{ and } C)$
$= \frac{1378}{2850} + \frac{1536}{2850} - \frac{741}{2850} \approx 0.762$

(c) $P(D \text{ or } A) = P(D) + P(A) - P(D \text{ and } A)$
$$= \frac{1472}{2850} + \frac{428}{2850} - \frac{221}{2850} \approx 0.589$$
(d) $P(D' \text{ or } A') = P(D') + P(A') - P(D' \text{ and } A')$
$$= \frac{1378}{2850} + \frac{2422}{2850} - \frac{1171}{2850} \approx 0.922$$
(e) Not mutually exclusive. A female can be frequently involved in charity work.

27. Answers will vary.
Conclusion: If two events, A and B, are independent, $P(A \text{ and } B) = P(A) \cdot P(B)$. If two events are mutually exclusive, $P(A \text{ and } B) = 0$. The only scenario when two events can be independent and mutually exclusive is if $P(A) = 0$ or $P(B) = 0$.

29. $P(A \text{ or } B \text{ or } C) = P(A) + P(B) + P(C) - P(A \text{ and } B) - P(A \text{ and } C) - P(B \text{ and } C) + P(A \text{ and } B \text{ and } C)$
$= 0.38 + 0.26 + 0.14 - 0.12 - 0.03 - 0.09 + 0.01$
$= 0.55$

3.4 ADDITIONAL TOPICS IN PROBABILITY AND COUNTING

3.4 Try It Yourself Solutions

1a. $n = 8$ teams **b.** $8! = 40{,}320$

2a. $${}_8P_3 = \frac{8!}{(8-3)!} = \frac{8!}{5!} = \frac{8\cdot7\cdot6\cdot5\cdot4\cdot3\cdot2\cdot1}{5\cdot4\cdot3\cdot2\cdot1} = 8\cdot7\cdot6 = 336$$

b. There are 336 possible ways that the subject can pick a first, second, and third activity.

3a. $n = 12$, $r = 4$

b. $${}_{12}P_4 = \frac{12!}{(12-4)!} = \frac{12!}{8!} = 12\cdot11\cdot10\cdot9 = 11{,}880$$

4a. $n = 20$, $n_1 = 6$, $n_2 = 9$, $n_3 = 5$

b. $$\frac{n!}{n_1!\cdot n_2!\cdot n_3!} = \frac{20!}{6!\cdot9!\cdot5!} = 77{,}597{,}520$$

5a. $n = 20$, $r = 3$

b. $${}_{20}C_3 = \frac{20!}{(20-3)!3!} = \frac{20!}{17!3!} = \frac{20\cdot19\cdot18\cdot17!}{17!\cdot3!} = 1140$$

c. There are 1140 different possible three-person committees that can be selected from 20 employees.

6a. $_{20}P_2 = \dfrac{20!}{(20-2)!} = \dfrac{20!}{18!} = \dfrac{20 \cdot 19 \cdot 18!}{18!} = 20 \cdot 19 = 380$

b. $P(\text{selecting the two members}) = \dfrac{1}{380} \approx 0.003$

7a. One favorable outcome and $\dfrac{6!}{1! \cdot 2! \cdot 2! \cdot 1!} = 180$ distinguishable permutations.

b. $P(\text{letter}) = \dfrac{1}{180} \approx 0.006$

8a. $_{15}C_5 = 3003$ **b.** $_{54}C_5 = 3{,}162{,}510$ **c.** $\dfrac{3003}{3{,}162{,}510} \approx 0.0009$

9a. $_5C_3 \cdot {}_7C_0 = 10 \cdot 1 = 10$ **b.** $_{12}C_3 = 220$ **c.** $\dfrac{10}{220} \approx 0.045$

3.4 EXERCISE SOLUTIONS

1. The number of ordered arrangements of *n* objects taken *r* at a time. An example of a permutation is the number of seating arrangements of you and three friends.

3. False. A permutation is an ordered arrangement of objects.

5. True

7. $_9P_5 = \dfrac{9!}{(9-5)!} = \dfrac{9!}{4!} = 9 \cdot 8 \cdot 7 \cdot 6 \cdot 5 = 15{,}120$

9. $_8C_3 = \dfrac{8!}{(8-3)!3!} = \dfrac{8!}{5!3!} = \dfrac{8 \cdot 7 \cdot 6 \cdot 5!}{5! \cdot 3!} = 56$

11. $_{21}C_8 = \dfrac{21!}{(21-8)!8!} = \dfrac{21!}{13!8!} = \dfrac{21 \cdot 20 \cdot 19 \cdot 18 \cdot 17 \cdot 16 \cdot 15 \cdot 14 \cdot 13!}{13! \cdot 8!} = 203{,}490$

13. $\dfrac{_6P_2}{_{11}P_3} = \dfrac{\frac{6!}{(6-2)!}}{\frac{11!}{(11-3)!}} = \dfrac{\frac{6!}{4!}}{\frac{11!}{8!}} = \dfrac{6 \cdot 5}{11 \cdot 10 \cdot 9} \approx 0.030$

15. Permutation, because the order of the 8 cars in line matters.

17. Combination, because the order of the captains does not matter.

19. $7! = 5040$

21. $6! = 720$

23. ${}_{52}C_6 = 20,358,520$

25. $\dfrac{22!}{4! \cdot 10! \cdot 8!} = 320,089,770$

27. 3 S's, 3 T's, 1 A, 2 I's, 1 C

$$\frac{10!}{3! \cdot 3! \cdot 1! \cdot 2! \cdot 1!} = 50,400$$

29. $10 \cdot 8 \cdot {}_{13}C_2 = 6240$

31. ${}_5C_1 \cdot {}_{75}C_5 = 86,296,950$

33. (a) $6! = 720$

(b) sample

(c) $\dfrac{1}{720} \approx 0.001$

The event can be considered unusual because its probability is less than or equal to 0.05.

35. (a) 2 E's, 1 T, 1 R

$$\frac{4!}{2! \cdot 1! \cdot 1!} = 12$$

(b) tree

(c) $\dfrac{1}{12} \approx 0.083$

The event cannot be considered unusual because its probability is not less than or equal to 0.05.

37. (a) 1 U, 1 N, 2 O's, 2 P's, 1 L, 1 A, 1 T, 1 I

$$\frac{10!}{1! \cdot 1! \cdot 2! \cdot 2! \cdot 1! \cdot 1! \cdot 1! \cdot 1!} = 907,200$$

(b) population

(c) $\dfrac{1}{907,200} \approx 0.000001$

The event can be considered unusual because its probability is less than or equal to 0.05.

39. $\dfrac{1}{{}_{12}C_3} = \dfrac{1}{220} \approx 0.005$

41. (a) $\dfrac{{}_{15}C_3}{{}_{56}C_3} = \dfrac{455}{27,720} \approx 0.0164$ (b) $\dfrac{{}_{41}C_3}{{}_{56}C_3} = \dfrac{10,660}{27,720} \approx 0.385$

43. (a) ${}_8C_4 = 70$

(b) $({}_2C_1)\cdot({}_2C_1)\cdot({}_2C_1)\cdot({}_2C_1) = 2\cdot2\cdot2\cdot2 = 16$

(c) ${}_4C_2\left[\dfrac{({}_2C_0)\cdot({}_2C_0)\cdot({}_2C_2)\cdot({}_2C_2)}{{}_8C_4}\right] \approx 0.086$

45. (a) $(26)(26)(10)(10)(10)(10)(10) = 67{,}600{,}000$

(b) $(26)(25)(10)(9)(8)(7)(6) = 19{,}656{,}000$

(c) $\dfrac{1}{67{,}600{,}000} \approx 0.000000015$

47. (a) $5! = 120$ (b) $2!\,3! = 12$ (c) $3!\,2! = 12$ (d) $\dfrac{{}_2C_1}{{}_5C_1} = \dfrac{2}{5} = 0.4$

49. (7%)(1200) = (0.07)(1200) = 84 of the 1200 rate financial shape as excellent.

$P(\text{all four rate excellent}) = \dfrac{{}_{84}C_4}{{}_{1200}C_4} = \dfrac{1{,}929{,}501}{85{,}968{,}659{,}700} \approx 0.000022$

51. (39%)(500) = (0.39)(500) = 195 of the 500 rate financial shape as fair $\Rightarrow 500 - 195 = 305$ rate financial shape as not fair.

$P(\text{none of 80 rate fair}) = \dfrac{{}_{305}C_{80}}{{}_{500}C_{80}} \approx 6.00\times10^{-20}$

53. (a) ${}_{40}C_5 = 658{,}008$ (b) $P(\text{win}) = \dfrac{1}{658{,}008} \approx 0.00000152$

55. (a) $\dfrac{({}_{13}C_1)({}_4C_4)({}_{12}C_1)({}_4C_1)}{{}_{52}C_5} = \dfrac{(13)(1)(12)(4)}{2{,}598{,}960} \approx 0.0002$

(b) $\dfrac{({}_{13}C_1)({}_4C_3)({}_{12}C_1)({}_4C_2)}{{}_{52}C_5} = \dfrac{(13)(4)(12)(6)}{2{,}598{,}960} \approx 0.0014$

(c) $\dfrac{({}_{13}C_1)({}_4C_3)({}_{12}C_2)({}_4C_1)({}_4C_1)}{{}_{52}C_5} = \dfrac{(13)(4)(66)(4)(4)}{2{,}598{,}960} \approx 0.0211$

(d) $\dfrac{({}_{13}C_2)({}_{13}C_1)({}_{13}C_1)({}_{13}C_1)}{{}_{52}C_5} = \dfrac{(78)(13)(13)(13)}{2{,}598{,}960} \approx 0.0659$

57. ${}_{14}C_4 = 1001$ possible 4 digit arrangements if order is not important.
Assign 1000 of the 4 digit arrangements to the 13 teams since 1 arrangement is excluded.

59. $P(1\text{st}) = \dfrac{250}{1000} = 0.250$

$P(2\text{nd}) = \dfrac{199}{1000} = 0.199$

$P(3\text{rd}) = \dfrac{156}{1000} = 0.156$

$P(4\text{th}) = \dfrac{119}{1000} = 0.119$

$P(5\text{th}) = \dfrac{88}{1000} = 0.088$

$P(6\text{th}) = \dfrac{63}{1000} = 0.063$

$P(7\text{th}) = \dfrac{43}{1000} = 0.043$

$P(8\text{th}) = \dfrac{28}{1000} = 0.028$

$P(9\text{th}) = \dfrac{17}{1000} = 0.017$

$P(10\text{th}) = \dfrac{11}{1000} = 0.011$

$P(11\text{th}) = \dfrac{8}{1000} = 0.008$

$P(12\text{th}) = \dfrac{7}{1000} = 0.007$

$P(13\text{th}) = \dfrac{6}{1000} = 0.006$

$P(14\text{th}) = \dfrac{5}{1000} = 0.005$

Events in which any of Teams 7-14 would win the first pick would be considered unusual because the probabilities are all less than or equal to 0.05.

61. Let A = {team with the worst record wins third pick} and
B = {team with the best record, ranked 14th, wins first pick} and
C = {team ranked 2nd wins the second pick}.

$$P(A|B \text{ and } C) = \frac{250}{1000 - 199 - 5} = \frac{250}{796} \approx 0.314$$

CHAPTER 3 REVIEW EXERCISE SOLUTIONS

1. Sample space:
{HHHH, HHHT, HHTH, HHTT, HTHH, HTHT, HTTH, HTTT, THHH, THHT, THTH, THTT, TTHH, TTHT, TTTH, TTTT}

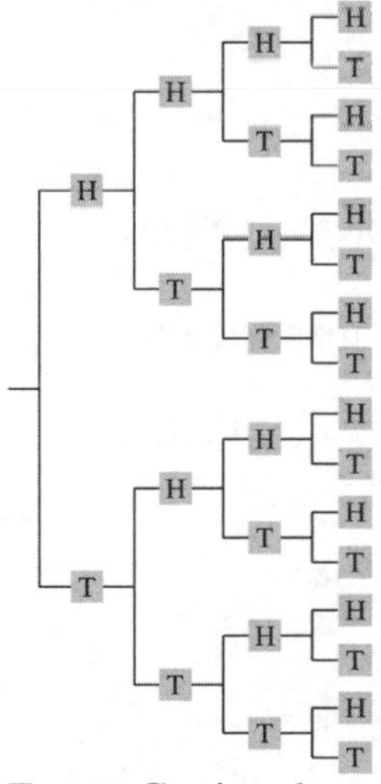

Event: Getting three heads
{HHHT, HHTH, HTHH, THHH}
There are 4 outcomes.

3. Sample space: {January, February, March, April, May, June, July, August, September, October, November, December}
Event: {January, June, July}
There are 3 outcomes.

5. $(7)(4)(3) = 84$

7. Empirical probability because it is based on observations obtained from probability experiments.

9. Subjective probability because it is based on opinion.

11. Classical probability because each outcome in the sample space is equally likely to occur.

13. $P(\text{at least } 10) = 0.107 + 0.090 + 0.018 = 0.215$

15. $\dfrac{1}{(8)(10)(10)(10)(10)(10)(10)} = 1.25 \times 10^{-7}$

17. $P(\text{undergrad} \mid +) = 0.92$

19. Independent. The outcomes of the first four coin tosses do not affect the outcome of the fifth coin toss.

21. Dependent. The outcome of getting high grades affects the outcome of being awarded an academic scholarship.

23.
$$\begin{aligned} P(\text{correct toothpaste and correct dental rinse}) &= P(\text{correct toothpaste}) \cdot P(\text{correct dental rinse}) \\ &= \frac{1}{8} \cdot \frac{1}{5} \\ &= \frac{1}{40} \\ &= 0.025 \end{aligned}$$
The event is unusual because its probability is less than or equal to 0.05.

25. Mutually exclusive. A jelly bean cannot be both completely red and completely yellow.

27. Mutually exclusive. A person cannot be registered legally to vote in more than one state.

29. $P(\text{home or work}) = P(\text{home}) + P(\text{work}) - P(\text{home and work}) = 0.44 + 0.37 - 0.21 = 0.6$

31. $P(\text{4-8 or club}) = P(\text{4-8}) + P(\text{club}) - P(\text{4-8 and club}) = \dfrac{20}{52} + \dfrac{13}{52} - \dfrac{5}{52} \approx 0.538$

33. $P(\text{odd or less than } 4) = P(\text{odd}) + P(\text{less than } 4) - P(\text{odd and less than } 4) = \dfrac{6}{12} + \dfrac{3}{12} - \dfrac{2}{12} \approx 0.583$

35.
$$\begin{aligned} P(\text{600 or more}) &= P(\text{600-999}) + P(\text{1000 or more}) \\ &= 0.244 + 0.047 = 0.291 \end{aligned}$$

37. $P(\text{action or horror}) = P(\text{action}) + P(\text{horror})$

$$= \frac{112}{874} + \frac{52}{874} = \frac{164}{874} \approx 0.188$$

39. $P(\text{not comedy}) = 1 - P(\text{comedy})$

$$= 1 - \frac{260}{874} = \frac{614}{874} \approx 0.703$$

41. $_{11}P_2 = \dfrac{11!}{(11-2)!} = \dfrac{11!}{9!} = 11 \cdot 10 = 110$

43. $_7C_4 = \dfrac{7!}{(7-4)!4!} = \dfrac{7!}{3!4!} = 35$

45. $_{50}P_5 = 254{,}251{,}200$

47. Order is important: $_{15}P_3 = 2730$

49. Order is not important: $_{17}C_4 = 2380$

51. $P(\text{3 kings and 2 queens}) = \dfrac{({}_4C_3)\cdot({}_4C_2)}{{}_{52}C_5} = \dfrac{4 \cdot 6}{2{,}598{,}960} \approx 0.00000923$

The event is unusual because its probability is less than or equal to 0.05.

53. (a) $P(\text{no defective}) = \dfrac{{}_{197}C_3}{{}_{200}C_3} = \dfrac{1{,}254{,}890}{1{,}313{,}400} \approx 0.955$

The event is not unusual because its probability is not less than or equal to 0.05.

(b) $P(\text{all defective}) = \dfrac{{}_3C_3}{{}_{200}C_3} = \dfrac{1}{1{,}313{,}400} \approx 0.000000761$

The event is unusual because its probability is less than or equal to 0.05.

(c) $P(\text{at least one defective}) = 1 - P(\text{no defective}) \approx 1 - 0.955 = 0.045$

The event is unusual because its probability is less than or equal to 0.05.

(d) $P(\text{at least one non-defective}) = 1 - P(\text{all defective}) \approx 1 - 0.000000761 \approx 0.999999239$

The event is not unusual because its probability is not less than or equal to 0.05.

55. (a) $P(\text{4 men}) = \dfrac{{}_6C_4}{{}_{10}C_4} = \dfrac{15}{210} \approx 0.071$

The event is not unusual because its probability is not less than or equal to 0.05.

(b) $P(\text{4 women}) = \dfrac{{}_4C_4}{{}_{10}C_4} = \dfrac{1}{210} \approx 0.005$

The event is unusual because its probability is less than or equal to 0.05.

(c) $P(\text{2 men and 2 women}) = \dfrac{({}_6C_2)\cdot({}_4C_2)}{{}_{10}C_4} = \dfrac{15 \cdot 6}{210} \approx 0.429$

The event is not unusual because its probability is not less than or equal to 0.05.

(d) $P(\text{1 man and 3 women}) = \dfrac{({}_6C_1)\cdot({}_4C_3)}{{}_{10}C_4} = \dfrac{6\cdot 4}{210} \approx 0.114$

The event is not unusual because its probability is not less than or equal to 0.05.

CHAPTER 3 QUIZ SOLUTIONS

1. (a) $P(\text{bachelor's degree}) = \dfrac{1525}{2917} \approx 0.523$

(b) $P(\text{bachelor's degree}|\text{female}) = \dfrac{875}{1724} \approx 0.508$

(c) $P(\text{bachelor's degree}|\text{male}) = \dfrac{650}{1193} \approx 0.545$

(d) $P(\text{associate's degree or bachelor's degree}) = P(\text{associate's degree}) + P(\text{bachelor's degree})$

$$= \frac{728}{2917} + \frac{1252}{2917}$$
$$\approx 0.772$$

(e) $P(\text{doctorate}|\text{male}) = \dfrac{30}{1193} \approx 0.025$

(f) $P(\text{master's degree or female}) = P(\text{master's degree}) + P(\text{female}) - P(\text{master's degree and female})$

$$= \frac{604}{2917} + \frac{1724}{2917} - \frac{366}{2917}$$
$$\approx 0.673$$

(g) $P(\text{associate's degree and male}) = P(\text{associate's degree}) \cdot P(\text{male}|\text{associate's degree})$

$$= \frac{728}{2917} \cdot \frac{275}{728}$$
$$\approx 0.094$$

(h) $P(\text{female}|\text{bachelor's degree}) = \dfrac{875}{1525} \approx 0.574$

2. The event in part (e) is unusual because its probability is less than or equal to 0.05.

3. Not mutually exclusive. A golfer can score the best round in a four-round tournament and still lose the tournament.
Dependent. The outcome of scoring the best round in a four-round tournament affects the outcome of losing the golf tournament.

4. (a) ${}_{247}C_3 = 2{,}481{,}115$

(b) ${}_3C_3 = 1$

(c) ${}_{250}C_3 - {}_3C_3 = 2{,}573{,}000 - 1 = 2{,}572{,}999$

5. (a) $\dfrac{{}_{247}C_3}{{}_{250}C_3} = \dfrac{2{,}481{,}115}{2{,}573{,}000} \approx 0.96$

(b) $\dfrac{{}_3C_3}{{}_{250}C_3} = \dfrac{1}{2{,}573{,}000} \approx 3.88 \times 10^{-7}$

(c) $\dfrac{{}_{250}C_3 - {}_3C_3}{{}_{250}C_3} = \dfrac{2{,}572{,}999}{2{,}573{,}000} \approx 0.9999996$

6. $9 \cdot 10 \cdot 10 \cdot 10 \cdot 10 \cdot 5 = 450{,}000$

7. ${}_{30}P_4 = 657{,}720$

8. There are 27 ways of getting a sum of 11 and 3 ways of getting a sum of 17, out of 216 different outcomes when you roll three dice. Thus the corresponding probabilities $P(\text{sum} = 11) = \dfrac{27}{216} = 0.125$ and $P(\text{sum} = 17) = \dfrac{3}{216} = 0.014$. Since getting the sum of 17 has a probability less than or equal to 0.05, it is unusual.

CHAPTER 4

Discrete Probability Distributions

4.1 PROBABILITY DISTRIBUTIONS

4.1 Try It Yourself Solutions

1a. (1) measured (2) counted

b. (1) The random variable is continuous because x can be any speed up to the maximum speed of a space shuttle.

(2) The random variable is discrete because the number of calves born on a farm in one year is countable.

2ab.

x	f	$P(x)$
0	16	0.16
1	19	0.19
2	15	0.15
3	21	0.21
4	9	0.09
5	10	0.10
6	8	0.08
7	2	0.02
	$n = 100$	$\sum P(x) = 1$

c.

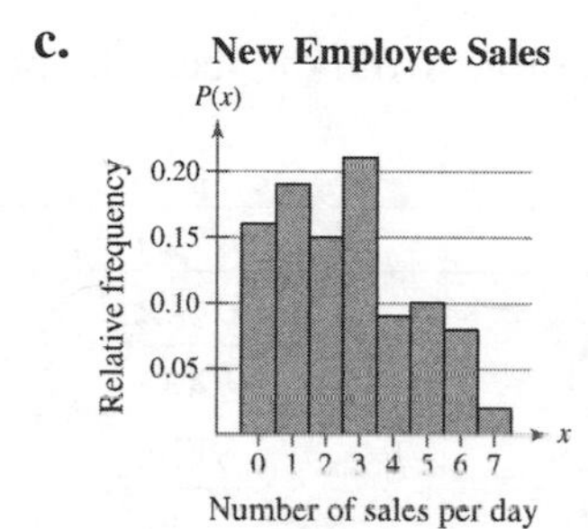

3a. Each $P(x)$ is between 0 and 1.

b. $\sum P(x) = 1$

c. Because both conditions are met, the distribution is a probability distribution.

4a. (1) Yes, each outcome is between 0 and 1. (2) Yes, each outcome is between 0 and 1.

b. (1) Yes, $\sum P(x) - 1$. (2) Yes, $\sum P(x) = 1$.

c. (1) Is a probability distribution (2) Is a probability distribution.

5ab.

x	$P(x)$	$xP(x)$
0	0.16	(0)(0.16) = 0.00
1	0.19	(1)(0.19) = 0.19
2	0.15	(2)(0.15) = 0.30
3	0.21	(3)(0.21) = 0.63
4	0.09	(4)(0.09) = 0.36
5	0.10	(5)(0.10) = 0.50
6	0.08	(6)0.08 = 0.48
7	0.02	(7)(0.02) = 0.14
	$\sum P(x) = 1$	$\sum xP(x) = 2.60$

c. $\mu = \sum xP(x) = 2.6$

On average, a new employee makes 2.6 sales per day.

6ab. From 5, $\mu - 2.6$,

x	f	$x - \mu$	$(x - \mu)^2$	$P(x)(x - \mu)^2$
0	0.16	−2.6	6.76	(0.16)(6.76) = 1.0816
1	0.19	−1.6	2.56	(0.19)(2.56) = 0.4864
2	0.15	−0.6	0.36	(0.15)(0.36) = 0.0540
3	0.21	0.4	0.16	(0.21)(0.16) = 0.0336
4	0.09	1.4	1.96	(0.09)(1.96) = 0.1764
5	0.10	2.4	5.76	(0.10)(5.76) = 0.5760
6	0.08	3.4	11.56	(0.08)(11.56) = 0.9248
7	0.02	4.4	19.36	(0.02)(19.36) = 0.3872
	$\sum P(x) = 1$			$\sum P(x)(x - \mu)^2 = 3.72$

c. $\sigma = \sqrt{\sigma^2} = \sqrt{3.72} \approx 1.9$

d. Most of the data valves differ from the mean by no more than 1.9 sales per day.

7ab.

Gain, x	$P(x)$	$xP(x)$
\$1995	$\frac{1}{2000}$	$\frac{1995}{2000}$
\$ 995	$\frac{1}{2000}$	$\frac{995}{2000}$
\$ 495	$\frac{1}{2000}$	$\frac{495}{2000}$
\$ 245	$\frac{1}{2000}$	$\frac{245}{2000}$
\$ 95	$\frac{1}{2000}$	$\frac{95}{2000}$
\$ −5	$\frac{1995}{2000}$	$-\frac{9975}{2000}$
	$\sum P(x) = 1$	$\sum xP(x) \approx -3.08$

c. $E(x) = \sum xP(x) = -\$3.08$

d. Because the expected value is negative, you can expect to lose an average of \$3.08 for each ticket you buy.

4.1 EXERCISE SOLUTIONS

1. A random variable represents a numerical value associated with each outcome of a probability experiment. Examples: Answers will vary.

3. No; Expected value may not be a possible value of x for one trial, but it represents the average value of x over a large number of trials.

5. False. In most applications, discrete random variables represent counted data, while continuous random variables represent measured data.

7. True

9. Discrete, because attendance is a random variable that is countable.

11. Continuous, because the distance a baseball travels after being hit is a random variable that must be measured.

13. Discrete, because the number of books in a university library is a random variable that is countable.

15. Continuous, because the volume of blood drawn for a blood test is a random variable that must be measured.

17. Discrete, because the number of measures that is posted each month on a social networking site is a random variable that is countable.

19. Continuous, because the amount of snow that fell in Nome, Alaska last winter is a random variable that cannot be counted.

21. (a) $P(x > 2) = 0.25 + 0.10 = 0.35$
(b) $P(x < 4) = 1 - P(4) = 1 - 0.10 = 0.90$

23. $\sum P(x) = 1 \rightarrow P(3) = 0.22$

25. Because each $P(x)$ is between 0 and 1, and $\sum P(x) = 1,$ the distribution is a probability distribution.

27. (a)

x	f	$P(x)$
0	1491	0.686
1	425	0.195
2	168	0.077
3	48	0.022
4	29	0.013
5	14	0.006
	$n = 2175$	$\sum P(x) \approx 1$

(b)

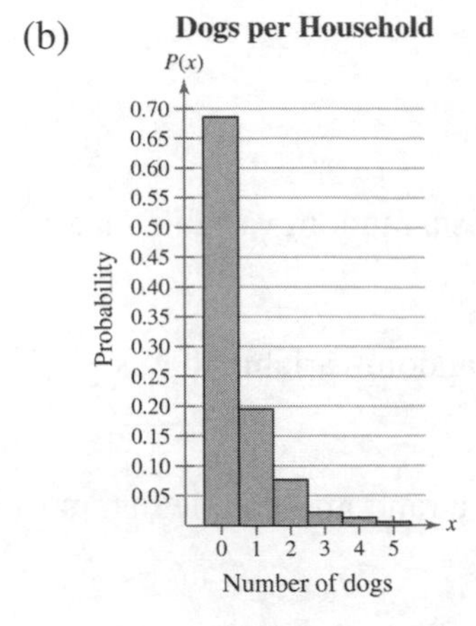

Skewed right

(c)

$xP(x)$	$(x-\mu)$	$(x-\mu)^2$	$(x-\mu)^2P(x)$
0	−0.497	0.247	0.169
0.195	0.503	0.253	0.049
0.154	1.503	2.259	0.174
0.066	2.503	6.265	0.138
0.052	3.503	12.271	0.160
0.030	4.503	20.277	0.122
$\sum xP(x) = 0.497$			$\sum (x-\mu)^2 P(x) = 0.812$

$\mu = \sum xP(x) = 0.497 \approx 0.5$

$\sigma^2 = \sum (x-\mu)^2 P(x) = 0.812 \approx 0.8$

$\sigma = \sqrt{\sigma^2} = \sqrt{0.812} \approx 0.9$

(d) The mean is 0.5, so the average number of dogs per household is about 0 or 1 dog. The standard deviation is 0.9, so most of the households differ from the mean by no more than 1 dog.

29. (a)

x	f	$P(x)$
0	26	0.01
1	442	0.17
2	728	0.28
3	1404	0.54
	$n = 2600$	$\sum P(x) = 1$

(b)

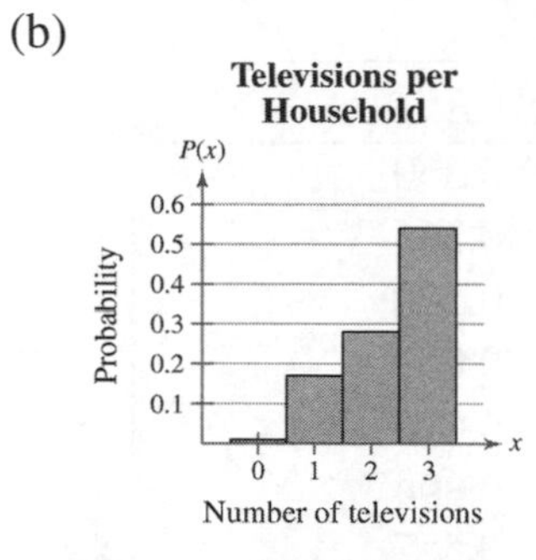

Skewed left

(c)

$xP(x)$	$(x-\mu)$	$(x-\mu)^2$	$(x-\mu)^2P(x)$
0.00	−2.35	5.523	0.055
0.17	−1.35	1.823	0.310
0.56	−0.35	0.123	0.034
1.62	0.65	0.423	0.228
$\sum xP(x) = 2.35$			$\sum (x-\mu)^2 P(x) = 0.627$

$\mu = \sum xP(x) = 2.35 \approx 2.4$

$\sigma^2 = \sum (x-\mu)^2 P(x) = 0.627 \approx 0.6$

$\sigma = \sqrt{\sigma^2} = \sqrt{0.627} \approx 0.8$

(d) The mean is 2.4, so the average household in the town has about 2 televisions. The standard deviation is 0.8, so most of the households differ from the mean by no more than 1 television.

31. (a)

x	f	$P(x)$
0	6	0.031
1	12	0.063
2	29	0.151
3	57	0.297
4	42	0.219
5	30	0.156
6	16	0.083
	$n = 192$	$\sum P(x) = 1$

(b)

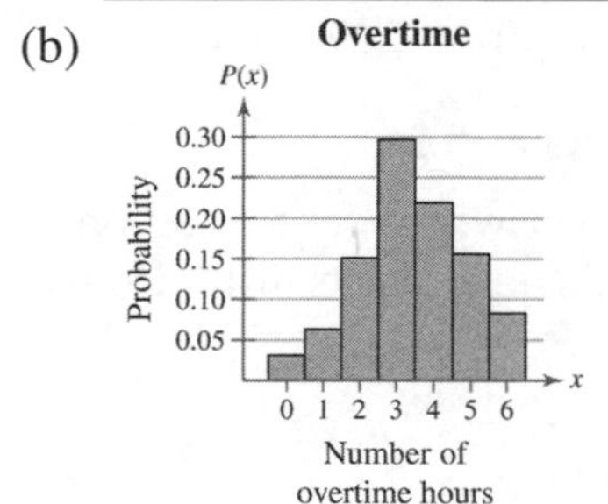

Approximately symmetric

(c)

$xP(x)$	$(x - \mu)$	$(x - \mu)^2$	$(x - \mu)^2 P(x)$
0	−3.41	11.628	0.360
0.063	−2.41	5.808	0.366
0.302	−1.41	1.988	0.300
0.891	−0.41	0.168	0.050
0.876	−0.59	0.348	0.076
0.780	1.59	2.528	0.394
0.498	2.59	6.708	0.557
$\sum xP(x) = 3.410$			$\sum (x - \mu)^2 P(x) = 2.103$

$\mu = \sum xP(x) = 3.410 \approx 3.4$

$\sigma^2 = \sum (x - \mu)^2 P(x) = 2.103 \approx 2.1$

$\sigma = \sqrt{\sigma^2} = \sqrt{2.103} \approx 1.5$

(d) The mean is 3.4, so the average employee worked 3.4 hours of overtime. The standard deviation is 1.5, so the overtime worked by most of the employees differed from the mean by no more than 1.5 hours.

33. An expected value of 0 means that the money gained is equal to the spent, representing the breakeven point.

35.

x	$P(x)$	$xP(x)$	$(x - \mu)$	$(x - \mu)^2$	$(x - \mu)^2 P(x)$
0	0.02	0.00	−5.30	28.09	0.562
1	0.02	0.02	−4.30	18.49	0.370
2	0.06	0.12	−3.30	10.89	0.653
3	0.06	0.18	−2.30	5.29	0.317
4	0.08	0.32	−1.30	1.69	0.135
5	0.22	1.10	−0.30	0.09	0.020
6	0.30	1.80	0.70	0.49	0.147
7	0.16	1.12	1.70	2.89	0.462
8	0.08	0.64	2.70	7.29	0.583
		$\sum xP(x) = 5.30$			$\sum (x - \mu)^2 P(x) = 3.249$

(a) $\mu = \sum xP(x) = 5.3$

(b) $\sigma^2 = \sum (x - \mu)^2 P(x) = 3.249 \approx 3.3$

(c) $\sigma = \sqrt{\sigma^2} = \sqrt{3.249} \approx 1.9$

(d) $E(x) = \mu = \sum xP(x) = 5.3$

(e) The expected value is 5.3, so an average student is expected to answer about 5 questions correctly. The standard deviation is 1.9, so most of the student's quiz results differ from the expected value by no more than about 2 questions.

37.

x	$P(x)$	$xP(x)$	$(x-\mu)$	$(x-\mu)^2$	$(x-\mu)^2P(x)$
1	0.392	0.392	−1.04	1.082	0.424
2	0.265	0.530	−0.04	0.002	0.001
3	0.269	0.807	0.96	0.922	0.248
4	0.064	0.256	1.96	3.842	0.246
5	0.011	0.055	2.96	8.762	0.096
		$\sum xP(x)=2.04$			$\sum(x-\mu)^2P(x)=1.015$

(a) $\mu=\sum xP(x)=2.04\approx 2.0$

(b) $\sigma^2=\sum(x-\mu)^2P(x)=1.015\approx 1.0$

(c) $\sigma=\sqrt{\sigma^2}=\sqrt{1.015}\approx 1.0$

(d) $E(x)=\mu=2.0$

(e) The expected value is 2.0, so an average hurricane that hits the U.S. mainland is expected to be a category 2 hurricane. The standard deviation is 1.0, so most of the hurricanes differ from the expected value by no more than 1 category level.

39.

x	$P(x)$	$xP(x)$	$(x-\mu)$	$(x-\mu)^2$	$(x-\mu)^2P(x)$
1	0.275	0.275	−1.491	2.223	0.611
2	0.332	0.664	−0.491	0.241	0.080
3	0.159	0.447	0.509	0.259	0.041
4	0.136	0.544	1.509	2.277	0.310
5	0.063	0.315	2.509	6.295	0.397
6+	0.036	0.216	3.509	12.313	0.443
		$\sum xP(x)=2.491$			$\sum(x-\mu)^2P(x)=1.882$

(a) $\mu=\sum xP(x)=2.491\approx 2.5$

(b) $\sigma^2=\sum(x-\mu)^2P(x)=1.882\approx 1.9$

(c) $\sigma=\sqrt{\sigma^2}=\sqrt{1.882}\approx 1.4$

(d) $E(x)=\mu=2.5$

(e) The expected value is 2.5, so an average household is expected to have either 2 or 3 people. The standard deviation is 1.4, so most of the household sites differ from the expected value by no more than 1 or 2 people.

41. (a) $P(x<2)=0.686+0.195=0.881$

(b) $P(x\geq 1)=1-P(x=0)=1-0.686=0.314$

(c) $P(1\leq x\leq 3)=0.195+0.077+0.022=0.294$

43. A household with three dogs is unusual because the probability of this event is 0.022, which is less than or equal to 0.05.

45. $E(x)=\mu=\sum xP(x)=(-\$1)\cdot\left(\frac{37}{38}\right)+(\$35)\cdot\left(\frac{1}{38}\right)\approx -\$0.05$

47. (a)

x	$P(x)$
0	0.432
1	0.403
2	0.137
3	0.029
	$\sum xP(x) = 1$

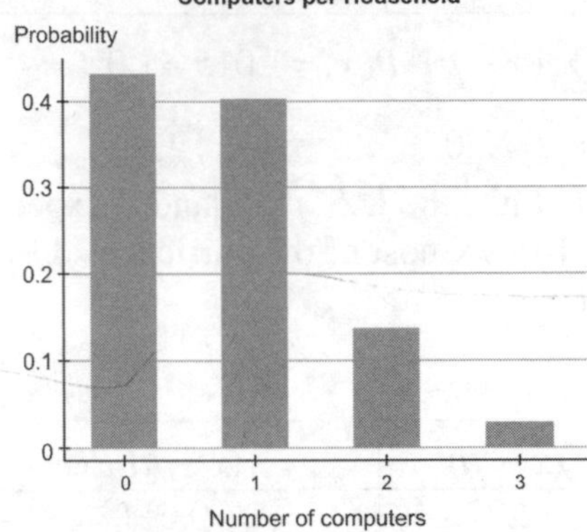

(b) Skewed right

49. $\mu_x = a + b\mu_x = 1000 + 1.05(36,000) = \$38,800$

51. $\mu_{x+y} = \mu_x + \mu_y = 1524 + 1496 = 3020$

$\mu_{x-y} = \mu_x - \mu_y = 1524 - 1496 = 28$

4.2 BINOMIAL DISTRIBUTIONS

4.2 Try It Yourself Solutions

1a. Trial answering a question
Success: the question answered correctly

b. Yes, the experiment satisfies the four conditions of a binomial experiment.

c. It is a binomial experiment.
$n = 10, p = 0.25, q = 0.75, x = 0, 1, 2, 3, 4, 5, 6, 7, 8, 9, 10$

2a. Trial: drawing a card with replacement
Success: card drawn is a club
Failure: card drawn is not a club

b. $n = 5, p = 0.25, q = 0.75, x = 3$

c. $P(3) = \dfrac{5!}{(5-3)!3!}(0.25)^3(0.75)^2 \approx 0.088$

3a. Trial: Selecting an adult and asking a question
Success: selecting an adult who likes texting because it works where talking won't do
Failure: selecting an adult who does not like texting because it works where talking won't do

b. $n = 7, p = 0.75, q = 0.25, x = 0, 1, 2, 3, 4, 5, 6, 7$

c. $P(0) = {}_7C_0(0.75)^0(0.25)^7 \approx 0.00006$

$P(1) = {}_7C_1(0.75)^1(0.25)^6 \approx 0.00128$

$P(2) = {}_7C_2(0.75)^2(0.25)^5 \approx 0.01154$

$P(3) = {}_7C_3(0.75)^3(0.25)^4 \approx 0.05768$

$P(4) = {}_7C_4(0.75)^4(0.25)^3 \approx 0.17303$

$P(5) = {}_7C_5(0.75)^5(0.25)^2 \approx 0.31146$

$P(6) = {}_7C_6(0.75)^6(0.25)^1 \approx 0.31146$

$P(7) = {}_7C_7(0.75)^7(0.25)^0 \approx 0.13348$

d.

x	$P(x)$
0	0.00006
1	0.00128
2	0.01154
3	0.05768
4	0.17303
5	0.31146
6	0.31146
7	0.13348
	$\sum xP(x) = 1$

4a. $n = 250, p = 0.71, x = 178$

b. $P(178) \approx 0.056$

c. The probability that exactly 178 people from a random sample of 250 people in the United States will use more than one topping on their hot dog is about 0.056.

d. Because 0.056 is not less than or equal to 0.05, this event is not unusual.

5a. (1) $x = 2$ (2) $x = 2, 3, 4,$ or 5 (3) $x = 0$ or 1

b. (1) $P(2) = {}_5C_2(0.21)^2(0.79)^3 \approx 0.217$

(2) $P(1) = {}_5C_1(0.21)^1(0.79)^4 \approx 0.409$

$P(x \geq 2) = 1 - P(0) - P(1) \approx 1 - 0.308 - 0.409 = 0.283$

or

$P(x \geq 2) = P(2) + P(3) + P(4) + P(5)$

$\approx 0.217 + 0.058 + 0.008 + 0.0004$

≈ 0.283

(3) $P(x < 2) = P(0) + P(1) \approx 0.308 + 0.409 = 0.717$

c. (1) The probability that exactly two of the five men consider fishing their favorite leisure-time activity is about 0.217.

(2) The probability that at least two of the five men consider fishing their favorite leisure-time activity is about 0.283.

(3) The probability that fewer than two of the five men consider fishing their favorite leisure-time activity is about 0.717.

6a. Trial: Selecting a business and asking if it has a website
Success: Selecting a business with a website
Failure: Selecting a business without a website

b. $n = 10, p = 0.55, x = 4$

c. $P(4) \approx 0.160$

d. The probability of randomly selecting 10 small businesses and finding exactly 4 that have a website is 0.160.

e. Because 0.160 is not less than or equal to 0.05, this event is not unusual.

7a. $P(0) = {}_4C_0(0.81)^0(0.19)^4 \approx 0.001$

$P(1) = {}_4C_1(0.81)^1(0.19)^3 \approx 0.022$

$P(2) = {}_4C_2(0.81)^2(0.19)^2 \approx 0.142$

$P(3) = {}_4C_3(0.81)^3(0.91)^1 \approx 0.404$

$P(4) = {}_4C_4(0.81)^4(0.19)^0 \approx 0.430$

b.

x	$P(x)$
0	0.001
1	0.022
2	0.142
3	0.404
4	0.430

c.

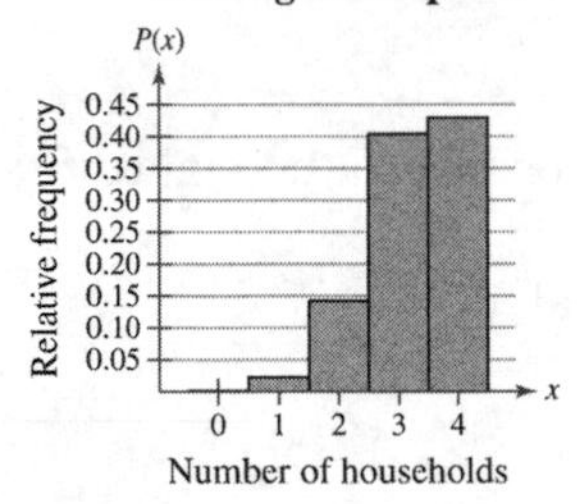

Skewed left

d. Yes, it would be unusual if exactly zero or exactly one of the four households owned a computer, because each of these events has a probability that is less than 0.05.

8a. Success: selecting a clear day
$n = 31, p = 0.44, q = 0.56$

b. $\mu = np = (31)(0.44) \approx 13.6$

c. $\sigma^2 = npq = (31)(0.44)(0.56) \approx 7.6$

d. $\sigma = \sqrt{npq} = \sqrt{(31)(0.44)(0.56)} \approx 2.8$

e. On average, there are about 14 clear days during the month of May. The standard deviation is about 3 days.

f. Values that are more than 2 standard deviations from the mean are considered unusual. Because $13.6 - 2(2.8) = 8$ and $13.6 + 2(2.8) = 19.2$, a May with fewer than 8 clear days, or more than 19 clear days would be unusual.

4.2 EXERCISE SOLUTIONS

1. Each trial is independent of the other trials if the outcome of one trial does not affect the outcome of any of the other trials.

3. (a) $p = 0.50$ (graph is symmetric)
(b) $p = 0.20$ (graph is skewed right $\rightarrow p < 0.5$)
(c) $p = 0.80$ (graph is skewed left $\rightarrow p > 0.5$)

5. (a) $n = 12$ ($x = 0, 1, 2, 3, 4, 5, 6, 7, 8, 9, 10, 11, 12$)
(b) $n = 4$ ($x = 0, 1, 2, 3, 4$)
(c) $n = 8$ ($x = 0, 1, 2, 3, 4, 5, 6, 7, 8$)
As n increases, the distribution becomes more symmetric.

7. (a) 0, 1, 2, 3, 4, 11, 12 (b) 0 (c) 0, 1, 2, 8

9. It is a binomial experiment.
Success: baby recovers
$n = 5, p = 0.80, q = 0.20, x = 0, 1, 2, 3, 4, 5$

11. It is a binomial experiment.
Success: Selecting an officer who is postponing or reducing the amount of vacation
$n = 20, p = 0.31, q = 0.69, x = 0, 1, 2, 3, 4, 5, 6, 7, 8, 9, 10, 11, 12, 13, 14, 15, 16, 17, 18, 19, 20$

13. $\mu = np = (50)(0.4) = 20$

$\sigma^2 = npq = (50)(0.4)(0.6) = 12$

$\sigma = \sqrt{npq} = \sqrt{(50)(0.4)(0.6)} \approx 3.5$

15. $\mu = np = (124)(0.26) = 32.2$

$\sigma^2 = npq = (124)(0.26)(0.74) = 23.9$

$\sigma = \sqrt{npq} = \sqrt{(124)(0.26)(0.74)} \approx 4.9$

17. $n = 5, p = 0.25$
(a) $P(3) \approx 0.088$
(b) $P(x \geq 3) = P(3) + P(4) + P(5) \approx 0.088 + 0.015 + 0.001 = 0.104$
(c) $P(x < 3) = 1 - P(x \geq 3) \approx 1 - 0.104 = 0.896$

19. $n = 10, p = 0.59$
(a) $P(8) \approx 0.111$
(b) $P(x \geq 8) = P(8) + P(9) + P(10) \approx 0.111 + 0.036 + 0.005 = 0.152$
(c) $P(x < 8) = 1 - P(x \geq 8) \approx 1 - 0.152 = 0.848$

21. $n = 8, p = 0.55$
(a) $P(5) \approx 0.257$
(b) $P(x > 5) = P(6) + P(7) + P(8) \approx 0.157 + 0.055 + 0.008 = 0.220$
(c) $P(x \leq 5) = 1 - P(x > 5) \approx 1 - 0.220 = 0.780$

23. $n = 14, p = 0.43$

(a) $P(5) \approx 0.187$

(b) $P(x \geq 6) = 1 - P(x \leq 5)$

$= 1 - (P(0) + P(1) + P(2) + P(3) + P(4) + P(5)$

$\approx 1 - (0.0004 + 0.004 + 0.020 + 0.060) + 0.124 + 0.187)$

≈ 0.605

(c) $P(x \leq 3) = P(0) + P(1) + P(2) + P(3) \approx 0.0004 + 0.004\ 0.020 + 0.060 \approx 0.084$

25. $n = 10, p = 0.28$

(a) $P(2) \approx 0.255$

(b) $P(x > 2) = 1 - P(x \leq 2) = 1 - (P(0) + P(1) + P(2))$

$\approx 1 - (0.037 + 0.146 + 0.255)$

≈ 0.562

(c) $P(x \leq 2 \leq 5) = P(2) + P(3) + P(4) + P(5)$

$\approx 0.255 + 0.264 + 0.180 + 0.084$

$= 0.783$

27. (a) $n = 6, p = 0.63$

x	$P(x)$
0	0.003
1	0.026
2	0.112
3	0.253
4	0.323
5	0.220
6	0.063

(b)

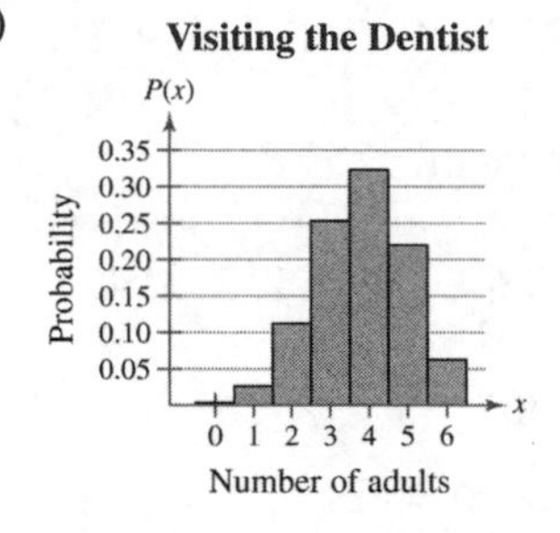

Skewed left

(c) $\mu = np = (6)(0.63) \approx 3.8$

$\sigma^2 = npq = (6)(0.63)(0.37) \approx 1.4$

$\sigma = \sqrt{npq} = \sqrt{(6)(0.63)(0.37)} \approx 1.2$

(d) On average, 3.8 out of 6 adults are visiting the dentist less because of the economy. The standard deviation is 1.2, so most samples of 6 adults would differ from the mean by no more than 1.2 people The values $x = 0$ and $x = 1$ would be unusual because their probabilities are less an 0.05.

29. (a) $n = 4, p = 0.05$

x	$P(x)$
0	0.814506
1	0.171475
2	0.013538
3	0.000475
4	0.000006

(b)

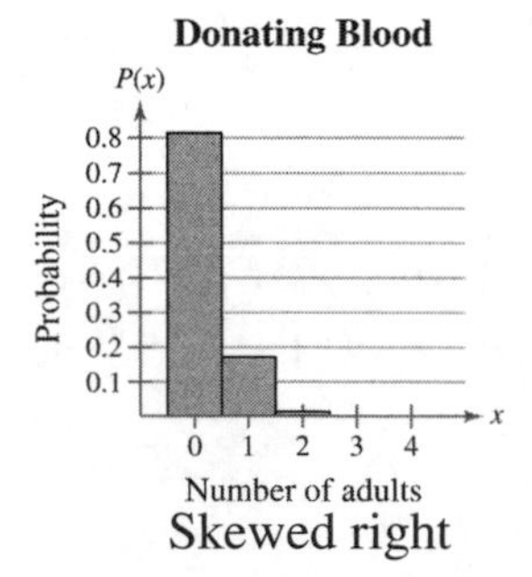

Skewed right

(c) $\mu = np = (4)(0.05) \approx 0.2$

$\sigma^2 = npq = (4)(0.05)(0.95) \approx 0.2$

$\sigma = \sqrt{npq} = \sqrt{(4)(0.05)(0.95)} \approx 0.4$

(d) On average, 0.2 eligible adults out of every 4 gives blood. The standard deviation is 0.4, so most sample of 4 eligible adults would differ from the mean by at most 0.4 adult. The values $x = 2$, $x = 3$, or $x = 4$ would be unusual because their probabilities are less than 0.05.

31. (a) $n = 6, p = 0.37$

x	$P(x)$
0	0.063
1	0.220
2	0.323
3	0.253
4	0.112
5	0.026
6	0.003

(b) $P(x = 2) \approx 0.323$

(c) $P(x \geq 5) = P(x = 5) + P(x = 6) \approx 0.026 + 0.003 = 0.029$

33. $n = 6, p = 0.37, q = 0.63$

$\mu = np = (6)(0.37) \approx 2.2$

$\sigma = \sqrt{npq} = \sqrt{(6)(0.37)(0.63)} \approx 1.2$

On average, 2.2 out of 6 travelers would name "crying kids" as the most annoying. The standard deviation is 1.2, so most samples of 6 travelers would differ from the mean by at most 1.2 travelers. The values $x = 5$ and $x = 6$ would be unusual because their probabilities are less than 0.05.

35. (a) $P(x = 9) \approx 0.081$

(b) $P(x \geq 7) \approx 0.541$

(c) $P(x \leq 3) \approx 0.022$; This event is unusual because its probability is less than 0.05.

37. $P(5, 2, 2, 1) = \dfrac{10!}{5!2!2!1!}\left(\dfrac{9}{16}\right)^5\left(\dfrac{3}{16}\right)^2\left(\dfrac{3}{16}\right)^2\left(\dfrac{1}{16}\right)^1 = 0.033$

4.3 MORE DISCRETE PROBABILITY DISTRIBUTIONS

4.3 Try It Yourself Solutions

1a. $P(1) = (0.74)(0.26)^0 \approx 0.74$

$P(2) = (0.74)(0.26)^0 \approx 0.192$

b. $P(\text{shot made before third attempt}) = P(1) + P(2) \approx 0.932$

c. The probability that LeBron makes his first free throw shot before his third attempt is 0.932.

2a. $P(0) \approx \dfrac{3^0(2.71828)^{-3}}{0!} \approx 0.050$ $\quad P(1) \approx \dfrac{3^1(2.71828)^{-3}}{1!} \approx 0.149$

$P(2) \approx \dfrac{3^2(2.71828)^{-3}}{2!} \approx 0.224$ $\quad P(3) \approx \dfrac{3^3(2.71828)^{-3}}{3!} \approx 0.224$

$P(4) \approx \dfrac{3^4(2.71828)^{-3}}{4!} \approx 0.168$

b. $P(0) + P(1) + P(2) + P(3) + P(4) \approx 0.050 + 0.149 + 0.224 + 0.224 + 0.168 = 0.815$

c. $1 - 0.815 = 0185$

d. The probability that more than four accidents will occur in any given month at the intersection is 0.185.

3a. $\mu = \dfrac{2000}{20,000} = 0.10$

b. $\mu = 0.10, x = 3$

c. $P(3) = 0.0002$

d. The probability of finding three brown trout in any given cubic meter of the lake is 0.0002.

e. Because 0.0002 is less than 0.05, this can be considered an unusual event.

4.3 EXERCISE SOLUTIONS

1. $P(3) = (0.65)(0.35)^2 = 0.08$

3. $P(5) = (0.09)(0.91)^4 = 0.062$

5. $P(4) \approx \dfrac{(5)^4(2.71828)^{-5}}{4!} = 0.175$

7. $P(2) \approx \dfrac{(1.5)^2(2.71828)^{-1.5}}{2!} = 0.251$

9. In a binomial distribution, the value of x represents the number of success in n trials, and in a geometric distribution the value of x represents the first trial that results in a success.

11. Geometric, You are interested in counting the number of trials until the first success.

13. Binomial. You are interested in counting the number of successes out of n trials.

15. $p = 0.19$

(a) $P(5) = (0.19)(0.81)^4 \approx 0.082$

(b) $P(\text{sale on 1st, 2nd, or 3}^{\text{rd}}\text{ call}) = P(1) + P(2) + P(3)$

$= (0.19)(0.81)^0 + (0.19)(0.81)^1 + (0.19)(0.81)^2 \approx 0.469$

(c) $P(\text{no sale of first 3 calls}) = 1 - P(\text{sale on 1}^{\text{st}}, 2^{\text{nd}}, \text{or 3}^{\text{rd}}\text{ call}) \approx 1 - 0.469 = 0.531$

17. $\mu = 4$

(a) $P(3) = \dfrac{(4)^3(2.71828)^{-4}}{3!} = 0.195$

(b) $P(x \geq 3) = 1 - P(0) + P(1) + P(2) + P(3)$
$\approx 0.018 + 0.073 + 0.147 + 0.195$
$= 0.433$

(c) $P(x > 3) = 1 - (P(x \leq 3) \approx 1 - 0.433 = 0.567$

19. $\mu = 0.6$

(a) $P(1) \approx \dfrac{(0.6)^1(2.71828)^{-0.6}}{1!} = 0.329$

(b) $P(x \leq 1) = P(0) + P(1) \approx 0.549 + 0.329 = 0.878$

(c) $P(x > 1) = 1 - (P(x \leq 1) \approx 1 - 0.878 = 0.122$

21. $p = \dfrac{1}{4} = 0.25$

(a) $P(4) = (0.25)(0.75)^3 \approx 0.105$

(b) P(prize with 1st, 2nd, or 3rd purchase)
$= P(1) + P(2) + P(3) = (0.25)(0.75)^0 + (0.25)(0.75)^1 + (0.25)(0.75)^2 = 0.578$

(c) $P(x > 4) = 1 - P(x \leq 4) = 1 - [P(1) + P(2) + P(3) + P(4)] = 1 - 0.684 = 0.316$

23. $\mu = 8$

(a) $P(8) \approx 0.140$

(b) $P(x \leq 3) \approx 0.042$
This event is unusual because its probability is less than 0.005.

(c) $P(x > 12) \approx 0.064$

25. (a) $n = 6000, p = \dfrac{1}{2500} = 0.0004$

$P(4) = \dfrac{6000!}{5996!4!}(0.0004)^4(0.9996)^{5996} \approx 0.1254235482$

(b) $\mu = \dfrac{6000}{2500} = 2.4$ cars with defects per 6000.

$P(4) \approx 0.1254084986$

The results are approximately the same.

27. $p = \frac{1}{1000} = 0.001$

(a) $\mu = \frac{1}{p} = \frac{1}{0.001} = 1000$

$\sigma^2 = \frac{q}{p^2} = \frac{0.999}{(0.001)^2} = 999{,}000$

$\sigma = \sqrt{\sigma^2} \approx 999.5$

On average, you would have to play 1000 times in order to win the lottery. The standard deviation is 999.5 times.

(b) 1000 times, because it is the mean.

You would expect to lose money, because, on average, you would win \$500 every 1000 times you play the lottery and pay \$1000 to play it. So, the net gain would be –\$500.

29. $\mu = 3.9$

(a) $\sigma^2 = \mu = 3.9$

$\sigma = \sqrt{\sigma^2} \approx 2.0$

The standard deviation is 2.0 strokes, so most of Phil's scores per hole differ from the mean by no more than 2.0 strokes.

(b) For 18 holes, Phil's average would be (18)(3.9) = 70.2 strokes.

So, $\mu = 70.2$.

$P(x > 72) = 1 - P(x \leq 72) \approx 1 - 0.615 = 0.385$

CHAPTER 4 REVIEW EXERCISE SOLUTIONS

1. Continuous, because the length of time spent sleeping is a random variable that cannot be counted.

3. Discrete

5. Continuous

7. No, $\sum P(x) \neq 1$.

9. Yes

11. (a)

x	f	$P(x)$
2	3	0.005
3	12	0.018
4	72	0.111
5	115	0.177
6	169	0.260
7	120	0.185
8	83	0.128
9	48	0.074
10	22	0.034
11	6	0.009
	$n = 650$	$\sum P(x) \approx 1$

(b)

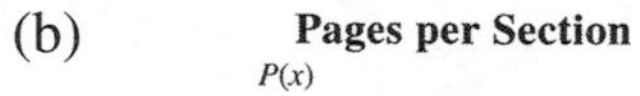

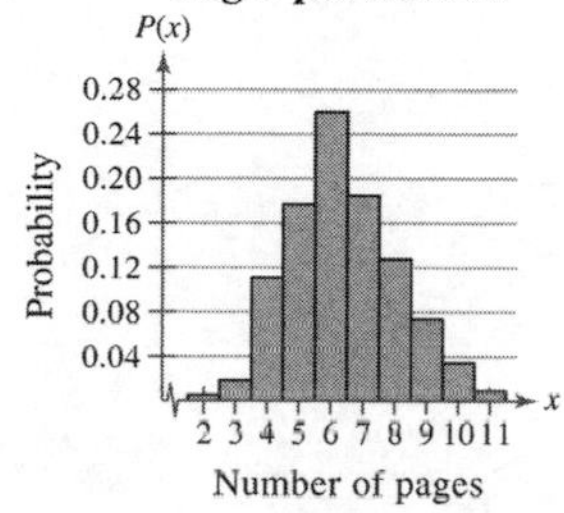

Approximately symmetric

(c)

$xP(x)$	$(x-\mu)$	$(x-\mu)^2$	$(x-\mu)^2P(x)$
0.010	−4.377	19.158	0.096
0.054	−3.377	11.404	0.205
0.444	−2.377	5.650	0.627
0.885	−1.377	1.896	0.336
1.560	−0.377	0.142	0.037
1.295	0.623	0.388	0.072
1.024	1.623	2.634	0.337
0.666	2.623	6.880	0.509
0.340	3.623	13.126	0.446
0.099	4.623	21.372	0.192
$\sum xP(x) = 6.377$			$\sum (x-\mu)^2 P(x) = 2.857$

$\mu = \sum xP(x) = 6.377 \approx 6.4$

$\sigma^2 = \sum (x-\mu)^2 P(x) = 2.857 \approx 2.9$

$\sigma = \sqrt{\sigma^2} = \sqrt{2.857} \approx 1.7$

(d) The mean is 6.4, so the average number of pages per section is about 6 pages. The standard deviation is 1.7, so most of the sections differ from the mean by no more than about 2 pages.

13. (a)

x	f	$P(x)$
0	5	0.020
1	35	0.140
2	68	0.272
3	73	0.292
4	42	0.168
5	19	0.076
6	8	0.032
	$n = 250$	$\sum P(x) = 1$

(b)

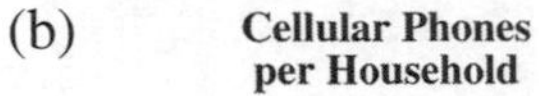

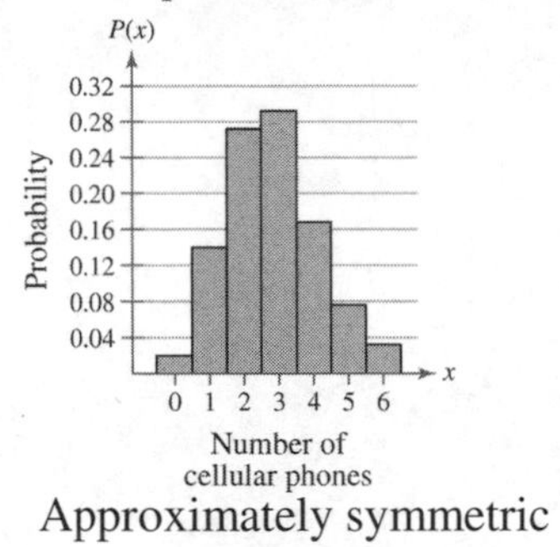

Approximately symmetric

(c)

$xP(x)$	$(x-\mu)$	$(x-\mu)^2$	$(x-\mu)^2P(x)$
0	−2.804	7.862	0.157
0.140	−1.804	3.254	0.456
0.544	−0.804	0.646	0.176
0.876	0.196	0.038	0.011
0.672	1.196	1.430	0.240
0.380	2.196	4.822	0.366
0.192	3.196	10.214	0.327
$\sum xP(x) = 2.804$			$\sum (x-\mu)^2 P(x) = 1.733$

$\mu = \sum xP(x) = 2.804 \approx 2.8$

$\sigma^2 = \sum (x-\mu)^2 P(x) = 1.733 \approx 1.7$

$\sigma = \sqrt{\sigma^2} = \sqrt{1.733} \approx 1.3$

(d) The mean is 2.8, so the average number of cellular phones per household is about 3. The standard deviation is 1.3, so most of the households differ from the mean by no more than about 1 cellular phone.

15. $E(x) = \mu = \sum xP(x) \approx 3.4$

17. No; In a binomial experiment, there are only two possible outcomes: success or failure.

19. It is a binomial experiment.
$n = 12, p = 0.24, q = 0.76, x = 0, 1, 2, 3, 4, 5, 6, 7, 8, 9, 10, 11, 12$

21. $n = 8, p = 0.25$
(a) $P(3) \approx 0.208$
(b) $P(x \geq 3) = 1 - P(x < 3) = 1 - [P(0) + P(1) + P(2)] = 1 - [0.100 + 0.267 + 0.311] = 0.322$
(c) $P(x > 3) = 1 - P(x \leq 3) =$
$1 - [P(0) + P(1) + P(2) + P(3)] = 1 - [0.100 + 0.267 + 0.311 + 0.208] = 0.114$

23. $n = 9, p = 0.43$
(a) $P(5) \approx 0.196$
(b) $P(x \geq 5) = P(5) + P(6) + P(7) + P(8) + P(9)$
$\approx 0.196 + 0.098 + 0.032 + 0.006 + 0.001$
$= 0.333$
(c) $P(x > 5) = P(6) + P(7) + P(8) + P(9)$
$\approx 0.098 + 0.032 + 0.006 + 0.001$
$= 0.137$

25. (a) $n = 5, p = 0.34$

x	$P(x)$
0	0.125
1	0.323
2	0.332
3	0.171
4	0.044
5	0.005

(b)

Help with Chores
P(x)
Probability
0.35
0.30
0.25
0.20
0.15
0.10
0.05
0 1 2 3 4 5
x
Number of women

Skewed right

(c) $\mu = np = (5)(0.34) = 1.7$
$\sigma^2 = npq = (5)(0.34)(0.66) \approx 1.1$
$\sigma = \sqrt{npq} = \sqrt{(5)(0.34)(0.66)} \approx 1.1$
The mean is 1.7 , so an average of 1.7 out of 5 women have spouses who never help with household chores. The standard deviation is 1.1, so most samples of 5 women differ from the mean by no more than 1.1 women.
(d) The values $x = 4$ and $x = 5$ are unusual because their probabilities are less than 0.05.

27. (a) $n = 4, p = 0.4$

x	$P(x)$
0	0.130
1	0.346
2	0.346
3	0.154
4	0.025

(b) Skewed right

(c) $\mu = np = (4)(0.4) = 1.6$

$\sigma^2 = npq = (4)(0.4)(0.6) \approx 1.0$

$\sigma = \sqrt{npq} = \sqrt{(4)(0.4)(0.6)} \approx 1.0$

The mean is 1.6, so an average of 1.6 out of 4 tracks have diesel engines. The standard deviation is 1.0, so most samples of 4 trucks differ from the mean by no more than 1 truck.

(d) The value $x = 4$ is unusual because its probability is less than 0.05.

29. $p = 0.22$

(a) $P(3) = (0.22)(0.78)^2 \approx 0.134$

(b) $P(4 \text{ or } 5) = P(4) + P(5) = (0.22)(0.78)^2 + (0.22)(0.78)^4 \approx 0.186$

(c) $P(x > 7) = 1 - P(x \le 7)$

$= 1 - [P(1) + P(2) + P(3) + P(4) + P(5) + P(6) + P(7)]$

$= 1 - [0.220 + 0.172 + 0.134 + 0.104 + 0.081 + 0.064 + 0.050]$

$= 0.175$

31. $\mu = \frac{6755}{69} \approx 97.9$ tornado deaths/year $\rightarrow \mu = \frac{97.9}{365} \approx 0.268$ deaths/day

(a) $P(0) \approx \frac{(0.268)^0(2.71828)^{-0.268}}{0!} \approx 0.765$

(b) $P(1) \approx \frac{(0.268)^1(2.71828)^{-0.268}}{1!} \approx 0.205$

(c) $P(x \le 2) = P(0) + P(1) + P(2)$

$\approx 0.765 + 0.205 + 0.027$

$= 0.997$

(d) $P(x > 1) = 1 - P(x \le 1) = 1 - [P(0) + P(1)] = 1 - [0.765 + 0.205] = 0.030$

This event is unusual because its probability is less than 0.05.

33. The probability increases as the rate increases, and decreases as the rate decreases.

CHAPTER 4 QUIZ SOLUTIONS

1. (a) Discrete because the number of lightning strikes that occur in Wyoming during the month of June is a random variable that is countable.

(b) Continuous because the fuel (in gallons) used by the Space Shuttle during takeoff is a random variable that has an infinite number of possible outcomes and cannot be counted.

2. (a)

x	f	$P(x)$
1	114	0.400
2	74	0.260
3	76	0.267
4	18	0.063
5	3	0.011
	$n = 285$	$\sum P(x) \approx 1$

(b)

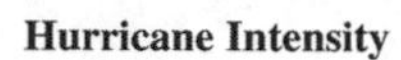

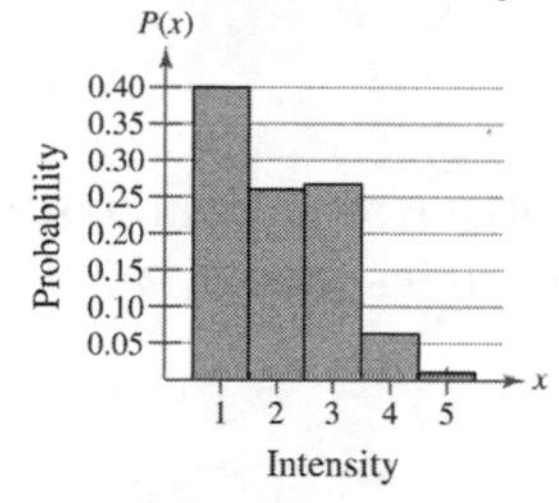

Skewed right

(c)

$xP(x)$	$(x - \mu)$	$(x - \mu)^2$	$(x - \mu)^2 P(x)$
0.400	−1.028	1.057	0.423
0.520	−0.028	0.001	0.000
0.801	0.972	0.945	0.252
0.252	1.972	3.889	0.245
0.055	2.972	8.833	0.097
$\sum xP(x) = 2.028$			$\sum (x-\mu)^2 P(x) = 1.017$

$\mu = \sum xP(x) = 2.028 \approx 2.0$

$\sigma^2 = \sum \leq (x-\mu)^2 P(x) = 1.017 = 1.0$

$\sigma = \sqrt{\sigma^2} = \sqrt{1.017} = 1.0$

On average, the intensity of a hurricane will be 2.0. The standard deviation is 1.0, so most hurricane intensities will differ from the mean by no more than 1.0.

(d) $P(x \geq 4) = P(4) + P(5) = \dfrac{18}{285} + \dfrac{3}{285} = 0.074$

3. (a) $n = 6, p = 0.85$

x	$P(x)$
0	0.00001
1	0.00039
2	0.00549
3	0.04145
4	0.17618
5	0.39933
6	0.37715

(b)

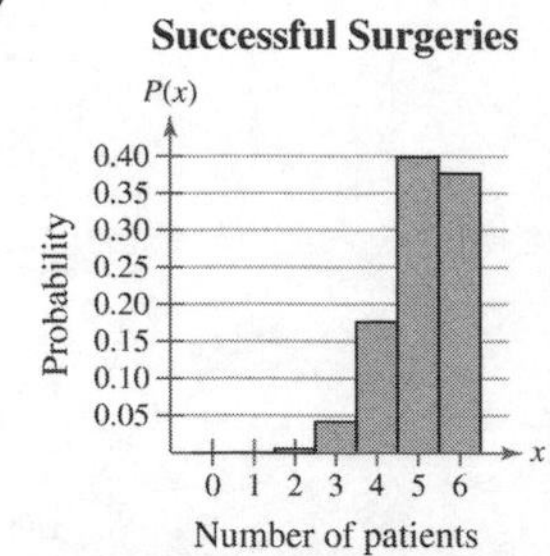

(c) $\mu = np = (6)(0.85) = 5.1$

$\sigma^2 = npq = (6)(0.85)(0.15) = 0.8$

$\sigma = \sqrt{npq} = \sqrt{(16)(0.85)(0.15)} = 0.9$

The average number of successful surgeries is 5.1 out of 6. The standard deviation is 0.9, so most samples of 6 surgeries differ from the mean by no more than 0.9 surgery.

(d) $P(3) \approx 0.041$

This event is unusual because its probability is less than 0.05.

(e) $P(x < 4) = P(0) + P(1) + P(2) + P(3) \approx 0.047$

This event is unusual because its probability is less than 0.05.

4. $\mu = 5$

(a) $P(5) \approx \dfrac{(5)^5(2.71828)^{-5}}{5!} \approx 0.175$

(b) $P(x < 5) = P(0) + P(1) + P(2) + P(3) + P(4)$

$\approx 0.007 + 0.034 + 0.084 + 0.140 + 0.175$

$= 0.440$

(c) $P(0) \approx \dfrac{(5)^0(2.71828)^{-5}}{0!} \approx 0.007$

5. $p = 0.602$

$P(4) = (0.602)(0.398)^3 \approx 0.038$

This event is unusual because its probability is less than 0.05.

6. $p = 0.602$

$P(2 \text{ or } 3) = P(2) + P(3) = (0.602)(0.398)^1 + (0.602)(0.398)^2 \approx 0.335$

This event is not unusual because its probability is greater than 0.05.

CHAPTER

5

Normal Probability Distributions

5.1 INTRODUCTION TO NORMAL DISTRIBUTIONS AND THE STANDARD NORMAL DISTRIBUTION

5.1 Try It Yourself Solutions

1a. A: $x = 45$, B: $x = 60$, C: $x = 45$ (B has the greatest mean.)
b. Curve C is more spread out, so curve C has the greatest standard deviation.

2a. Mean = 660
b. Inflection points: 630 and 690
Standard deviation = 30

3a. (1) 0.0143 (2) 0.9850

4a.

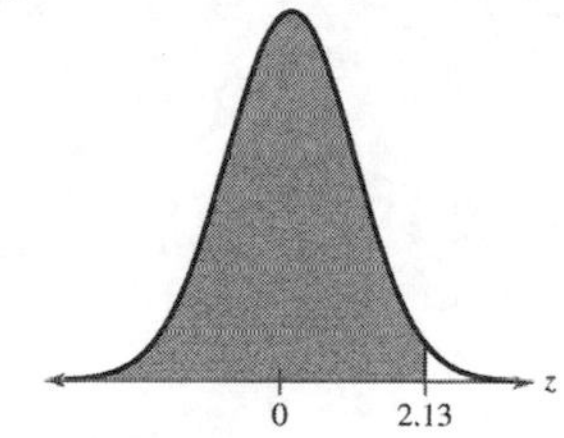

b. 0.9834

5a.

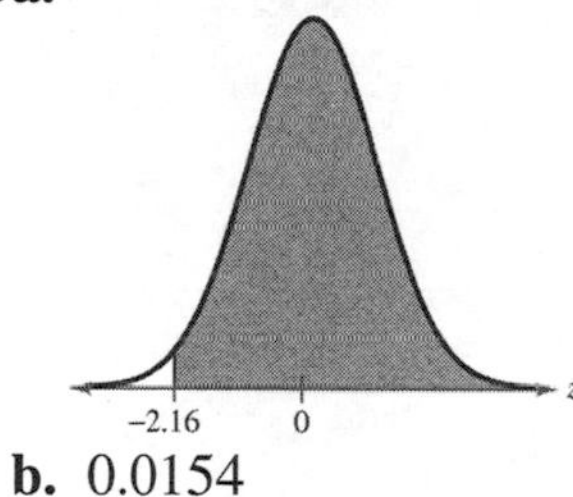

b. 0.0154

6a. 0.0885 **b.** 0.0152
c. Arca = 0.0885 − 0.0152 − 0.0733

5.1 EXERCISE SOLUTIONS

1. Answers will vary.

3. 1

5. Answers will vary.
Similarities: The two curves will have the same line of symmetry.
Differences: The curve within the larger standard deviation will be more spread out than the curve with the smaller standard deviation.

7. $\mu = 0$, $\sigma = 1$

9. "The" standard normal distribution is used to describe one specific normal distribution ($\mu = 0$, $\sigma = 1$). "a" normal distribution is used to describe a normal distribution with any mean and standard deviation.

11. No, the graph crosses the x-axis.

13. Yes, the graph fulfills the properties of the normal distribution.

15. No, the graph is skewed to the right.

17. The histogram represents data from a normal distribution because it is bell-shaped.

19. (Area left of $z = -1.3$) = 0.0968

21. (Area right of $z = 2$) = 1 – (Area left of $z = 2$)
= 1 – 0.9772
= 0.0228

23. (Area left of $z = 0$) – (Area left of $z = -2.25$) = 0.5 – 0.0122 = 0.4878

25. 0.5319

27. 0.005

29. 1 – 0.2578 = 0.7422

31. 1 – 0.3613 = 0.6387

33. 0.9979 – 0.5 = 0.4979

35. 0.9750 – 0.0250 = 0.950

37. 0.1003 + 0.1003 = 0.2006

39. (a)

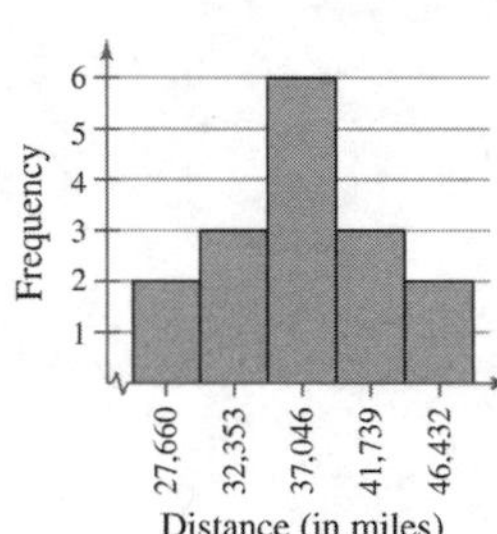

It is reasonable to assume that the life spans are normally distributed because the histogram is symmetric and bell-shaped.

(b) $\bar{x} = 37{,}234.7$

$s = 6259.2$

(c) The sample mean of 37,234.7 hours is less than the claimed mean, so, on average, the tires in the sample lasted for a shorter time. The sample standard deviation of 6259.2 is greater than the claimed standard deviation, so the tires in the sample had greater variation in life span than the manufacturer's claim.

41. (a) A = 105; B = 113; C = 121; D = 127

(b) $x = 105 \Rightarrow z = \dfrac{x-\mu}{\sigma} = \dfrac{105-115}{3.6} \approx -2.78$

$x = 113 \Rightarrow z = \dfrac{x-\mu}{\sigma} = \dfrac{113-115}{3.6} \approx -0.56$

$x = 121 \Rightarrow z = \dfrac{x-\mu}{\sigma} = \dfrac{121-115}{3.6} \approx 1.67$

$x = 127 \Rightarrow z = \dfrac{x-\mu}{\sigma} = \dfrac{127-115}{3.6} \approx 3.33$

(c) $x = 105$ is unusual because its corresponding z-score (-2.78) lies more than 2 standard deviations from the mean, and $x = 127$ is very unusual because its corresponding z-score (3.33) lies more than 3 standard deviations from the mean.

43. (a) A = 1241; B = 1392; C = 1924; D = 2202

(b) $x = 1241 \Rightarrow z = \frac{x-\mu}{\sigma} = \frac{1241-1509}{312} \approx -0.86$

$x = 1392 \Rightarrow z = \frac{x-\mu}{\sigma} = \frac{1392-1509}{312} = -0.375$

$x = 1924 \Rightarrow z = \frac{x-\mu}{\sigma} = \frac{1924-1509}{312} \approx 1.33$

$x = 2202 \Rightarrow z = \frac{x-\mu}{\sigma} = \frac{2202-1509}{312} \approx 2.22$

(c) $x = 2202$ is unusual because its corresponding z-score (2.22) lies more than 2 standard deviations from the mean.

45. 0.9750

47. $1 - 0.0225 = 0.9775$

49. $0.9987 - 0.1587 = 0.8400$

51. $P(z < 1.45) = 0.9265$

53. $P(z > 2.175) = 1 - P(z < 2.175) = 1 - 0.9852 = 0.0148$

55. $P(-0.89 < z < 0) = 0.5 - 0.1867 = 0.3133$

57. $P(-1.65 < z < 1.65) = 0.9505 - 0.0495 = 0.901$

59. $P(z < -2.58 \text{ or } z > 2.58) = 2(0.0049) = 0.0098$

61.

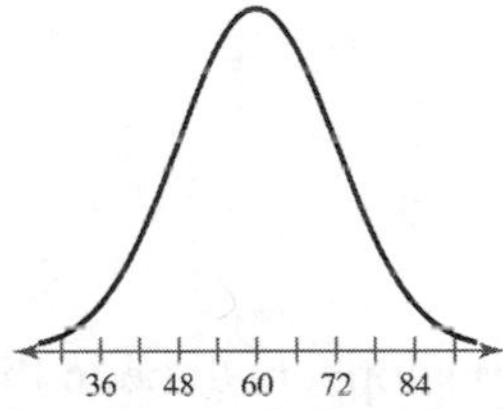

The normal distribution curve is centered at its mean (60) and has 2 points of inflection (48 and 72) representing $\mu \pm \sigma$.

63. (a) Area under curve = area of square = (base)(height) = (1)(1) = 1
(b) $P(0.25 < x\ 0.5)$ = (base)(height) = (0.25)(1) = 0.25
(c) $P(0.3 < x\ 0.7)$ = (base)(height) = (0.4)(1) = 0.4

5.2 NORMAL DISTRIBUTIONS: FINDING PROBABILITIES

5.2 Try It Yourself Solutions

1a.

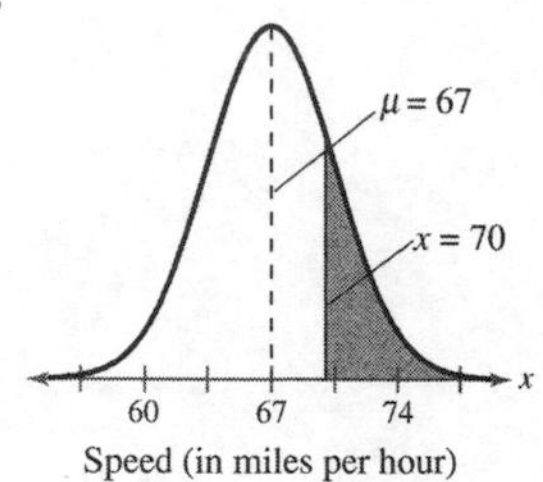

b. $z = \dfrac{x-\mu}{\sigma} = \dfrac{70-67}{3.5} \approx 0.86$

c. $P(z < 0.86) = 0.8051$

$P(z > 0.86) = 1 - 0.8051 = 0.1949$

d. The probability that a randomly selected vehicle is violating the 70 mile per hour speed limit is 0.1949.

2a.

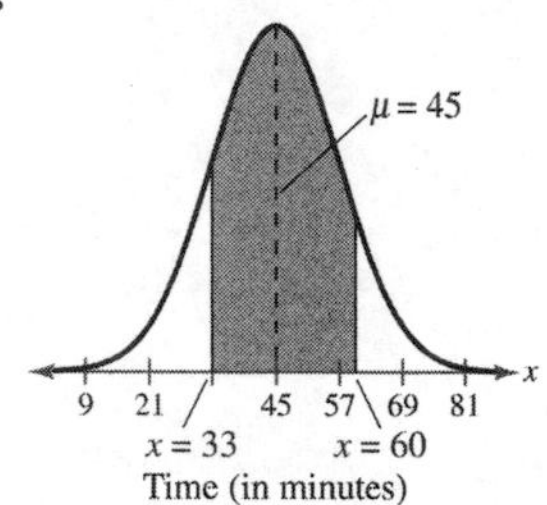

b. $z = \dfrac{x-\mu}{\sigma} = \dfrac{33-45}{12} = -1$

$z = \dfrac{x-\mu}{\sigma} = \dfrac{60-45}{12} = 1.25$

c. $P(z < -1) = 0.1587$

$P(z > 1.25) = 0.8944$

$0.8944 - 0.1587 = 0.7357$

d. If 150 shoppers enter the store, then you would expect $150(0.7357) \approx 110$ shoppers to be in the store between 33 and 60 minutes.

3a. Read user's guide for the technology tool.

b. Enter the data.

$P(100 < x < 150) = P(-0.97 < z < 0.46) = 0.5105$

c. The probability that a randomly selected U.S. person's triglyceride level is between 100 and 150 is 0.5105.

5.2 EXERCISE SOLUTIONS

1. $P(x < 170) = P(z < -0.2) = 0.4207$

3. $P(x > 182) = P(z > 0.4) = 1 - 0.6554 = 0.3446$

5. $P(160 < x < 170) = P(-0.7 < z < -0.2) = 0.4207 - 0.2420 = 0.1787$

7. $P(200 < x < 450) = P(-2.64 < z < -0.39) = 0.3483 - 0.0041 = 0.3442$

9. $P(220 < x < 255) = P(0.29 < z < 1.22) = 0.8888 - 0.6141 = 0.2747$

11. $P(145 < x < 155) = P(0.39 < z < 2.34) = 0.9904 - 0.6517 = 0.3387$

13. (a) $P(x < 66) = P(z < -1.3) = 0.0968$
(b) $P(66 < x < 72) = P(-1.3 < z < 0.7) = 0.7580 - 0.0968 = 0.6612$
(c) $P(x > 72) = P(z > 0.7) = 1 - P(z < 0.7) = 1 - 0.7580 = 0.2420$
(d) No, none of these events are unusual because their probabilities are greater than 0.05.

15. (a) $P(x < 15) = P(z < -0.89) = 0.1867$
(b) $P(18 < x < 25) = P(-0.41 < z < 0.70) = 0.7580 - 0.3409 = 0.4171$
(c) $P(x > 34) = P(z > 2.13) = 1 - P(z < 2.13) = 1 - 0.9834 = 0.0166$
(d) Yes, the event in part (c) is unusual because its probability is less than 0.05.

17. (a) $P(x < 5) = P(z < -2) = 0.0228$
(b) $P(5.5 < x < 9.5) = P(-1.5 < z < 2.5) = 0.9938 - 0.0668 = 0.927$
(c) $P(x > 10) = P(z > 3) = 1 - P(z < 3) = 1 - 0.9987 = 0.0013$

19. (a) $P(x < 4) = P(z < -2.44) = 0.0073$
(b) $P(5 < x < 7) = P(-1.33 < z < 0.89) = 0.8133 - 0.0918 = 0.7215$
(c) $P(x > 8) = P(z > 2) = 1 - 0.9722 = 0.0228$

21. (a) $P(x < 600) = P(z < 0.96) = 0.8315 \Rightarrow 83.15\%$
(b) $P(x > 550) = P(z > 0.51) = 1 - P(z < 0.51) = 1 - 0.6950 = 0.3050$
$(1000)(0.3050) = 305 \Rightarrow 305$ scores

23. (a) $P(x < 225) = P(z < 0.42) = 0.6628 \Rightarrow 66.28\%$
(b) $P(x > 260) = P(z > 1.35) = 1 - P(z < 1.35) = 1 - 0.9115 = 0.0885$
$(250)(0.0885) = 22.125 \Rightarrow 22$ men

25. (a) $P(x > 4) = P(z > -3) = 1 - P(z < -3) = 1 - 0.0013 = 0.9987 \Rightarrow 99.87\%$
(b) $P(x < 5) = P(z < -2) = 0.0228$
$(35)(0.0228) = 0.798 \Rightarrow 1$ adult

27. $P(x > 2065) = P(z > 2.17) = 1 - P(z < 2.17) = 1 - 0.9850 = 0.0150 \Rightarrow 1.5\%$
It is unusual for a battery to have a life span that is more than 2065 hours because the probability is less than 0.05.

29. (a) 0.3085
(b) 0.1499
(c) 0.0668
No, because $0.0668 > 0.05$, this event is not unusual.

31. Out of control, because the 10th observation plotted beyond 3 standard deviations.

33. Out of control, because the first nine observations lie below the mean and since two out of three consecutive points lie more than 2 standard deviations from the mean.

5.3 NORMAL DISTRIBUTIONS: FINDING VALUES

5.3 Try It Yourself Solutions

1ab. (1)

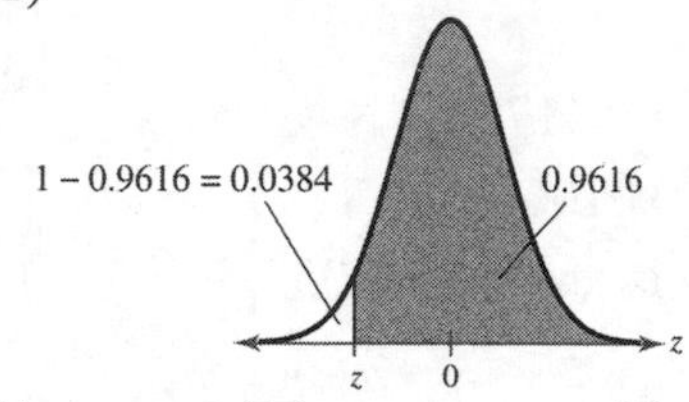

(2)

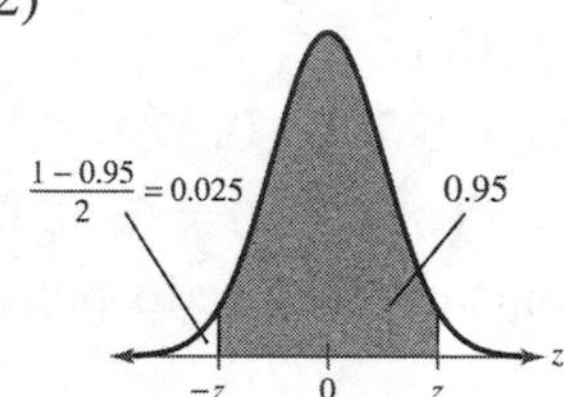

c. (1) $z = -1.77$ (2) $z = \pm 1.96$

2a. (1)

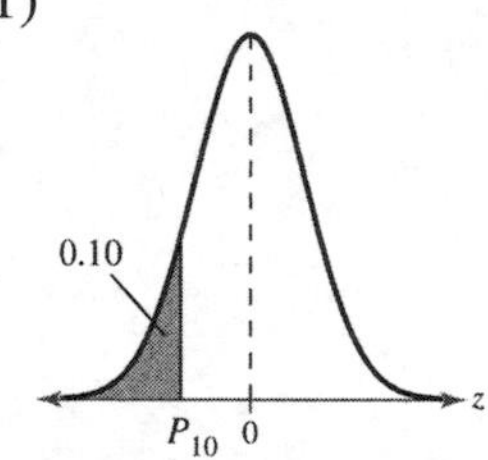

(2)

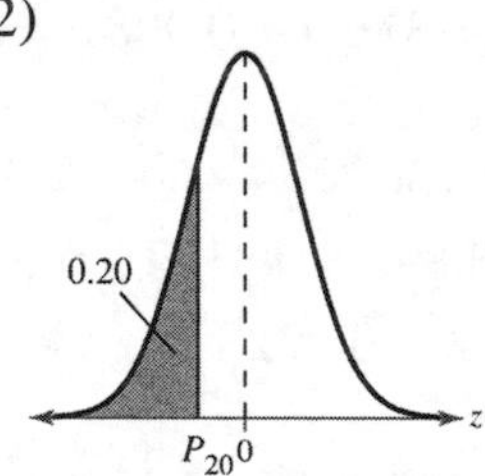

(3)

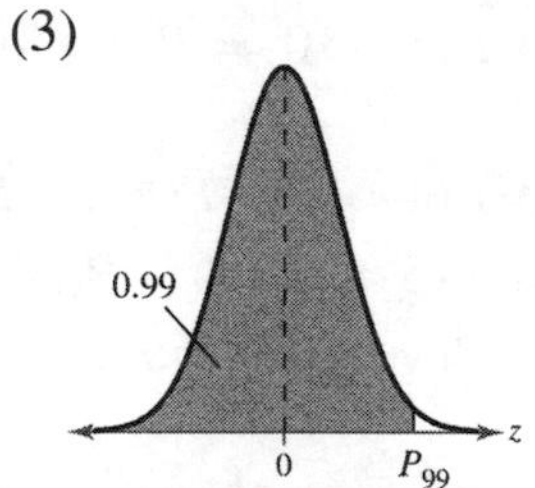

b. (1) use area = 0.1003 (2) use area = 0.2005 (3) use area = 0.9901
c. (1) $z = -1.28$ (2) $z = -0.84$ (3) $z = 2.33$

3a. $\mu = 52$, $\sigma = 15$

b. $z = -2.33 \Rightarrow x = \mu + z\sigma = 52 + (-2.33)(15) = 17.05$

$z = 3.10 \Rightarrow x = \mu + z\sigma = 52 + (3.10)(15) = 98.50$

$z = 0.58 \Rightarrow x = \mu + z\sigma = 52 + (0.58)(15) = 60.70$

c. 17.05 pounds is below the mean, 60.7 pounds and 98.5 pounds are above the mean.

4a.

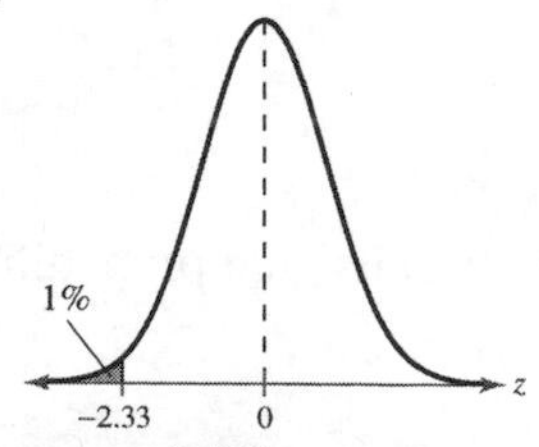

b. $z = -2.33$

c. $x = \mu + z\sigma = 129 + (-2.33)(5.18) \approx 116.93$

d. So, the longest braking distance a Nissan Altima could have and still be in the bottom 1% is about 117 feet.

5a.

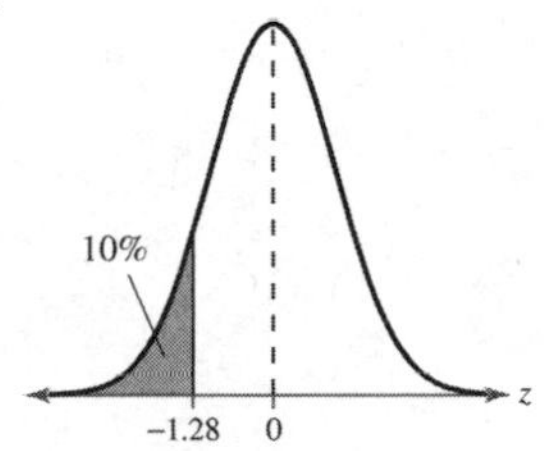

b. $z = -1.28$

c. $x = \mu + z\sigma = 11.2 + (-1.28)(2.1) = 8.512$

d. So, the maximum length of time an employee could have worked and still be laid off is about 8.5 years.

5.3 EXERCISE SOLUTIONS

1. $z = -0.81$

3. $z = 2.39$

5. $z = -1.645$

7. $z = 1.555$

9. $z = -1.04$

11. $z = 1.175$

13. $z = -0.67$

15. $z = 0.67$

17. $z = -0.38$

19. $z = -0.58$

21. $z = \pm 1.645$

23.

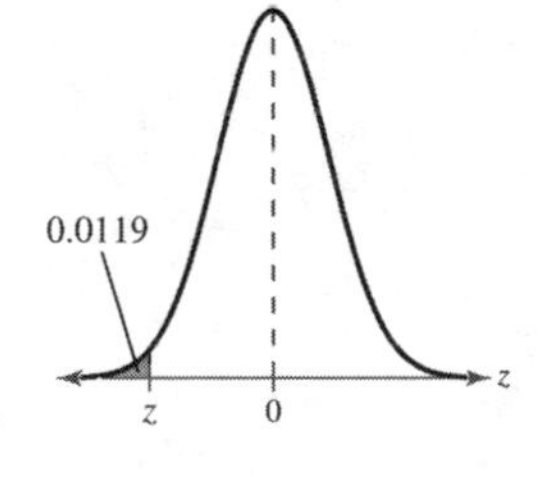

$\Rightarrow z = -1.18$

25.

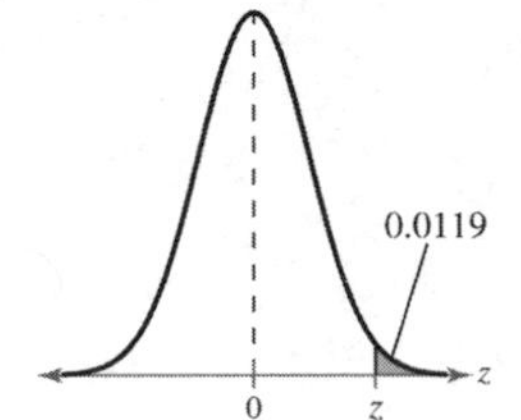

$\Rightarrow z = 1.18$

27.

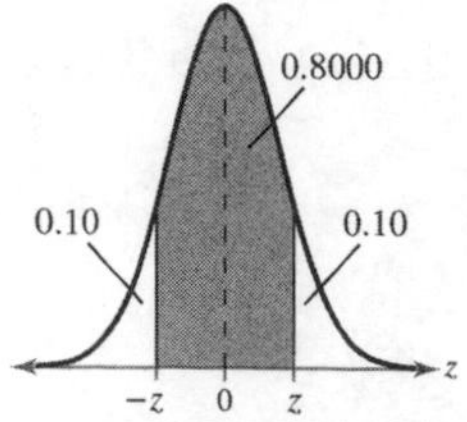

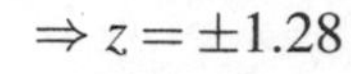
$\Rightarrow z = \pm 1.28$

29.

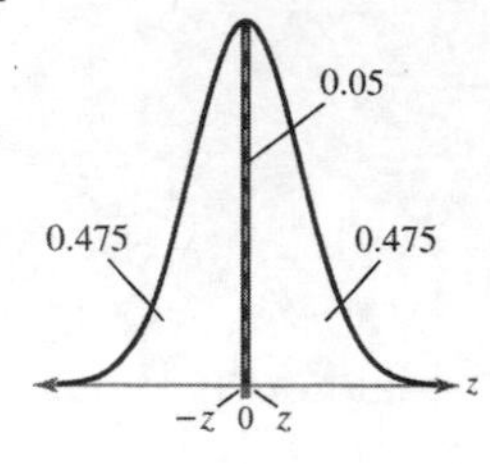

$\Rightarrow z = \pm 0.06$

31. (a) 95th percentile $\Rightarrow$ Area = 0.95 $\Rightarrow$ $z = 1.645$
$x = \mu + z\sigma = 64.3 + (1.645)(2.6) \approx 68.58$ inches
(b) 1st quartile $\Rightarrow$ Area = 0.25 $\Rightarrow$ $z = -0.67$
$x = \mu + z\sigma = 64.3 + (-0.67)(2.6) \approx 62.56$ inches

33. (a) 5th percentile $\Rightarrow$ Area = 0.05 $\Rightarrow$ $z = -1.645$
$x = \mu + z\sigma = 204 + (-1.645)(25.7) \approx 161.72$ days
(b) 3rd quartile $\Rightarrow$ Area = 0.75 $\Rightarrow$ $z = 0.67$
$x = \mu + z\sigma = 204 + (0.67)(25.7) \approx 221.22$ days

35. (a) Top 5% $\Rightarrow$ Area = 0.95 $\Rightarrow$ $z = 1.645$
$x = \mu + z\sigma = 6.1 + (1.645)(1.0) \approx 7.75$ hours
(b) Middle 50% $\Rightarrow$ Area = 0.25 to 0.75 $\Rightarrow$ $z = \pm 0.675$
$x = \mu + z\sigma = 6.1 + (\pm 0.67)(1.0) = 5.43$ to 6.77 hours

37. Upper 4.5% $\Rightarrow$ Area = 0.955 $\Rightarrow$ $z = 1.70$
$x = \mu + z\sigma = 32 + (1.70)(0.36) \approx 32.61$ ounces

39. (a) 18.89 pounds
(b) 12.05 pounds

41. Bottom 10% $\Rightarrow$ Area = 0.10 $\Rightarrow$ $z = -1.28$
$x = \mu + z\sigma = 30{,}000 + (-1.28)(2500) = 26{,}800$
Tires that wear out by 26,800 miles will be replaced free of charge.

5.4 SAMPLING DISTRIBUTIONS AND THE CENTRAL LIMIT THEOREM

5.4 Try It Yourself Solutions

1a.

Sample	Mean	Sample	Mean	Sample	Mean	Sample	Mean
1, 1, 1	1	3, 1, 1	1.67	5, 1, 1	2.33	7, 1, 1	3
1, 1, 3	1.67	3, 1, 3	2.33	5, 1, 3	3	7, 1, 3	3.67
1, 1, 5	2.33	3, 1, 5	3	5, 1, 5	3.67	7, 1, 5	4.33
1, 1, 7	3	3, 1, 7	3.67	5, 1, 7	4.33	7, 1, 7	5
1, 3, 1	1.67	3, 3, 1	2.33	5, 3, 1	3	7, 3, 1	3.67
1, 3, 3	2.33	3, 3, 3	3	5, 3, 3	3.67	7, 3, 3	4.33
1, 3, 5	3	3, 3, 5	3.67	5, 3, 5	4.33	7, 3, 5	5
1, 3, 7	3.67	3, 3, 7	4.33	5, 3, 7	5	7, 3, 7	5.67
1, 5, 1	2.33	3, 5, 1	3	5, 5, 1	3.67	7, 5, 1	4.33
1, 5, 3	3	3, 5, 3	3.67	5, 5, 3	4.33	7, 5, 3	5
1, 5, 5	3.67	3, 5, 5	4.33	5, 5, 5	5	7, 5, 5	5.67
1, 5, 7	4.33	3, 5, 7	5	5, 5, 7	5.67	7, 5, 7	6.33
1, 7, 1	3	3, 7, 1	3.67	5, 7, 1	4.33	7, 7, 1	5
1, 7, 3	3.67	3, 7, 3	4.33	5, 7, 3	5	7, 7, 3	5.67
1, 7, 5	4.33	3, 7, 5	5	5, 7, 5	5.67	7, 7, 5	6.33
1, 7, 7	5	3, 7, 7	5.67	5, 7, 7	6.33	7, 7, 7	7

b.

$\bar{x}$	f	Probability
1	1	0.0156
1.67	3	0.0469
2.33	6	0.0938
3	10	0.1562
3.67	12	0.1875
4.33	12	0.1875
5	10	0.1563
5.67	6	0.0938
6.33	3	0.0469
7	1	0.0156

$\mu_{\bar{x}} = 4,\ \sigma_{\bar{x}}^2 \approx 1.667,\ \sigma_{\bar{x}} \approx 1.291$

c. $\mu_{\bar{x}} = \mu = 4,$

$$\sigma_{\bar{x}}^2 = \frac{\sigma^2}{n} = \frac{5}{3} \approx 1.667,$$

$$\sigma_{\bar{x}} = \frac{\sigma}{\sqrt{n}} = \frac{\sqrt{5}}{\sqrt{3}} \approx 1.291$$

2a. $\mu_{\bar{x}} = \mu = 63,\ \sigma_{\bar{x}} = \dfrac{\sigma}{\sqrt{n}} = \dfrac{11}{\sqrt{64}} \approx 1.4$

b. $n = 64$

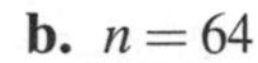

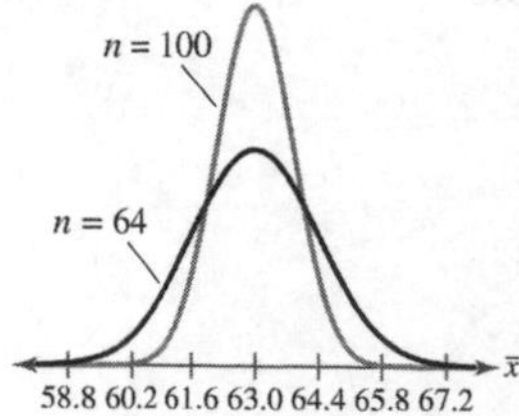

c. With a smaller sample size, the mean stays the same but the standard deviation increases.

3a. $\mu_{\bar{x}} = \mu = 3.5,\ \sigma_{\bar{x}} = \dfrac{\sigma}{\sqrt{n}} = \dfrac{0.2}{\sqrt{16}} = 0.05$

b.

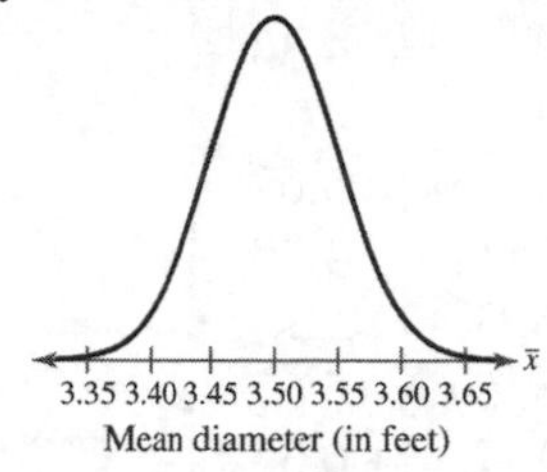

4a. $\mu_{\bar{x}} = \mu = 25,\ \sigma_{\bar{x}} = \dfrac{\sigma}{\sqrt{n}} = \dfrac{1.5}{\sqrt{100}} = 0.15$

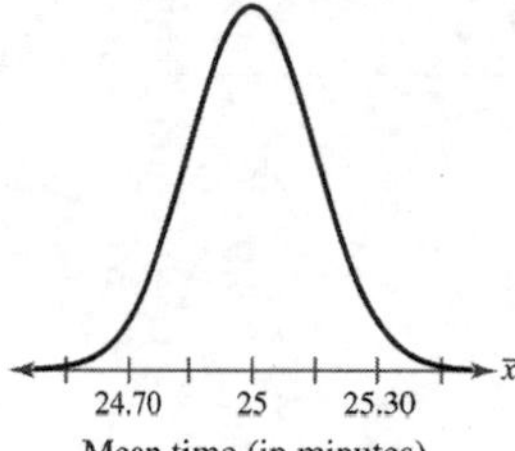

b. $\bar{x} = 24.7:\ z = \dfrac{\bar{x} - \mu}{\frac{\sigma}{\sqrt{n}}} = \dfrac{24.7 - 25}{\frac{1.5}{\sqrt{100}}} = -\dfrac{0.3}{0.15} = -2$

$\bar{x} = 25.5:\ z = \dfrac{\bar{x} - \mu}{\frac{\sigma}{\sqrt{n}}} = \dfrac{25.5 - 25}{\frac{1.5}{\sqrt{100}}} = -\dfrac{0.5}{0.15} \approx 3.33$

c. $P(z < -2) = 0.0228$

$P(z < 3.33) = 0.9996$

$P(24.7 < \bar{x} < 25.5) = P(-2 < z < 3.33) = 0.9996 - 0.0228 = 0.9768$

d. Of the samples of 100 drivers ages 15 to 19, 97.68% will have a mean driving time that is between 24.7 and 25.5 minutes.

5a. $\mu_{\bar{x}} = \mu = 290,600,\ \sigma_{\bar{x}} = \dfrac{\sigma}{\sqrt{n}} = \dfrac{36,000}{\sqrt{12}} \approx 10,392.30$

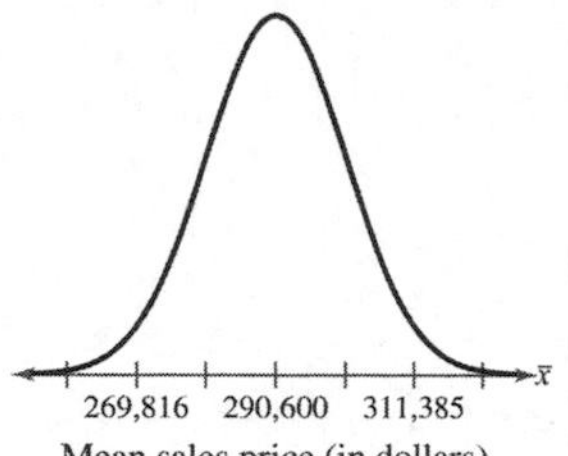

b. $\bar{x} = 265,000:\ z = \dfrac{\bar{x} - \mu}{\dfrac{\sigma}{\sqrt{n}}} = \dfrac{265,000 - 290,600}{\dfrac{36,000}{\sqrt{12}}} \approx \dfrac{-25,600}{10,392.30} \approx -2.46$

c. $P(\bar{x} > 265,000) = P(z > -2.46) = 1 - P(z < -2.46) = 1 - 0.0069 = 0.9931$

d. 99.31% of samples of 12 single-family houses will have a mean sales price greater than \$265,000.

6a. $x = 200:\ z = \dfrac{x - \mu}{\sigma} = \dfrac{200 - 190}{48} \approx 0.21$

$\bar{x} = 200:\ z = \dfrac{\bar{x} - \mu}{\dfrac{\sigma}{\sqrt{n}}} = \dfrac{200 - 190}{\dfrac{48}{\sqrt{10}}} \approx \dfrac{10}{15.18} \approx 0.66$

b. $P(z < 0.21) = 0.5832$

$P(z < 0.66) = 0.7454$

c. There is about a 58% chance that an LCD computer monitor will cost less than \$200. There is about a 75% chance that the mean of a sample of 10 LCD computer monitors is less than \$200.

5.4 EXERCISE SOLUTIONS

1. $\mu_{\bar{x}} = \mu = 150$

$\sigma_{\bar{x}} = \dfrac{\sigma}{\sqrt{n}} = \dfrac{25}{\sqrt{50}} \approx 3.536$

3. $\mu_{\bar{x}} = \mu = 150$

$\sigma_{\bar{x}} = \dfrac{\sigma}{\sqrt{n}} = \dfrac{25}{\sqrt{250}} \approx 1.581$

5. False. As the size of a sample increases, the mean of the distribution of sample means does not change.

7. False. A sampling distribution is normal if either $n \geq 30$ or the population is normal.

9. (c) Because $\mu_{\bar{x}} = 16.5,\ \sigma_{\bar{x}} = \dfrac{\sigma}{\sqrt{n}} = \dfrac{11.9}{\sqrt{100}} = 1.19,$ and the graph approximates a normal curve.

11.

Sample	Mean	Sample	Mean	Sample	Mean
2, 2, 2	2	4, 4, 8	5.33	8, 16, 2	8.67
2, 2, 4	2.67	4, 4, 16	8	8, 16, 4	9.33
2, 2, 8	4	4, 8, 2	4.67	8, 16, 8	10.67
2, 2, 16	6.67	4, 8, 4	5.33	8, 16, 16	13.33
2, 4, 2	2.67	4, 8, 8	6.67	16, 2, 2	6.67
2, 4, 4	3.33	4, 8, 16	9.33	16, 2, 4	7.33
2, 4, 8	4.67	4, 16, 2	7.33	16, 2, 8	8.67
2, 4, 16	7.33	4, 16, 4	8	16, 2, 16	11.33
2, 8, 2	4	4, 16, 8	9.33	16, 4, 2	7.33
2, 8, 4	4.67	4, 16, 16	12	16, 4, 4	8
2, 8, 8	6	8, 2, 2	4	16, 4, 8	9.33
2, 8, 16	8.67	8, 2, 4	4.67	16, 4, 16	12
2, 16, 2	6.67	8, 2, 8	6	16, 8, 2	8.67
2, 16, 4	7.33	8, 2, 16	8.67	16, 8, 4	9.33
2, 16, 8	8.67	8, 4, 2	4.67	16, 8, 8	10.67
2, 16, 16	11.33	8, 4, 4	5.33	16, 8, 16	13.33
4, 2, 2	2.67	8, 4, 8	6.67	16, 16, 2	11.33
4, 2, 4	3.33	8, 4, 16	9.33	16, 16, 4	12
4, 2, 8	4.67	8, 8, 2	6	16, 16, 8	13.33
4, 2, 16	7.33	8, 8, 4	6.67	16, 16, 16	16
4, 4, 2	3.33	8, 8, 8	8		
4, 4, 4	4	8, 8, 16	10.67		

$\mu = 7.5,\ \sigma \approx 5.36$

$$\mu_{\bar{x}} = 7.5,\ \sigma_{\bar{x}} \approx \frac{5.36}{\sqrt{3}} \approx 3.09$$

The means are equal but the standard deviation of the sampling distribution is smaller.

13. $$z = \frac{\bar{x}-\mu}{\frac{\sigma}{\sqrt{n}}} = \frac{24.3-24}{\frac{1.25}{\sqrt{64}}} \approx \frac{0.3}{0.156} \approx 1.92$$

$P(\bar{x} < 24.3) = P(z < 1.92) = 0.9726$

The probability is not unusual because it is greater than 0.05.

15. $$z = \frac{\bar{x}-\mu}{\frac{\sigma}{\sqrt{n}}} = \frac{551-550}{\frac{3.7}{\sqrt{45}}} \approx \frac{1}{0.552} \approx 1.81$$

$P(\bar{x} > 551) = P(z > 1.81) = 1 - P(z < 1.81) = 1 - 0.9649 = 0.0351$

The probability is unusual because it is less than 0.05.

17. $\mu_{\bar{x}} = 7.6$

$$\sigma_{\bar{x}} = \frac{\sigma}{\sqrt{n}} = \frac{0.35}{\sqrt{12}} \approx 0.101$$

7.3 7.4 7.5 7.6 7.7 7.8 7.9 $\bar{x}$

Mean time (in hours)

19. $\mu_{\bar{x}} = 235$

$$\sigma_{\bar{x}} = \frac{\sigma}{\sqrt{n}} = \frac{62}{\sqrt{20}} \approx 13.864$$

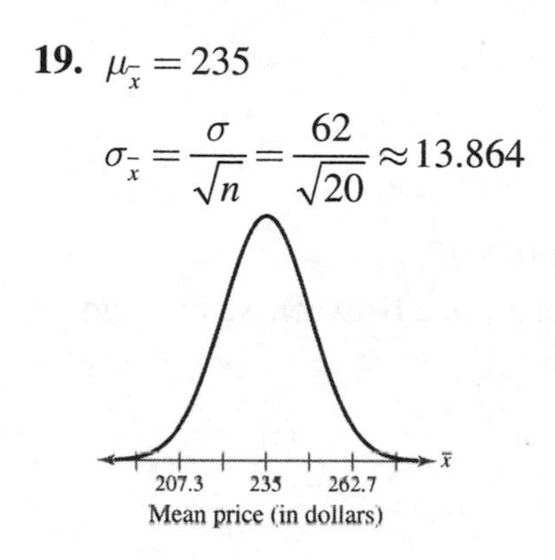

21. $\mu_{\bar{x}} = 188.4$

$$\sigma_{\bar{x}} = \frac{\sigma}{\sqrt{n}} = \frac{54.5}{\sqrt{25}} = 10.9$$

166.6 188.4 210.2 $\bar{x}$

Mean consumption of fresh vegetables (in pounds)

23. $n = 24$: $\mu_{\bar{x}} = 7.6$, $\sigma_{\bar{x}} = \frac{\sigma}{\sqrt{n}} = \frac{0.35}{\sqrt{24}} \approx 0.07$

$n = 36$: $\mu_{\bar{x}} = 7.6$, $\sigma_{\bar{x}} = \frac{\sigma}{\sqrt{n}} = \frac{0.35}{\sqrt{36}} \approx 0.06$

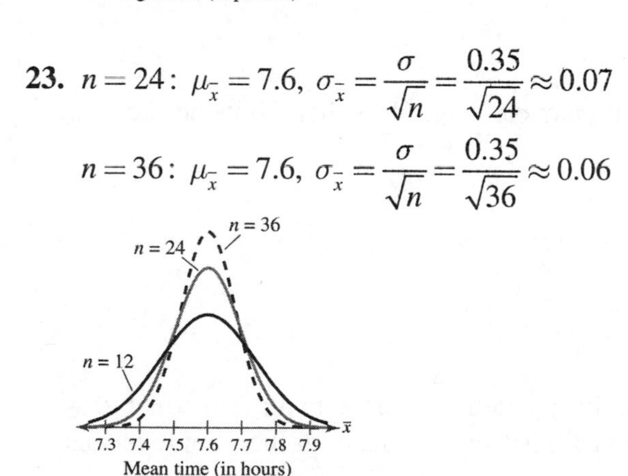

As the sample size increases, the standard error decreases, while the mean of the sample means remains constant.

25. $z=\dfrac{\bar{x}-\mu}{\frac{\sigma}{\sqrt{n}}}=\dfrac{60{,}000-63{,}500}{\frac{6100}{\sqrt{35}}}\approx\dfrac{-3500}{1031.09}\approx-3.39$

$P(\bar{x}<60{,}000)=P(z<-3.39)=0.0003$

Only 0.03% of samples of 35 specialists will have a mean salary less than \$60,000. This is an extremely unusual event.

27. $z=\dfrac{\bar{x}-\mu}{\frac{\sigma}{\sqrt{n}}}=\dfrac{2.695-2.714}{\frac{0.045}{\sqrt{32}}}\approx\dfrac{-0.019}{0.00795}\approx-2.39$

$z=\dfrac{\bar{x}-\mu}{\frac{\sigma}{\sqrt{n}}}=\dfrac{2.725-2.714}{\frac{0.045}{\sqrt{32}}}\approx\dfrac{0.011}{0.00795}\approx1.38$

$P(2.695<\bar{x}<2.725)=P(-2.39<z<1.38)\approx0.9162-0.0084=0.9078$

About 91% of samples of 32 gas stations that week will have a mean price between \$2.695 and \$2.725.

29. $z=\dfrac{\bar{x}-\mu}{\frac{\sigma}{\sqrt{n}}}=\dfrac{66-64.3}{\frac{2.6}{\sqrt{60}}}\approx\dfrac{1.7}{0.336}\approx5.06$

$P(\bar{x}>66)=P(z>5.06)\approx0$

There is almost no chance that a random sample of 60 women will have a mean height greater than 66 inches. This event is almost impossible.

31. $z=\dfrac{x-\mu}{\sigma}=\dfrac{70-64.3}{2.6}\approx2.19$

$P(x<70)=P(z<2.19)=0.9857$

$z=\dfrac{\bar{x}-\mu}{\frac{\sigma}{\sqrt{n}}}=\dfrac{70-64.3}{\frac{2.6}{\sqrt{20}}}\approx\dfrac{5.7}{0.581}\approx9.81$

$P(\bar{x}<70)=P(z<9.81)\approx1$

It is more likely to select a sample of 20 women with a mean height less than 70 inches because the sample of 20 has a higher probability.

33. $z=\dfrac{\bar{x}-\mu}{\frac{\sigma}{\sqrt{n}}}=\dfrac{127.9-128}{\frac{0.20}{\sqrt{40}}}\approx\dfrac{-0.1}{0.0316}\approx-3.16$

$P(\bar{x}<127.9)=P(z<-3.16)\approx0.0008$

Yes, it is very unlikely that you would have randomly sampled 40 cans with a mean equal to 127.9 ounces because it is more than 2 standard deviations from the mean of the sample means.

35. (a) $\mu = 96$

$\sigma = 0.5$

$$z = \frac{\bar{x} - \mu}{\frac{\sigma}{\sqrt{n}}} = \frac{96.25 - 96}{\frac{0.5}{\sqrt{40}}} \approx \frac{0.25}{0.079} \approx 3.16$$

$P(\bar{x} \geq 96.25) = P(z > 3.16) = 1 - P(z < 3.16) = 1 - 0.9992 = 0.0008$

(b) The seller's claim is inaccurate.

(c) Assuming the distribution is normally distributed:

$$z = \frac{x - \mu}{\sigma} = \frac{96.25 - 96}{0.5} = 0.5$$

$P(x > 96.25) = P(z > 0.5) = 1 - P(z < 0.5) = 1 - 0.6915 = 0.3085$

Assuming the manufacturer's claim is true, an individual board with a length of 96.25 would not be unusual. It is within 1 standard deviation of the mean for an individual board.

37. (a) $\mu = 50{,}000$

$\sigma = 800$

$$z = \frac{\bar{x} - \mu}{\frac{\sigma}{\sqrt{n}}} = \frac{49{,}721 - 50{,}000}{\frac{800}{\sqrt{100}}} = \frac{-279}{80} \approx -3.49$$

$P(\bar{x} \leq 49{,}721) = P(z \leq -3.49) = 0.0002$

(b) The manufacturer's claim is inaccurate.

(c) Assuming the distribution is normally distributed:

$$z = \frac{x - \mu}{\sigma} = \frac{49{,}721 - 50{,}000}{800} \approx -0.35$$

$P(x < 49{,}721) = P(z < -0.35) = 0.3632$

Assuming the manufacturer's claim is true, an individual tire with a life span of 49,721 miles is not unusual. It is within 1 standard deviation of the mean for an individual tire.

39. $\mu = 501$

$\sigma = 112$

$$z = \frac{\bar{x} - \mu}{\frac{\sigma}{\sqrt{n}}} = \frac{515 - 501}{\frac{112}{\sqrt{50}}} \approx \frac{14}{15.84} \approx 0.88$$

$P(\bar{x} \geq 515) = P(z \geq 0.88) = 1 - P(z \leq 0.88) = 1 - 0.8106 = 0.1894$

The high school's claim is not justified because it is not rare to find a sample mean as large as 515.

41. Use the finite correction factor since $n = 55 > 45 = 0.05N$.

$$z = \frac{\bar{x} - \mu}{\frac{\sigma}{\sqrt{n}}\sqrt{\frac{N-n}{N-1}}} = \frac{2.698 - 2.702}{\frac{0.009}{\sqrt{55}}\sqrt{\frac{900-55}{900-1}}} \approx \frac{-0.004}{(0.00121)\sqrt{0.9399}} \approx -3.41$$

$P(\bar{x} < 2.698) = P(z < -3.41) \approx 0.0003$

43.

Sample	Number of boys from 3 births	Proportion of boys from 3 births
bbb	3	1
bbg	2	$\frac{2}{3}$
bgb	2	$\frac{2}{3}$
gbb	2	$\frac{2}{3}$
bgg	1	$\frac{1}{3}$
gbg	1	$\frac{1}{3}$
ggb	1	$\frac{1}{3}$
ggg	0	0

45.

Sample	Numerical representation	Sample mean
bbb	111	1
bbg	110	$\frac{2}{3}$
bgb	101	$\frac{2}{3}$
gbb	011	$\frac{2}{3}$
bgg	100	$\frac{1}{3}$
gbg	010	$\frac{1}{3}$
ggb	001	$\frac{1}{3}$
ggg	000	0

The sample means are equal to the proportions.

47. $z = \dfrac{\hat{p} - p}{\sqrt{\dfrac{pq}{n}}} = \dfrac{0.70 - 0.77}{\sqrt{\dfrac{0.77(0.23)}{105}}} = \dfrac{-0.07}{0.0411} = -1.70$

$P(p < 0.70) = P(z < -1.70) = 0.0446$

About 4.5% of samples of 105 female heart transplant patients will have a mean 3-year survival rate of less than 70%.

5.5 NORMAL APPROXIMATIONS TO BINOMIAL DISTRIBUTIONS

5.5 Try It Yourself Solutions

1a. $n = 125, p = 0.05, q = 0.95$

b. $np = 6.25, nq = 118.75$

c. Because $np \geq 5$ and $nq \geq 5$, the normal distribution can be used.

d. $\mu = np = (125)(0.05) = 6.25$

$\sigma = \sqrt{npq} = \sqrt{(125)(0.05)(0.95)} \approx 2.44$

2a. (1) 57, 58, ..., 83 (2) ..., 52, 53, 54

b. (1) $56.5 < x < 83.5$ (2) $x < 54.5$

3a. $n = 125, p = 0.05$

$np = 6.25 \geq 5$ and $nq = 118.75 \geq 5$

The normal distribution can be used.

b. $\mu = np = 6.25$

$\sigma = \sqrt{npq} \approx 2.44$

c. $x > 9.5$

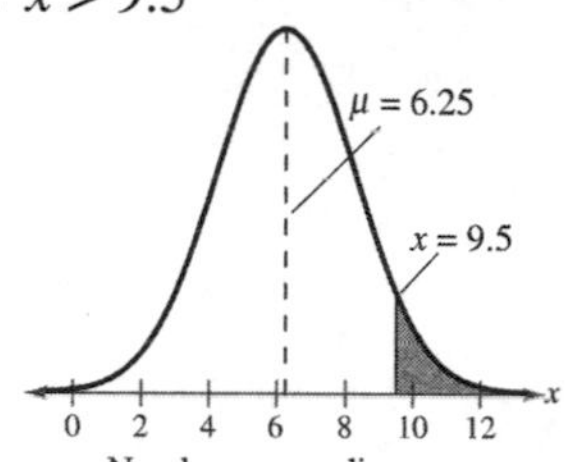

d. $z = \dfrac{x - \mu}{\sigma} = \dfrac{9.5 - 6.25}{2.44} \approx 1.33$

e. $P(z < 1.33) = 0.9082$

$P(x > 9.5) = P(z > 1.33) = 1 - P(z < 1.33) = 0.0918$

The probability that more than 9 respond yes is 0.0918.

4a. $n = 200, p = 0.58$

$np = 116 \geq 5$ and $nq = 84 \geq 5$

The normal distribution can be used.

b. $\mu = np = 116$

$\sigma = \sqrt{npq} \approx 6.98$

c. $P(x \leq 100.5)$

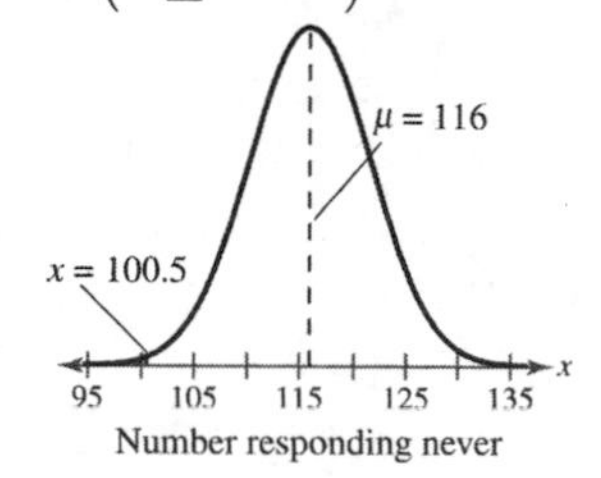

d. $z = \dfrac{x-\mu}{\sigma} \approx \dfrac{100.5-116}{6.98} \approx -2.22$

e. $P(x < 100.5) = P(z < -2.22) = 0.0132$

The probability that at most 100 people will say never is approximately 0.0132.

5a. $n = 150, p = 0.24$

$np = 36 \geq 5$ and $nq = 114 \geq 5$

The normal distribution can be used.

b. $\mu = np = 36$

$\sigma = \sqrt{npq} \approx 5.23$

c. $P(26.5 < x < 27.5)$

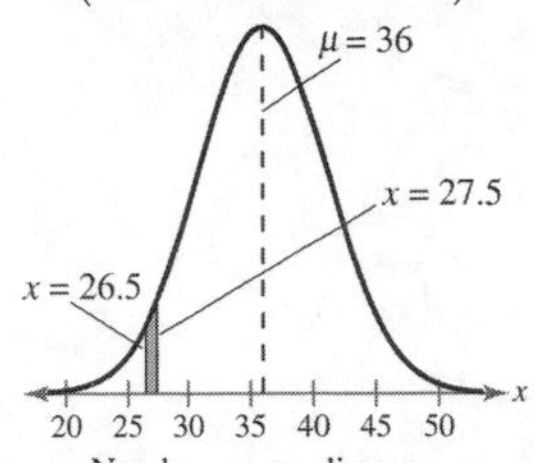

d. $z = \dfrac{x-\mu}{\sigma} = \dfrac{26.5-36}{5.23} \approx -1.82$

$z = \dfrac{x-\mu}{\sigma} = \dfrac{27.5-36}{5.23} \approx -1.625$

e. $P(z < -1.82) = 0.0344$

$P(z < -1.625) = 0.0521$

$P(-1.82 < z < -1.63) = 0.0521 - 0.0344 = 0.0177$

The probability that exactly 27 people will respond yes is 0.0177.

5.5 EXERCISE SOLUTIONS

1. Properties of a binomial experiment:

(1) The experiment is repeated for a fixed number of independent trials.
(2) There are two possible outcomes: success or failure.
(3) The probability of success is the same for each trial.
(4) The random variable x counts the number of successful trials.

3. $np = (24)(0.85) = 20.4 \geq 5$

$nq = (24)(0.15) = 3.6 < 5$

Cannot use normal distribution.

5. $np = (18)(0.90) = 16.2 \geq 5$

$nq = (18)(0.10) = 1.8 < 5$

Cannot use normal distribution.

7. $n = 10,\ p = 0.85,\ q = 0.15$

$np = 8.5 \geq 5,\ nq = 1.5 < 5$

Cannot use normal distribution because $nq < 5$.

9. $n = 50,\ p = 0.55,\ q = 0.45$

$np = 27.5 \geq 5,\ nq = 22.5 \geq 5$

Can use normal distribution.

$\mu = np = (50)(0.55) = 27.5$

$\sigma = \sqrt{npq} \approx 3.52$

11. $n = 20,\ p = 0.76,\ q = 0.24$

$np = 15.2 \geq 5,\ nq = 4.8 < 5$

Cannot use normal distribution because $nq < 5$.

13. a

15. c

17. The probability of getting fewer than 25 successes; $P(x < 24.5)$

19. The probability of getting exactly 33 successes; $P(32.5 < x < 33.5)$

21. The probability of getting at most 150 successes; $P(x < 150.5)$

23. $n = 100,\ p = 0.93$

$np = 93 \geq 5,\ nq = 7 \geq 5$

Can use normal distribution.

a. $z = \dfrac{x - \mu}{\sigma} \approx \dfrac{89.5 - 93}{2.55} \approx -1.37$

$z = \dfrac{x - \mu}{\sigma} \approx \dfrac{90.5 - 93}{2.55} \approx -0.98$

$$\begin{aligned} P(x = 90) &= P(89.5 < x < 90.5) \\ &= P(-1.37 < z < -0.98) \\ &= 0.1635 - 0.0835 \\ &= 0.0782 \end{aligned}$$

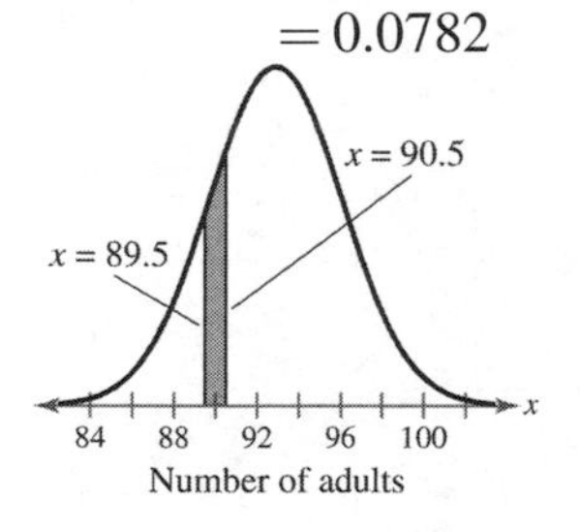

b. $P(x \geq 90) = P(x > 89.5)$
$= P(z > -1.37)$
$= 1 - P(z < -1.37)$
$= 1 - 0.0853$
$= 0.9147$

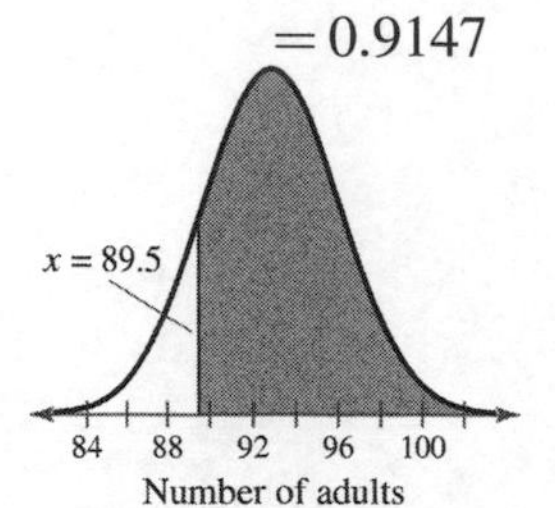

c. $P(x < 90) = P(x < 89.5)$
$= P(z < -1.37)$
$= 0.0853$

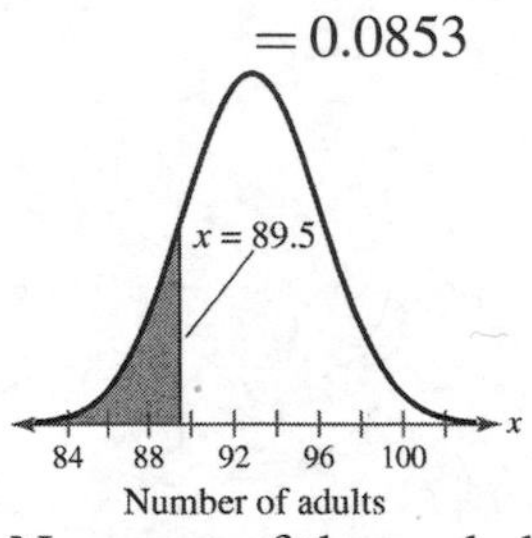

d. No, none of the probabilities are less than 0.05.

25. $n = 150,\ p = 0.35$

$np = 52.5 \geq 5,\ nq = 97.5 \geq 5$

Can use normal distribution.

a. $z = \dfrac{x - \mu}{\sigma} \approx \dfrac{75.5 - 52.5}{5.84} \approx 3.94$

$P(x \leq 75) \approx P(x < 75.5) = P(z < 3.94) \approx 1$

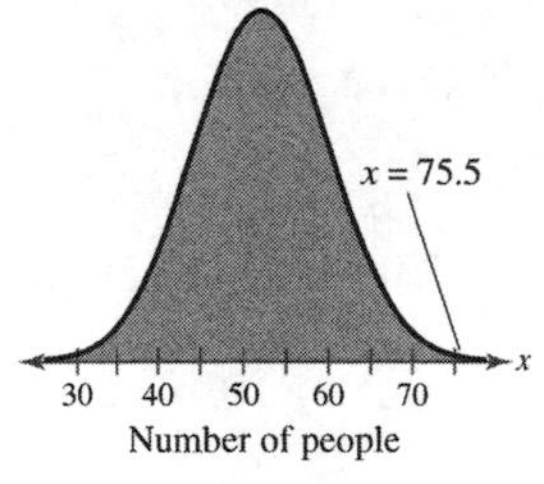

b. $z = \dfrac{x-\mu}{\sigma} \approx \dfrac{40.5-52.5}{5.84} \approx -2.05$

$P(x > 40) \approx P(x > 40.5)$
$= P(z > -2.05)$
$= 1 - P(z < -2.05)$
$= 1 - 0.0202$
$= 0.9798$

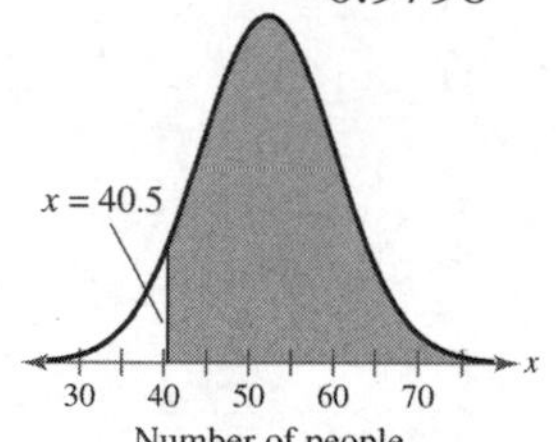

c. $z = \dfrac{x-\mu}{\sigma} \approx \dfrac{49.5-52.5}{5.84} \approx -0.51$

$z = \dfrac{x-\mu}{\sigma} \approx \dfrac{60.5-52.5}{5.84} \approx 1.37$

$P(50 \le x \le 60) \approx P(49.5 < x < 60.5)$
$= P(-0.51 < z < 1.37)$
$= 0.9147 - 0.3050$
$= 0.6097$

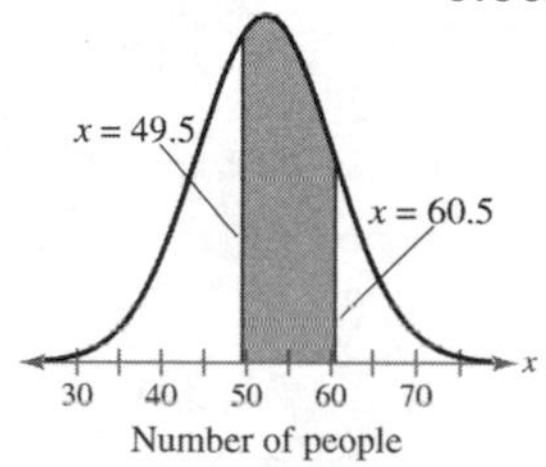

d. No, none of the probabilities are less than 0.05.

27. $n = 250,\ p = 0.05,\ q = 0.95$

$np = 12.5 \ge 5,\ nq = 237.5 \ge 5$

Can use normal distribution.

a. $z = \dfrac{x-\mu}{\sigma} \approx \dfrac{15.5-12.5}{3.45} \approx 0.87$

$z = \dfrac{x-\mu}{\sigma} \approx \dfrac{16.5-12.5}{3.45} \approx 1.16$

$P(x = 16) \approx P(15.5 \le x \le 16.5) = P(0.87 \le z \le 1.16)$
$= 0.8770 - 0.8078 = 0.0692$

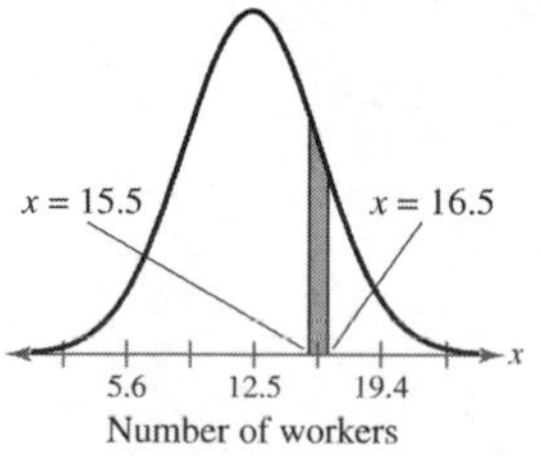

b. $P(x \geq 9) \approx P(x \geq 8.5) = P(z \geq -1.16) = 1 - P(z \leq -1.16) = 1 - 0.1230 = 0.8770$

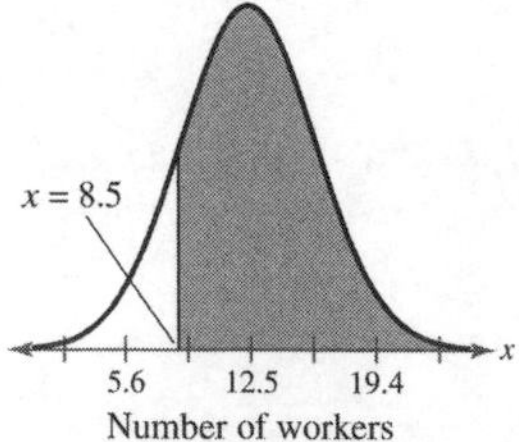

c. $P(x < 16) \approx P(x \leq 15.5) = P(z \leq 0.87) = 0.8078$

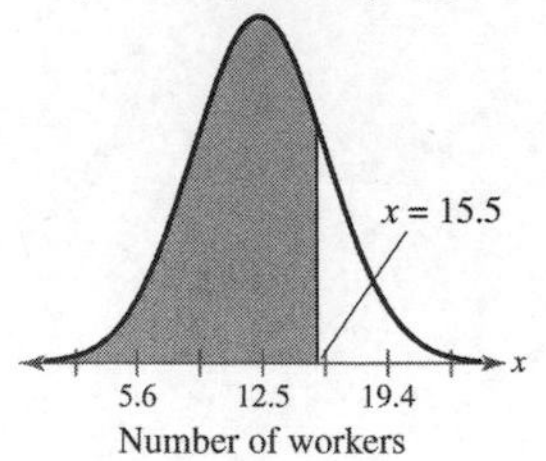

d. $n = 500,\ p = 0.05,\ q = 0.95$

$np = 25 \geq 5,\ nq = 475 \geq 5$

Can use normal distribution.

$$z = \frac{x - \mu}{\sigma} \approx \frac{29.5 - 25}{4.87} \approx 0.92$$

$P(x < 30) \approx P(x < 29.5) = P(z < 0.92) = 0.8212$

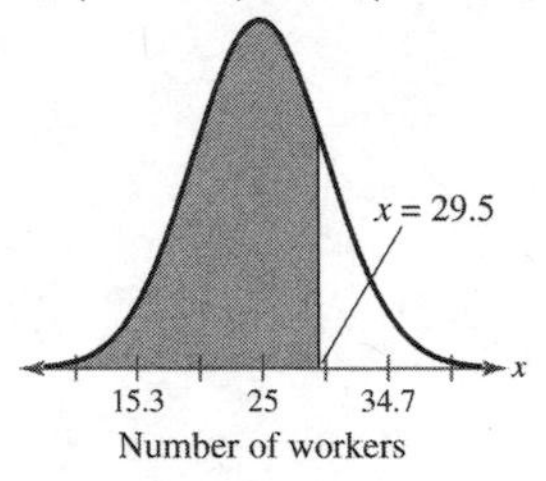

29. $n = 200,\ p = 0.34$

$np = 68 \geq 5,\ nq = 132 \geq 5$

Can use normal distribution.

a. $$z = \frac{x - \mu}{\sigma} \approx \frac{84.5 - 68}{6.70} \approx 2.46$$

$P(x \geq 85) \approx P(x > 84.5) = P(z > 2.46) = 1 - P(z < 2.46) = 1 - 0.9931 = 0.0069$

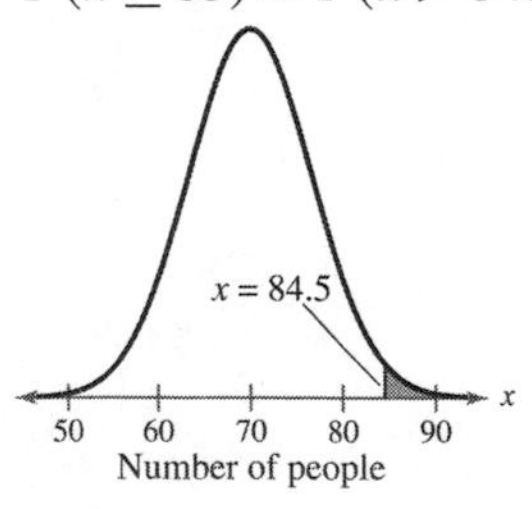

b. $z = \dfrac{x-\mu}{\sigma} \approx \dfrac{65.5-68}{6.70} \approx -0.37$

$P(x<66) \approx P(x<65.5) = P(z<-0.37) = 0.3557$

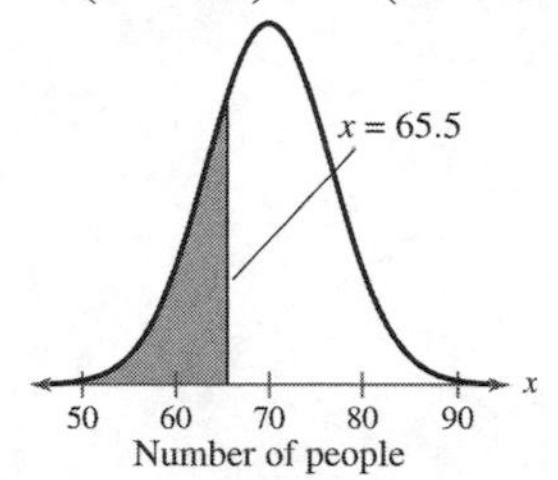

c. $z = \dfrac{x-\mu}{\sigma} \approx \dfrac{67.5-68}{6.70} \approx -0.07$

$z = \dfrac{x-\mu}{\sigma} \approx \dfrac{68.5-68}{6.70} \approx 0.07$

$P(x=68) \approx P(67.5<x<68.5) = P(-0.07<z<0.07) = 0.5279-0.4721 = 0.0558$

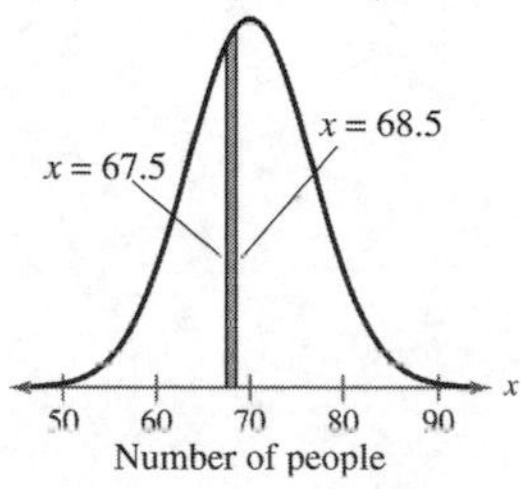

d. $n=6,\ p=0.34$

$np = 2.04 < 5,\ nq = 3.96 < 5$

Cannot use normal distribution because $np < 5$ and $nq < 5$.

$P(x=6) = {}_6C_6 (0.34)^6 (0.66)^0 \approx 0.002$

31. Binomial: $P(5 \le x \le 7) = P(x=5) + P(x=6) + P(x=7)$

$$= {}_{16}C_5 (0.4)^5 (0.6)^{11} + {}_{16}C_6 (0.4)^6 (0.6)^{10} + {}_{16}C_7 (0.4)^7 (0.6)^9$$

$$\approx 0.549$$

Normal: $\mu = np = 6.4,\ \sigma = \sqrt{npq} \approx 1.96$

$z = \dfrac{x-\mu}{\sigma} \approx \dfrac{4.5-6.4}{1.96} \approx -0.97$

$z = \dfrac{x-\mu}{\sigma} \approx \dfrac{7.5-6.4}{1.96} \approx 0.56$

$P(5 \le x \le 7) \approx P(4.5 \le x \le 7.5) = P(-0.97 \le z \le 0.56)$

$= 0.7123 - 0.1660$

$= 0.5463$

33. $n = 250,\ p = 0.70$

60% say no $\rightarrow$ 250(0.6) = 150 say no while 100 say yes.

$z = \dfrac{x-\mu}{\sigma} \approx \dfrac{100.5-175}{7.25} \approx -10.28$

P(less than or equal to 100 say yes) $= P(x \le 100) = P(x < 100.5) = P(z < -10.28) \approx 0$

It is highly unlikely that 60% responded no. Answers will vary.

35. $n=100,\ p=0.75$

$$z=\frac{x-\mu}{\sigma}\approx\frac{69.5-75}{4.33}\approx-1.27$$

$P(\text{reject claim})=P(x<70)=P(x<69.5)=P(z<-1.27)=0.1020$

CHAPTER 5 REVIEW EXERCISE SOLUTIONS

1. $\mu=15,\ \sigma=3$

3. Curve B has the greatest mean because its line of symmetry occurs the farthest to the right.

5. $x=1.32$: $z=\dfrac{x-\mu}{\sigma}=\dfrac{1.32-1.5}{0.08}=-2.25$

$x=1.54$: $z=\dfrac{x-\mu}{\sigma}=\dfrac{1.54-1.5}{0.08}=0.5$

$x=1.66$: $z=\dfrac{x-\mu}{\sigma}=\dfrac{1.66-1.5}{0.08}=2$

$x=1.78$: $z=\dfrac{x-\mu}{\sigma}=\dfrac{1.78-1.5}{0.08}=3.5$

7. 0.6772

9. 0.6293

11. 1 – 0.2843 = 0.7157

13. 0.00235

15. 0.5 – 0.0505 = 0.4495

17. 0.9564 – 0.5199 = 0.4365

19. 0.0668 + 0.0668 = 0.1336

21. A: 8
B: 17
C: 23
D: 29

23. $P(z<1.28)=0.8997$

25. $P(-2.15<x<1.55)=0.9394-0.0158=0.9236$

27. $P(z<-2.50 \text{ or } z>2.50)=2(0.0062)=0.0124$

29. $z=\dfrac{x-\mu}{\sigma}=\dfrac{84-74}{8}=1.25$

$P(x<84)=P(z<1.25)=0.8944$

31. $z=\dfrac{x-\mu}{\sigma}=\dfrac{80-74}{8}=0.75$

$P(x>80)=P(z>0.75)=1-P(z<0.75)=1-0.7734=0.2266$

33. $z = \dfrac{x-\mu}{\sigma} = \dfrac{60-74}{8} = -1.75$

$z = \dfrac{x-\mu}{\sigma} = \dfrac{70-74}{8} = -0.5$

$P(60 < x < 70) = P(-1.75 < z < -0.5) = 0.3085 - 0.0401 = 0.2684$

35. (a) $z = \dfrac{x-\mu}{\sigma} = \dfrac{1900-2200}{625} = -0.48$

$P(x < 1900) = P(z < -0.48) = 0.3156$

(b) $z = \dfrac{x-\mu}{\sigma} = \dfrac{2000-2200}{625} = -0.32$

$z = \dfrac{x-\mu}{\sigma} = \dfrac{2500-2200}{625} = 0.48$

$P(2000 < x < 2500) = P(-0.32 < z < 0.48) = 0.6844 - 0.3745 = 0.3099$

(c) $z = \dfrac{x-\mu}{\sigma} = \dfrac{2450-2200}{625} = 0.4$

$P(x > 2450) = P(z > 0.4) = 0.3446$

37. No, none of the events are unusual because their probabilities are greater than 0.05.

39. $z = -0.07$

41. $z = 1.13$

43. $z = 1.04$

45. 0.51

47. $x = \mu + z\sigma = 48 + (-2.5)(2.2) = 42.5$ meters

49. 95th percentile $\Rightarrow$ Area = 0.95 $\Rightarrow$ $z = 1.645$

$x = \mu + z\sigma = 48 + (1.645)(2.2) \approx 51.6$ meters

51. Top 10% $\Rightarrow$ Area = 0.90 $\Rightarrow$ $z = 1.28$

$x = \mu + z\sigma = 48 + (1.28)(2.2) \approx 50.8$ meters

53. {90 90 90, 90 90 120, 90 90 160, 90 90 210, 90 120 90, 90 120 120, 90 120 160, 90 120 210, 90 160 90, 90 160 120, 90 160 160, 90 160 210, 90 210 90, 90 210 120, 90 210 160, 90 210 210, 120 90 90, 120 90 120, 120 90 160, 120 90 210, 120 120 90, 120 120 120, 120 120 160, 120 120 210, 120 160 90, 120 160 120, 120 160 160, 120 160 210, 120 210 90, 120 210 120, 120 210 160, 120 210 210, 160 90 90, 160 90 120, 160 90 160, 160 90 210, 160 120 90, 160 120 120, 160 120 160, 160 120 210, 160 160 90, 160 160 120, 160 160 160, 160 160 210, 160 210 90, 160 210 120, 160 210 160, 160 210 210, 210 90 90, 210 90 120, 210 90 160, 210 90 210, 210 120 90, 210 120 120, 210 120 160, 210 120 210, 210 160 90, 210 160 120, 210 160 160, 210 160 210, 210 210 90, 210 210 120, 210 210 160, 210 210 210}

$\mu = 145,\ \sigma = 45$

$\mu_{\bar{x}} = 145,\ \sigma_{\bar{x}} = \dfrac{45}{\sqrt{3}} \approx 25.98$

The means are the same, but $\sigma_{\bar{x}}$ is less than σ.

55. $\mu_{\bar{x}} = 76,\ \sigma_{\bar{x}} = \dfrac{\sigma}{\sqrt{n}} = \dfrac{20.5}{\sqrt{35}} \approx 3.465$

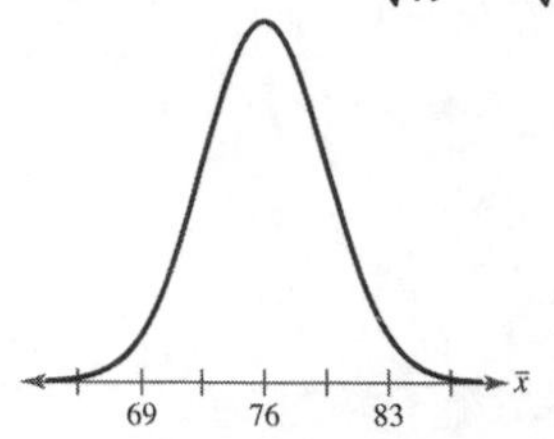

57. (a) $z = \dfrac{\bar{x}-\mu}{\frac{\sigma}{\sqrt{n}}} = \dfrac{1900-2200}{\frac{625}{\sqrt{12}}} \approx \dfrac{-300}{180.42} \approx -1.66$

$P(\bar{x} < 1900) = P(z < -1.66) = 0.0485$

(b) $z = \dfrac{\bar{x}-\mu}{\frac{\sigma}{\sqrt{n}}} = \dfrac{2000-2200}{\frac{625}{\sqrt{12}}} \approx \dfrac{-200}{180.42} \approx -1.11$

$z = \dfrac{\bar{x}-\mu}{\frac{\sigma}{\sqrt{n}}} = \dfrac{2500-2200}{\frac{625}{\sqrt{12}}} \approx \dfrac{300}{180.42} \approx 1.66$

$P(2000 < \bar{x} < 2500) = P(-1.11 < z < 1.66) = 0.9515 - 0.1335 = 0.8180$

(c) $z = \dfrac{\bar{x}-\mu}{\frac{\sigma}{\sqrt{n}}} = \dfrac{2450-2200}{\frac{625}{\sqrt{12}}} \approx \dfrac{250}{180.42} \approx 1.39$

$P(\bar{x} > 2450) = P(z > 1.39) = 0.0823$

(a) and (c) are smaller, (b) is larger. This is to be expected because the standard error of the sample mean is smaller.

59. (a) $z = \dfrac{\bar{x}-\mu}{\frac{\sigma}{\sqrt{n}}} = \dfrac{29{,}000-29{,}200}{\frac{1500}{\sqrt{45}}} \approx \dfrac{-200}{223.61} \approx -0.89$

$P(\bar{x} < 29{,}000) = P(z < -0.89) = 0.1867$

(b) $z = \dfrac{\bar{x}-\mu}{\frac{\sigma}{\sqrt{n}}} = \dfrac{31{,}000-29{,}200}{\frac{1500}{\sqrt{45}}} \approx \dfrac{1800}{223.61} \approx 8.05$

$P(\bar{x} > 31{,}000) = P(z > 8.05) \approx 0$

61. Assuming the distribution is normally distributed:

$z = \dfrac{\bar{x}-\mu}{\frac{\sigma}{\sqrt{n}}} = \dfrac{1.125-1.5}{\frac{0.5}{\sqrt{15}}} \approx \dfrac{-0.375}{0.1291} \approx -2.90$

$P(\bar{x} < 1.125) = P(z < -2.90) = 0.0019$

63. $n = 12,\ p = 0.96,\ q = 0.04$

$np = 11.52 \geq 5$, but $nq = 0.48 < 5$

Cannot use the normal distribution because $nq < 5$.

65. $P(x \geq 25) = P(x > 24.5)$

67. $P(x = 45) = P(44.5 < x < 45.5)$

69. $n = 45,\ p = 0.70$

$np = 31.5 \geq 5,\ nq = 13.5 \geq 5$

Can use the normal distribution.

$\mu = np = 31.5,\ \sigma = \sqrt{npq} = \sqrt{45(0.70)(0.30)} \approx 3.07$

$z = \dfrac{x - \mu}{\sigma} \approx \dfrac{20.5 - 31.5}{3.07} \approx -3.58$

$P(x \leq 20) \approx P(x < 20.5) = P(z < -3.58) \approx 0$

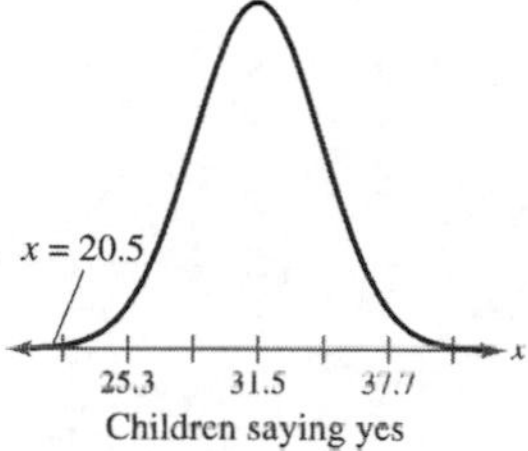

CHAPTER 5 QUIZ SOLUTIONS

1. (a) $P(z > -2.54) = 0.9945$

(b) $P(z < 3.09) = 0.9990$

(c) $P(-0.88 < z < 0.88) = 0.8106 - 0.1894 = 0.6212$

(d) $P(z < -1.445 \text{ or } z > -0.715) = 0.0742 + 0.76265 = 0.83685$

2. (a) $z = \dfrac{x - \mu}{\sigma} = \dfrac{5.36 - 5.5}{0.08} = -1.75$

$z = \dfrac{x - \mu}{\sigma} = \dfrac{5.64 - 5.5}{0.08} = 1.75$

$P(5.36 < x < 5.64) = P(-1.75 < z < 1.75) = 0.9599 - 0.0401 = 0.9198$

(b) $z = \dfrac{x - \mu}{\sigma} = \dfrac{-5.00 - (-8.2)}{7.84} \approx 0.41$

$z = \dfrac{x - \mu}{\sigma} = \dfrac{0 - (-8.2)}{7.84} \approx 1.05$

$P(-5.00 < x < 0) = P(0.41 < z < 1.05) = 0.8531 - 0.6591 = 0.1940$

(c) $z = \dfrac{x - \mu}{\sigma} = \dfrac{0 - 18.5}{9.25} = -2$

$z = \dfrac{x - \mu}{\sigma} = \dfrac{37 - 18.5}{9.25} = 2$

$P(x < 0 \text{ or } x > 37) = P(z < -2 \text{ or } z > 2) = 2(0.0228) = 0.0456$

3. $z = \dfrac{x-\mu}{\sigma} = \dfrac{125-100}{15} \approx 1.67$

$P(x > 125) = P(z > 1.67) = 0.0475$

Yes, the event is unusual because its probability is less than 0.05.

4. $z = \dfrac{x-\mu}{\sigma} = \dfrac{95-100}{15} \approx -0.33$

$z = \dfrac{x-\mu}{\sigma} = \dfrac{105-100}{15} \approx 0.33$

$P(95 < x < 105) = P(-0.33 < z < 0.33) = 0.6293 - 0.3707 = 0.2586$

No, the event is not unusual because its probability is greater than 0.05.

5. $P(x > 112) = P(z > 0.80) = 0.2119 \rightarrow 21.19\%$

6. $z = \dfrac{x-\mu}{\sigma} = \dfrac{90-100}{15} \approx -0.67$

$P(x < 90) = P(z < -0.67) = 0.2514$

(2000)(0.2514)=502.8→503 students

7. Top 5% → z=1.645

$\mu + z\sigma = 100 + (1.645)(15) \approx 124.7 \rightarrow 125$

8. Bottom 10% → $z = -1.28$

$\mu + z\sigma = 100 + (-1.28)(15) = 80.8 \rightarrow 80$

(Because you are finding the highest score that would still place you in the bottom 10%, round down, because if you rounded up you would be outside the bottom 10%.)

9. $z = \dfrac{\bar{x}-\mu}{\frac{\sigma}{\sqrt{n}}} = \dfrac{105-100}{\frac{15}{\sqrt{60}}} \approx \dfrac{5}{1.936} \approx 2.58$

$P(\bar{x} > 105) = P(z > 2.58) \approx 0.0049$

About 0.5% of samples of 60 students will have a mean IQ score greater than 105. This is a very unusual event.

10. $z = \dfrac{x-\mu}{\sigma} = \dfrac{105-100}{15} \approx 0.33$

$P(x > 105) = P(z > 0.33) = 0.3707$

$z = \dfrac{\bar{x}-\mu}{\frac{\sigma}{\sqrt{n}}} = \dfrac{105-100}{\frac{15}{\sqrt{15}}} \approx \dfrac{5}{3.873} \approx 1.29$

$P(\bar{x} > 105) = P(z > 1.29) \approx 0.0985$

You are more likely to select one student with a test score greater than 105 because the standard error of the mean is less than the standard deviation.

11. $n = 35,\ p = 0.81 \rightarrow np = 28.35,\ nq = 6.65$

Can use normal distribution.

$\mu = np = 28.35,\ \sigma = \sqrt{npq} \approx 2.321$

12. $z = \dfrac{x-\mu}{\sigma} \approx \dfrac{20.5 - 28.35}{2.321} \approx -3.38$

$P(x \leq 20) = P(x < 20.5) = P(z < -3.38) = 0.0004$

This event is extremely unusual because its probability is much less than 0.05.

CUMULATIVE REVIEW, CHAPTERS 3-5

1. (a) $np = 50(0.15) = 7.5 \geq 5$

$nq = 50(0.85) = 42.5 \geq 5$

Can use normal distribution.

$\mu = np = 50(0.15) = 7.5$

(b) $\sigma = \sqrt{npq} = \sqrt{50(0.15)(0.85)} \approx 2.52$

$P(x \leq 14) \approx P(x \leq 14.5)$

$= P\left(z \leq \dfrac{14.5 - 7.5}{2.52}\right)$

$= P(z \leq 2.78)$

$= 0.9973$

(c) It is unusual for 14 out of 50 voters to rate the U.S. health care system as excellent because the probability is less than 0.05.

$P(x = 14) \approx P(13.5 \leq x \leq 14.5)$

$= P\left(\dfrac{13.5 - 7.5}{2.52} \leq z \leq \dfrac{14.5 - 7.5}{2.52}\right)$

$= P(2.38 \leq z \leq 2.78)$

$= 0.9973 - 0.9913$

$= 0.0060$

2.

x	$P(x)$	$xP(x)$	$x-\mu$	$(x-\mu)^2$	$(x-\mu)^2 P(x)$
2	0.427	0.854	–1.13	1.277	0.545
3	0.227	0.681	–0.13	0.017	0.004
4	0.200	0.800	0.87	0.757	0.151
5	0.093	0.465	1.87	3.497	0.325
6	0.034	0.204	2.87	8.237	0.280
7	0.018	0.126	3.87	14.977	0.270
		$\sum xP(x) = 3.13$			$\sum (x-\mu)^2 P(x) \approx 1.575$

(a) $\mu = \sum xP(x) \approx 3.1$

(b) $\sigma^2 = \sum (x-\mu)^2 P(x) \approx 1.6$

(c) $\sigma = \sqrt{\sigma^2} \approx 1.3$

(d) $E(x) = \sum xP(x) \approx 3.1$

(e) The size of a family household on average is about 3 persons. The standard deviation is 1.3, so most households differ from the mean by no more than about 1 person.

3.

x	$P(x)$	$xP(x)$	$x-\mu$	$(x-\mu)^2$	$(x-\mu)^2 P(x)$
0	0.012	0.000	–3.596	12.931	0.155
1	0.049	0.049	–2.596	6.739	0.330
2	0.159	0.318	–1.596	2.547	0.405
3	0.256	0.768	–0.596	0.355	0.091
4	0.244	0.976	0.404	0.163	0.040
5	0.195	0.975	1.404	1.971	0.384
6	0.085	0.510	2.404	5.779	0.491
		$\sum xP(x) = 3.596$			$\sum (x-\mu)^2 P(x) \approx 1.896$

(a) $\mu = \sum xP(x) \approx 3.6$

(b) $\sigma^2 = \sum (x-\mu)^2 P(x) \approx 1.9$

(c) $\sigma = \sqrt{\sigma^2} \approx 1.4$

(d) $E(x) = \mu \approx 3.6$

(e) The number of fouls for a player in a game on average is bout 4 fouls. The standard deviation is 1.4, so games differ from the mean by no more than about 1 or 2 fouls.

4. (a) $P(x<4) = 0.012 + 0.049 + 0.159 + 0.256 = 0.476$

(b) $P(x \geq 3) = 1 - P(x \leq 2)$
$= 1 - (0.012 + 0.049 + 0.159)$
$= 0.78$

(c) $P(2 \leq x \leq 4) = 0.159 + 0.256 + 0.244 = 0.659$

5. (a) (16)(15)(14)(13) = 43,680

(b) $\dfrac{(7)(6)(5)(4)}{(16)(15)(14)(13)} = \dfrac{840}{43,680} \approx 0.0192$

6. 0.7642

7. 0.0010

8. $1 - 0.2005 = 0.7995$

9. $0.9984 - 0.500 = 0.4984$

10. $0.3974 - 0.1112 = 0.2862$

11. $0.5478 + (1 - 0.9573) = 0.5905$

12. $n = 11,\ p = 0.45$

(a) $P(8) = 0.0462$; unusual because the probability is less than 0.05.

(b) $P(x \geq 5) = 0.6029$

(c) $P(x < 2) = 0.0139$; unusual because the probability is less than 0.05.

13. $p = \dfrac{1}{200} = 0.005$

(a) $P(x = 10) = (0.005)(0.995)^9 \approx 0.0048$

(b) $P(x \le 3) \approx 0.0149$

(c) $P(x > 10) = 1 - P(x \le 10) \approx 1 - 0.0489 = 0.9511$

14. (a) 0.2777

(b) 0.8657

(c) Dependent. *P*(being a public school teacher | having 20 years or more of full-time teaching experience) ≠ *P*(being a public school teacher)

(d) $0.8799 + 0.0612 - 0.0530 = 0.8881$

(e) $0.3438 + 0.1201 - 0.0462 = 0.4177$

15. (a) $\mu_{\bar{x}} = 70$

$\sigma_{\bar{x}} = \dfrac{\sigma}{\sqrt{n}} = \dfrac{1.2}{\sqrt{40}} \approx 0.1897$

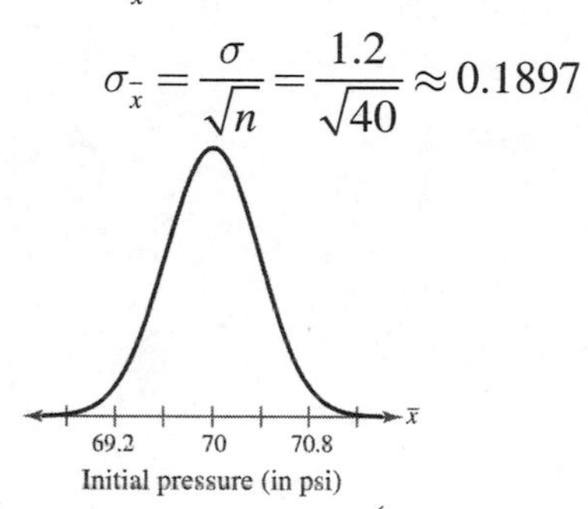

(b) $P(\bar{x} \le 69) = P\left(z \le \dfrac{69 - 70}{\frac{1.2}{\sqrt{15}}}\right) = P(z < -3.23) = 0.0006$

16. (a) $P(x < 36) = P\left(z < \dfrac{36 - 44}{5}\right) = P(z < -1.6) = 0.0548$

(b) $P(42 < x < 60) = P\left(\dfrac{42 - 44}{5} < z < \dfrac{60 - 44}{5}\right)$

$= P(-0.4 < z < 3.2)$

$= 0.9993 - 0.3446 = 0.6547$

(c) Top 5% $\Rightarrow z = 1.645$

$x = \mu + z\sigma = 44 + (1.645)(5) \approx 52.2$ months

17. (a) ${}_{12}C_4 = 495$

(b) $\dfrac{(1)(1)(1)(1)}{{}_{12}C_4} = 0.0020$

18. $n = 16$, $p = 0.50$

(a) $P(12) = 0.0278$; unusual because the probability is less than 0.05.

(b) $P(x \le 6) = 0.2272$

(c) $P(x > 7) = 1 - P(x \le 7) = 1 - 0.4018 = 0.5982$

CHAPTER 6

Confidence Intervals

6.1 CONFIDENCE INTERVALS FOR THE MEAN (LARGE SAMPLES)

6.1 Try It Yourself Solutions

1a. $\bar{x} = \dfrac{\sum x}{n} = \dfrac{4155}{30} = 138.5$

b. A point estimate for the population mean number of friends is 138.5.

2a. $z_c = 1.96,\ n = 30,\ s \approx 51.0$

b. $E = z_c \dfrac{s}{\sqrt{n}} \approx 1.96 \dfrac{51.0}{\sqrt{30}} \approx 18.3$

c. You are 95% confident that the maximum error of the estimate is about 18.3 friends.

3a. $\bar{x} = 138.5,\ E \approx 18.3$

b. $\bar{x} = E \approx 138.5 - 18.3 = 120.2$

$\bar{x} + E \approx 138.5 + 18.3 = 156.8$

c. With 95% confidence, you can say that the population mean number of friends is between 120.2 and 156.8. This confidence interval is wider than the one found in Example 3.

4b. 75% CI: (121.2, 140.4)

85% CI: (118.7, 142.9)

99% CI: (109.2, 152.4)

c. As the confidence level increases, so does the width of the interval.

5a. $n = 30,\ \bar{x} = 22.9,\ \sigma = 1.5,\ z_c = 1.645$

b. $E = z_c \dfrac{\sigma}{\sqrt{n}} = 1.645 \dfrac{1.5}{\sqrt{30}} \approx 0.5$

$\bar{x} - E \approx 22.9 - 0.5 - 22.4$

$\bar{x} + E \approx 22.9 + 0.5 = 23.4$

c. With 90% confidence, you can way that the mean age of the students is between 22.4 and 23.4 years. Because of the larger sample size, the confidence interval is slightly narrower.

6a. $z_c = 1.96,\ E = 10,\ \sigma \approx 53.0$

b. $n = \left(\dfrac{z_c \sigma}{E}\right)^2 \approx \left(\dfrac{1.96 \cdot 53.0}{10}\right)^2 \approx 107.91 \rightarrow 108$

c. You should have at least 108 users in you sample. Because of the larger margin of error, the sample size needed is much smaller.

6.1 EXERCISE SOLUTIONS

1. You are more likely to be correct using an interval estimate because it is unlikely that a point estimate will equal the population mean exactly.

3. d; As the level of confidence increases, z_c increases therefore creating wider intervals.

5. 1.28

7. 1.15

9. $\bar{x} - \mu = 3.8 - 4.27 = -0.47$

11. $\bar{x} - \mu = 26.43 - 24.67 = 1.76$

13. $E = z_c \dfrac{s}{\sqrt{n}} = 1.96 \dfrac{5.2}{\sqrt{30}} \approx 1.861$

15. $E = z_c \dfrac{s}{\sqrt{n}} = 1.28 \dfrac{1.3}{\sqrt{75}} \approx 0.192$

17. $c = 0.88 \Rightarrow z_c = 1.555$

$\bar{x} = 57.2,\ s = 7.1,\ n = 50$

$\bar{x} \pm z_c \dfrac{s}{\sqrt{n}} = 57.2 \pm 1.555 \dfrac{7.1}{\sqrt{50}} \approx 57.2 \pm 1.561 \approx (55.6,\ 58.8)$

Answer: (c)

19. $c = 0.95 \Rightarrow z_c = 1.96$

$\bar{x} = 57.2,\ s = 7.1,\ n = 50$

$\bar{x} \pm z_c \dfrac{s}{\sqrt{n}} = 57.2 \pm 1.96 \dfrac{7.1}{\sqrt{50}} \approx 57.2 \pm 1.968 \approx (55.2,\ 59.2)$

Answer: (b)

21. $\bar{x} \pm z_c \dfrac{s}{\sqrt{n}} = 12.3 \pm 1.645 \dfrac{1.5}{\sqrt{50}} \approx 12.3 \pm 0.349 \approx (12.0,\ 12.6)$

23. $\bar{x} \pm z_c \dfrac{s}{\sqrt{n}} = 10.5 \pm 2.575 \dfrac{2.14}{\sqrt{45}} \approx 10.5 \pm 0.821 \approx (9.7,\ 11.3)$

25. $(12.0, 14.8) \Rightarrow 13.4 \pm 1.4 \Rightarrow \bar{x} = 13.4,\ E = 1.4$

27. $(1.71, 2.05) \Rightarrow 1.88 \pm 0.17 \Rightarrow \bar{x} = 1.88,\ E = 0.17$

29. $c = 0.90 \Rightarrow z_c = 1.645$

$$n = \left(\frac{z_c \sigma}{E}\right)^2 = \left(\frac{(1.645)(6.8)}{1}\right)^2 \approx 125.13 \Rightarrow 126$$

31. $c = 0.80 \Rightarrow z_c = 1.28$

$$n = \left(\frac{z_c \sigma}{E}\right)^2 = \left(\frac{(1.28)(4.1)}{2}\right)^2 \approx 6.89 \Rightarrow 7$$

33. $(26.2, 30.1) \Rightarrow 2E = 30.1 - 26.2 = 3.9 \Rightarrow E = 1.95$ and $\bar{x} = 26.2 + E$
$= 26.2 + 1.95 = 28.15$

35. 90% CI: $\bar{x} \pm z_c \dfrac{s}{\sqrt{n}} = 452.80 \pm 1.645 \dfrac{85.50}{\sqrt{34}} \approx 452.80 \pm 24.12 \approx (428.68,\ 476.92)$

95% CI: $\bar{x} \pm z_c \dfrac{s}{\sqrt{n}} = 452.80 \pm 1.96 \dfrac{85.50}{\sqrt{34}} \approx 452.80 \pm 28.74 \approx (424.06,\ 481.54)$

With 90% confidence, you can say that the population mean price is between \$428.68 and \$476.92 with 95% confidence, you can say that the population mean price is between \$424.06 and \$481.54.

The 95% CI is wider.

37. 90% CI: $\bar{x} \pm z_c \dfrac{s}{\sqrt{n}} = 99.3 \pm 1.645 \dfrac{41.5}{\sqrt{31}} \approx 99.3 \pm 12.26 \approx (87.0,\ 111.6)$

95% CI: $\bar{x} \pm z_c \dfrac{s}{\sqrt{n}} = 99.3 \pm 1.96 \dfrac{41.5}{\sqrt{31}} \approx 99.3 \pm 14.61 \approx (84.7,\ 113.9)$

With 90% confidence, you can say that the population mean is between 87.0 and 111.6 with 95% confidence, you can say that the population mean is between 84.7 and 113.9 calories.

The 95% CI is wider.

39. $\bar{x} \pm z_c \dfrac{s}{\sqrt{n}} = 2650 \pm 1.96 \dfrac{425}{\sqrt{50}} \approx 2650 \pm 117.80 \approx (2532.20,\ 2767.80)$

With 95% confidence, you can say that the population mean cost is between \$2532.20 and \$2767.80.

41. $\bar{x} \pm z_c \dfrac{s}{\sqrt{n}} = 2650 \pm 1.96 \dfrac{425}{\sqrt{80}} \approx 2650 \pm 93.13 \approx (2556.87,\ 2743.13)$

The $n = 50$ CI is wider because a smaller sample is taken, giving less information about the population.

43. $\bar{x} \pm z_c \dfrac{s}{\sqrt{n}} = 3.12 \pm 1.96 \dfrac{0.09}{\sqrt{48}} \approx 3.12 \pm 0.03 \approx (3.09,\ 3.15)$

With 95% confidence, you can say that the population mean time is between 3.09 and 3.15 minutes.

45. $\bar{x} \pm z_c \dfrac{s}{\sqrt{n}} = 3.12 \pm 1.96 \dfrac{0.06}{\sqrt{48}} \approx 3.12 \pm 0.02 \approx (3.10,\ 3.14)$

The $s = 0.09$ CI is wider because of the increased variability within the sample.

47. (a) An increase in the level of confidence will widen the confidence interval.
(b) An increase in the sample size will narrow the confidence interval.
(c) An increase in the standard deviation will widen the confidence interval.

49. $\bar{x} = \frac{\sum x}{n} = \frac{302}{20} \approx 15.1$

90% CI: $\bar{x} \pm z_c \frac{\sigma}{\sqrt{n}} = 15.1 \pm 1.645 \frac{1.3}{\sqrt{20}} \approx 15.1 \pm 0.478 \approx (14.6,\ 15.6)$

99% CI: $\bar{x} \pm z_c \frac{\sigma}{\sqrt{n}} = 15.1 \pm 2.575 \frac{1.3}{\sqrt{20}} \approx 15.1 \pm 0.749 \approx (14.4,\ 15.8)$

With 90% confidence, you can say that the population mean length of time is between 14.6 and 15.6 minutes. With 99% confidence, you can say that the population mean length of time is between 14.4 and 15.8 minutes.

The 99% CI is wider.

51. $n = \left(\frac{z_c \sigma}{E}\right)^2 = \left(\frac{1.96 \cdot 4.8}{1}\right)^2 \approx 88.510 \to 89$

53. (a) $n = \left(\frac{z_c \sigma}{E}\right)^2 = \left(\frac{1.96 \cdot 2.8}{0.5}\right)^2 \approx 120.473 \to 121$ servings

(b) $n = \left(\frac{z_c \sigma}{E}\right)^2 = \left(\frac{2.575 \cdot 2.8}{0.5}\right)^2 \approx 207.936 \to 208$ servings

The 99% CI requires a larger sample because more information is needed from the population to be 99% confident.

55. (a) $n = \left(\frac{z_c \sigma}{E}\right)^2 = \left(\frac{1.645 \cdot 0.85}{0.25}\right)^2 \approx 31.282 \to 32$ cans

(b) $n = \left(\frac{z_c \sigma}{E}\right)^2 = \left(\frac{1.645 \cdot 0.85}{0.15}\right)^2 \approx 86.893 \to 87$ cans

$E = 0.15$ requires a larger sample size. As the error size decreases, a larger sample must be taken to obtain enough information from the population to ensure the desired accuracy.

57. (a) $n = \left(\frac{z_c \sigma}{E}\right)^2 = \left(\frac{1.96 \cdot 0.25}{0.125}\right)^2 \approx 15.3664 \to 16$ sheets

(b) $n = \left(\frac{z_c \sigma}{E}\right)^2 = \left(\frac{1.96 \cdot 0.25}{0.0625}\right)^2 \approx 61.4656 \to 62$ sheets

$E = 0.0625$ requires a larger sample size. As the error size decreases, a larger sample must be taken to obtain enough information from the population to ensure the desired accuracy.

59. (a) $n = \left(\frac{z_c \sigma}{E}\right)^2 = \left(\frac{2.575 \cdot 0.25}{0.1}\right)^2 \approx 41.441 \to 42$ soccer balls

(b) $n = \left(\frac{z_c \sigma}{E}\right)^2 = \left(\frac{2.575 \cdot 0.30}{0.1}\right)^2 \approx 59.676 \to 60$ soccer balls

$\sigma = 0.3$ requires a larger sample size. Due to the increased variability in the population, a larger sample is needed to ensure the desired accuracy.

61. (a) An increase in the level of confidence will increase the minimum sample size required.
(b) An increase (larger *E*) in the error tolerance will decrease the minimum sample size required.
(c) An increase in the population standard deviation will increase the minimum sample size required.

63. (212.74, 221.51)
With 95% confidence, you can say that the population mean airfare price is between $212.74 and $221.51.

65. 80% confidence interval results:
μ: population mean
standard deviation = 344.9

mean	n	Sample mean	Std. err.	L. Limit	U Limit
μ	30	1042.7	62.969837	962.0009	1123.399

90% confidence interval results:
μ: population mean
standard deviation = 344.9

mean	n	Sample mean	Std. err.	L. Limit	U Limit
μ	30	1042.7	62.969837	939.12384	1146.2761

95% confidence interval results:
μ: population mean
standard deviation = 344.9

mean	n	Sample mean	Std. err.	L. Limit	U Limit
μ	30	1042.7	62.969837	919.2814	1166.1187

With 80% confidence, you can say that the population mean sodium content is between 962.0 and 1123.4 milligrams. With 90% confidence, you can say it is between 929.1 and 1146.3 milligrams. With 95% confidence, you can say it is between 919.3 and 1166.1 milligrams.

67. (a) $\sqrt{\frac{N-n}{N-1}} = \sqrt{\frac{1000-500}{1000-1}} \approx 0.707$

(b) $\sqrt{\frac{N-n}{N-1}} = \sqrt{\frac{1000-100}{1000-1}} \approx 0.949$

(c) $\sqrt{\frac{N-n}{N-1}} = \sqrt{\frac{1000-75}{1000-1}} \approx 0.962$

(d) $\sqrt{\frac{N-n}{N-1}} = \sqrt{\frac{1000-50}{1000-1}} \approx 0.975$

(e) The finite population correction factor approaches 1 as the sample size decreases and the population size remains the same.

69. *Sample answer:*

$E = \frac{z_c \sigma}{\sqrt{n}}$ Write original equation.

$E\sqrt{n} = z_c \sigma$ Multiply each side by $\sqrt{n}$.

$\sqrt{n} = \frac{z_c \sigma}{E}$ Divide each side by E.

$n = \left(\frac{z_c \sigma}{E}\right)^2$ Square each side.

6.2 CONFIDENCE INTERVALS FOR THE MEAN (SMALL SAMPLES)

6.2 Try It Yourself Solutions

1a. d.f. $= n - 1 = 22 - 1 = 21$

b. $c = 0.90$

c. $t_c = 1.721$

2a. 90% CI: $t_c = 1.753$

$$E = t_c \frac{s}{\sqrt{n}} = 1.753 \frac{10}{\sqrt{16}} \approx 4.4$$

99% CI: $t_c = 2.947$

$$E = t_c \frac{s}{\sqrt{n}} - 2.947 \frac{10}{\sqrt{16}} \approx 7.4$$

b. 90% CI: $\bar{x} \pm E \approx 162 \pm 4.4 = (157.6, 166.4)$

99% CI: $\bar{x} \pm E \approx 162 \pm 7.4 = (154.6, 169.4)$

c. With 90% confidence, you can say that the population mean temperature of coffee sold is between 157.6°F and 166.4°F.

With 99% confidence, you can say that the population mean temperature of coffee sold is between 154.6°F and 169.4°F.

3a. 90% CI: $t_c = 1.729$

$$E = t_c \frac{s}{\sqrt{n}} = 1.729 \frac{2.39}{\sqrt{20}} \approx 0.92$$

95% CI: $t_c = 2.093$

$$E = t_c \frac{s}{\sqrt{n}} - 2.093 \frac{2.39}{\sqrt{20}} \approx 1.12$$

b. 90% CI: $\bar{x} \pm E \approx 9.75 \pm 0.92 = (8.83,\ 10.67)$

95% CI: $\bar{x} \pm E \approx 9.75 \pm 1.12 = (8.63,\ 10.87)$

c. With 90% confidence, you can say that the population mean number of days the car model sits on the lot is between 8.83 and 10.67 days.
With 95% confidence, you can say that the population mean number of days the car model sits on the lot is between 8.63 and 10.87 days. The 90% confidence interval is slightly narrower.

4a. Is $n \geq 30$? No
Is the population normally distributed? Yes
Is σ known? No
Use the t-distribution to construct the 90% CI.

6.2 EXERCISE SOLUTIONS

1. $t_c = 1.833$

3. $t_c = 2.947$

5. $E = t_c \frac{s}{\sqrt{n}} = 2.131 \frac{5}{\sqrt{16}} \approx 2.7$

7. $E = t_c \frac{s}{\sqrt{n}} = 1.796 \frac{2.4}{\sqrt{12}} \approx 1.2$

9. (a) $\bar{x} \pm t_c \frac{s}{\sqrt{n}} = 12.5 \pm 2.015 \frac{2.0}{\sqrt{6}} \approx 12.5 \pm 1.645 \approx (10.9,\ 14.1)$

(b) $\bar{x} \pm z_c \frac{s}{\sqrt{n}} = 12.5 \pm 1.645 \frac{2.0}{\sqrt{6}} \approx 12.5 \pm 1.343 \approx (11.2,\ 13.8)$

The t-CI is wider.

11. (a) $\bar{x} \pm t_c \frac{s}{\sqrt{n}} = 4.3 \pm 2.650 \frac{0.34}{\sqrt{14}} \approx 4.3 \pm 0.241 \approx (4.1,\ 4.5)$

(b) $\bar{x} \pm z_c \frac{s}{\sqrt{n}} = 4.3 \pm 2.326 \frac{0.34}{\sqrt{14}} \approx 4.3 \pm 0.211 \approx (4.1,\ 4.5)$

When rounded to the nearest tenth, the normal CI and the t-CI have the same width.

13. $(14.7,\ 22.1) \Rightarrow \bar{x} = 18.4 \Rightarrow E = 22.1 - 18.4 = 3.7$

15. $(64.6,\ 83.6) \Rightarrow \bar{x} = 74.1 \Rightarrow E = 83.6 - 74.1 = 9.5$

17. $E = t_c \frac{s}{\sqrt{n}} = 2.365 \frac{7.2}{\sqrt{8}} \approx 6.02$

$\bar{x} \pm E \approx 35.5 \pm 6.0 = (29.5,\ 41.5)$

With 95% confidence, you can say that the population mean commute time to work is between 29.5 and 41.5 minutes.

19. $E = z_c \frac{\sigma}{\sqrt{n}} = 1.96 \frac{9.3}{\sqrt{8}} \approx 6.4$

$\bar{x} \pm E \approx 35.5 \pm 6.4 = (29.1,\ 41.9)$

With 95% confidence, you can say that the population mean commute time to work is between 29.1 and 41.9 minutes. This confidence interval is slightly wider than the one found in Exercise 17.

21. (a) $\bar{x} \pm t_c \frac{s}{\sqrt{n}} = 4.50 \pm 1.833 \frac{1.21}{\sqrt{10}} \approx 4.50 \pm 0.701 \approx (3.80,\ 5.20)$

(b) $\bar{x} \pm z_c \frac{s}{\sqrt{n}} = 4.50 \pm 1.645 \frac{1.21}{\sqrt{500}} \approx 4.50 \pm 0.089 \approx (4.41,\ 4.59)$

The t-CI is wider.

23. (a) $\bar{x} \approx 90{,}182.9$ (b) $s \approx 3724.9$

(c) $\bar{x} \pm t_c \frac{s}{\sqrt{n}} = 90{,}182.9 \pm 2.947 \frac{3724.9}{\sqrt{16}} \approx 90{,}182.9 \pm 2744.3 \approx (87{,}438.6,\ 92{,}927.2)$

25. (a) $\bar{x} \approx 1767.7$

(b) $s \approx 252.2$

(c) $\bar{x} \pm t_c \frac{s}{\sqrt{n}} \approx 1767.7 \pm 3.106 \frac{252.23}{\sqrt{12}} \approx 1767.7 \pm 226.16 \approx (1541.5,\ 1993.8)$

27. $n \geq 30 \rightarrow$ use normal distribution

$\bar{x} \pm z_c \frac{s}{\sqrt{n}} = 27.7 \pm 1.96 \frac{6.12}{\sqrt{50}} \approx 27.7 \pm 1.70 = (26.00,\ 29.40)$

With 95% confidence, you can say that the population mean BMI is between 26.0 and 29.4.

29. $\bar{x} \approx 21.9$, $s = 3.46$, $n < 30$, σ known, and pop normally distributed $\rightarrow$ use t-distribution

$\bar{x} \pm t_c \frac{s}{\sqrt{n}} \approx 21.9 \pm 2.064 \frac{3.46}{\sqrt{25}} \approx 21.9 \pm 1.43 = (20.5,\ 23.3)$

With 95% confidence, you can say that the population mean per gallon is between 20.5 and 23.3 miles per gallon.

31. $n < 30$, σ unknown, and pop *not* normally distributed $\rightarrow$ cannot use either the normal or t-distributions.

33. 90% confidence interval results:
μ: mean of variable

Variable	Sample mean	Std. err.	DF	L. Limit	U Limit
Time (in hours)	12.194445	0.4136141	17	11.474918	12.91397

95% confidence interval results:
μ: mean of variable

Variable	Sample mean	Std. err.	DF	L. Limit	U Limit
Time (in hours)	12.194445	0.4136141	17	11.321795	13.067094

99% confidence interval results:
μ: mean of variable

Variable	Sample mean	Std. err.	DF	L. Limit	U Limit
Time (in hours)	12.194445	0.4136141	17	10.995695	13.393193

With 90% confidence, you can say the population mean time spent on homework is between 11.5 and 12.9 hours. With 95% confidence, you can say it is between 11.3 and 13.1 hours and with 99% confidence, you can say it is between 11.0 and 13.4 hours. As the level of confidence increases, the intervals get wider.

35. $n = 25,\ \bar{x} = 56.0,\ s = 0.25$

$\pm t_{0.99} \rightarrow 99\%$ t-CI

$$\bar{x} \pm t_c \frac{s}{\sqrt{n}} = 56.0 \pm 2.797 \frac{0.25}{\sqrt{25}} \approx 56.0 \pm 0.140 \approx (55.9,\ 56.1)$$

They are not making good tennis balls because desired bounce height of 55.5 inches is not contained between 55.9 and 56.1 inches.

6.3 CONFIDENCE INTERVALS FOR POPULATION PROPORTIONS

6.3 Try It Yourself Solutions

1a. $x = 181,\ n = 1006$

b. $\hat{p} = \dfrac{181}{1006} \approx 0.180$

2a. $\hat{p} \approx 0.180,\ \hat{q} \approx 0.820$

b. $n\hat{p} \approx (1006)(0.180) = 181.08 > 5$

$n\hat{q} \approx (1006)(0.820) = 824.92 > 5$

Distribution of $\hat{p}$ is approximately normal.

c. $z_c = 1.645$

$$E = z_c \sqrt{\frac{\hat{p}\hat{q}}{n}} \approx 1.645 \sqrt{\frac{0.180 \cdot 0.820}{1006}} \approx 0.020$$

d. $\hat{p} \pm E \approx 0.180 \pm 0.020 \approx (0.160,\ 0.200)$

e. With 90% confidence, you can say that the proportion of adults who think Abraham Lincoln was the greatest president is between 16.0% and 20.0%.

3a. $n = 498,\ \hat{p} = 0.25$

$\hat{q} = 1 = \hat{p} - 1 = 0.25 \approx 0.75$

b. $n\hat{p} = 498 \cdot 0.25 = 124.5 > 5$

$n\hat{q} = 498 \cdot 0.75 = 373.5 > 5$

Distribution of $\hat{p}$ is approximately normal.

c. $z_c = 2.575$

d. $\hat{p} \pm E \approx 0.25 \pm 0.050 = (0.20,\ 0.30)$

e. With 99% confidence, you can say that the proportion of adults who think that people over 65 are the more dangerous drivers is between 20% and 30%.

4a. (1) $\hat{p} = 0.5,\ \hat{q} = 0.5$

$z_c = 1.645,\ E = 0.02$

(2) $\hat{p} = 0.11,\ \hat{q} = 0.89$

$z_c = 1.645,\ E = 0.02$

b. (1) $n = \hat{p}\hat{q}\left(\dfrac{z_c}{E}\right)^2 = (0.5)(0.5)\left(\dfrac{1.645}{0.02}\right)^2 \approx 1691.266 \rightarrow 1692$

(2) $n = \hat{p}\hat{q}\left(\dfrac{z_c}{E}\right)^2 = (0.11)(0.89)\left(\dfrac{1.645}{0.02}\right)^2 \approx 662.3 \rightarrow 663$

c. (1) At least 1692 females should be included in the sample.
(2) At least 663 females should be included in the sample.

6.3 EXERCISE SOLUTIONS

1. False. To estimate the value of p, the population proportion of successes, use the point estimate $\hat{p} = \dfrac{x}{n}$.

3. $\hat{p} = \dfrac{x}{n} = \dfrac{752}{1002} \approx 0.750$

$\hat{q} = 1 - \hat{p} \approx 0.250$

5. $\hat{p} = \dfrac{x}{n} = \dfrac{4912}{11{,}605} \approx 0.423$

$\hat{q} = 1 - \hat{p} \approx 0.577$

7. $(0.905, 0.933) \rightarrow \hat{p} = 0.919 \Rightarrow E = 0.933 - 0.919 = 0.014$

9. $(0.512, 0.596) \rightarrow \hat{p} = 0.554 \Rightarrow E = 0.596 - 0.554 = 0.042$

11. $\hat{p} = \frac{x}{n} = \frac{396}{674} \approx 0.588$

$\hat{q} = 1 - \hat{p} \approx 0.412$

90% CI: $\hat{p} \pm z_c\sqrt{\frac{\hat{p}\hat{q}}{n}} \approx 0.588 \pm 1.645\sqrt{\frac{(0.588)(0.412)}{674}} \approx 0.588 \pm 0.031$

$= (0.557,\ 0.619)$

95% CI: $\hat{p} \pm z_c\sqrt{\frac{\hat{p}\hat{q}}{n}} \approx 0.588 \pm 1.96\sqrt{\frac{(0.588)(0.412)}{674}} \approx 0.588 \pm 0.037$

$= (0.551,\ 0.625)$

With 90% confidence, you can say that the population proportion of U.S. males ages 18-64 who say they have gone to the dentist in the past year is between 55.7% and 61.9%. With 95% confidence, you can say it is between 55.1% and 62.5%. The 95% confidence interval is slightly wider.

13. $\hat{p} = \frac{x}{n} = \frac{1435}{3110} \approx 0.461$

$\hat{q} = 1 - \hat{p} \approx 0.539$

$\hat{p} \pm z_c\sqrt{\frac{\hat{p}\hat{q}}{n}} \approx 0.461 \pm 2.575\sqrt{\frac{(0.461)(0.539)}{3110}} \approx 0.461 \pm 0.023$

$= (0.438,\ 0.484)$

With 99% confidence, you can say that the population proportion of U.S. adults who say they have started paying bills online in the past year is between 43.8% and 48.4%.

15. $\hat{p} = \frac{x}{n} = \frac{4431}{7000} \approx 0.633$

$\hat{q} = 1 - \hat{p} = 0.367$

$\hat{p} \pm z_c\sqrt{\frac{\hat{p}\hat{q}}{n}} \approx 0.633 \pm 1.96\sqrt{\frac{(0.633)(0.367)}{7000}} \approx 0.633 \pm 0.011$

$= (0.622,\ 0.644)$

17. (a) $n = \hat{p}\hat{q}\left(\frac{z_c}{E}\right)^2 = 0.5 \cdot 0.5\left(\frac{1.96}{0.04}\right)^2 \approx 600.25 \rightarrow 601$ adults

(b) $n = \hat{p}\hat{q}\left(\frac{z_c}{E}\right)^2 = 0.78 \cdot 0.22\left(\frac{1.96}{0.04}\right)^2 \approx 412.01 \rightarrow 413$ adults

(c) Having an estimate of the population proportion reduces the minimum sample size needed.

19. (a) $n = \hat{p}\hat{q}\left(\frac{z_c}{E}\right)^2 = 0.5 \cdot 0.5\left(\frac{1.645}{0.03}\right)^2 \approx 751.67 \to 752$ adults

(b) $n = \hat{p}\hat{q}\left(\frac{z_c}{E}\right)^2 = 0.201 \cdot 0.799\left(\frac{1.645}{0.03}\right)^2 \approx 482.87 \to 483$ adults

(c) Having an estimate of the population proportion reduces the minimum sample size needed.

21. (a) $\hat{p} = 0.27,\ \hat{q} = 0.73,\ n = 1017$

$$\hat{p} \pm z_c\sqrt{\frac{\hat{p}\hat{q}}{n}} = 0.27 \pm 2.575\sqrt{\frac{(0.27)(0.73)}{1017}} \approx 0.27 \pm 0.036 = (0.234,\ 0.306)$$

(b) $\hat{p} = 0.49,\ \hat{q} = 0.51,\ n = 1060$

$$\hat{p} \pm z_c\sqrt{\frac{\hat{p}\hat{q}}{n}} = 0.49 \pm 2.575\sqrt{\frac{(0.49)(0.51)}{1060}} \approx 0.49 \pm 0.040 = (0.450,\ 0.530)$$

(c) $\hat{p} = 0.31,\ \hat{q} = 0.69,\ n = 1126$

$$\hat{p} \pm z_c\sqrt{\frac{\hat{p}\hat{q}}{n}} = 0.31 \pm 2.575\sqrt{\frac{(0.31)(0.69)}{1126}} \approx 0.31 \pm 0.035 = (0.275,\ 0.345)$$

23. (a) $\hat{p} = 0.32$ $\qquad \hat{q} = 1 - \hat{p} = 0.68$

$$\hat{p} \pm z_c\sqrt{\frac{\hat{p}\hat{q}}{n}} = 0.32 \pm 1.96\sqrt{\frac{(0.32)(0.68)}{400}} \approx 0.32 \pm 0.046$$
$$= (0.274,\ 0.366)$$

(b) $\hat{p} = 0.56$ $\qquad \hat{q} = 1 - \hat{p} = 0.44$

$$\hat{p} \pm z_c\sqrt{\frac{\hat{p}\hat{q}}{n}} = 0.56 \pm 1.96\sqrt{\frac{(0.56)(0.44)}{400}} \approx 0.56 \pm 0.049$$
$$= (0.511,\ 0.609)$$

25. No, it is unlikely that the two proportions are equal because the confidence intervals estimating the proportions do not overlap. The 99% confidence intervals are (0.260, 0.380) and (0.496, 0.624). Although these intervals are wider, they still do not overlap.

27. 90% confidence interval results:
p: proportion of successes for population
method: Standard-Wald

Proportion	Count	Total	Sample Prop.	Std. err.	L. Limit	U Limit
p	802	1025	0.78243905	0.012887059	0.7612417	0.8036364

95% confidence interval results:
p: proportion of successes for population
method: Standard-Wald

Proportion	Count	Total	Sample Prop.	Std. err.	L. Limit	U Limit
p	802	1025	0.78243905	0.012887059	0.75718087	0.8076972

99% confidence interval results:
p: proportion of successes for population
method: Standard-Wald

Proportion	Count	Total	Sample Prop.	Std. err.	L. Limit	U Limit
p	802	1025	0.78243905	0.012887059	0.74924415	0.8156339

With 90% confidence, you can say the population proportion of U.S. adults who disapprove of the job Congress is doing is between 76.1% and 80.4%. With 95% confidence you can say it is between 75.7% and 80.8%. With 99% confidence you can say it is between 74.9% and 81.6%.

29. $31.4\% \pm 1\% \rightarrow (30.4\%,\ 32.4\%) \rightarrow (0.304,\ 0.324)$

$$E = z_c\sqrt{\frac{\hat{p}\hat{q}}{n}} \rightarrow z_c = E\sqrt{\frac{n}{\hat{p}\hat{q}}} = 0.01\sqrt{\frac{8451}{0.314 - 0.686}} \approx 1.981 \rightarrow z_c = 1.98 \rightarrow c = 0.9761$$

(30.4%, 32.4%) is approximately a 95.23% CI.

31. If $n\hat{p} < 5$ or $n\hat{q} < 5$, the sampling distribution of $\hat{p}$ may not be normally distributed, therefore preventing the use of z_c when calculating the confidence interval.

33.

$\hat{p}$	$\hat{q} = 1 - \hat{p}$	$\hat{p}\hat{q}$
0.0	1.0	0.00
0.1	0.9	0.09
0.2	0.8	0.16
0.3	0.7	0.21
0.4	0.6	0.24
0.5	0.5	0.25
0.6	0.4	0.24
0.7	0.3	0.21
0.8	0.2	0.16
0.9	0.1	0.09
1.0	0.0	0.00

$\hat{p}$	$\hat{q} = 1 - \hat{p}$	$\hat{p}\hat{q}$
0.45	0.55	0.2475
0.46	0.54	0.2484
0.47	0.53	0.2491
0.48	0.52	0.2496
0.49	0.51	0.2499
0.50	0.50	0.2500
0.51	0.49	0.2499
0.52	0.48	0.2496
0.53	0.47	0.2491
0.54	0.46	0.2484
0.55	0.45	0.2475

$\hat{p} = 0.5$ gives the maximum value of $\hat{p}\hat{q}$.

6.4 CONFIDENCE INTERVALS FOR VARIANCE AND STANDARD DEVIATION

6.4 Try It Yourself Solutions

1a. d.f. $= n - 1 = 29$
level of confidence $= 0.90$

b. Area to the right of χ_R^2 is 0.05.
Area to the right of χ_L^2 is 0.95.

c. $\chi_R^2 = 42.557,\ \chi_L^2 = 17.708$

d. So, 90% of the area under the curve lies between 17.708 and 42.557.

2a. 90% CI: $\chi_R^2 = 42.557,\ \chi_L^2 = 17.708$
95% CI: $\chi_R^2 = 42.722,\ \chi_L^2 = 16.047$

b. 90% CI for σ^2: $\left(\dfrac{(n-1)s^2}{\chi_R^2}, \dfrac{(n-1)s^2}{\chi_L^2}\right) = \left(\dfrac{29\cdot(1.2)^2}{42.557} \cdot \dfrac{29\cdot(1.2)^2}{17.708}\right) \approx (0.98,\ 2.36)$

95% CI for σ^2: $\left(\dfrac{(n-1)s^2}{\chi_R^2}, \dfrac{(n-1)s^2}{\chi_L^2}\right) = \left(\dfrac{29\cdot(1.2)^2}{42.722} \cdot \dfrac{29\cdot(1.2)^2}{16.047}\right) \approx (0.91,\ 2.60)$

c. 90% CI for σ: $\left(\sqrt{0.981},\ \sqrt{2.358}\right) = (0.99, 1.54)$
95% CI for σ: $\left(\sqrt{0.913},\ \sqrt{2.602}\right) = (0.96, 1.61)$

d. With 90% confidence, you can say that the population variance is between 0.98 and 2.36 and that the populations standard deviation is between 0.99 and 1.54. With 95% confidence, you can say that the population variance is between 0.91 and 2.60, and that the population standard deviation is between 0.96 and 1.61.

6.4 EXERCISE SOLUTIONS

1. Yes.

3. $\chi_R^2 = 14.067,\ \chi_L^2 = 2.167$

5. $\chi_R^2 = 32.852,\ \chi_L^2 = 8.907$

7. $\chi_R^2 = 52.336,\ \chi_L^2 = 13.121$

9. (a) $s \approx 0.00843$

$$\left(\frac{(n-1)s^2}{\chi_R^2}, \frac{(n-1)s^2}{\chi_L^2}\right) \approx \left(\frac{13\cdot(0.00843)^2}{22.362}, \frac{13\cdot(0.00843)^2}{5.892}\right) \approx (0.0000413,\ 0.000157)$$

(b) $\left(\sqrt{0.0000413},\ \sqrt{0.000157}\right) \approx (0.00643,\ 0.0125)$

With 90% confidence, you can say that the population variance is between 0.0000413 and 0.000157, and the population standard deviation is between 0.00643 and 0.0125 milligrams.

11. (a) $s \approx 0.253$

$$\left(\frac{(n-1)s^2}{\chi_R^2}, \frac{(n-1)s^2}{\chi_L^2}\right) \approx \left(\frac{17\cdot(0.253)^2}{35.718}, \frac{17\cdot(0.253)^2}{5.697}\right) \approx (0.0305,\ 0.191)$$

(b) $\left(\sqrt{0.0305},\ \sqrt{0.191}\right) \approx (0.175,\ 0.437)$

With 99% confidence, you can say that the population variance is between 0.0305 and 0.191, and the population standard deviation is between 0.175 and 0.437 hour.

13. (a) $$\left(\frac{(n-1)s^2}{\chi_R^2}, \frac{(n-1)s^2}{\chi_L^2}\right) \approx \left(\frac{13\cdot(3.90)^2}{29.819}, \frac{13\cdot(390)^2}{3.565}\right) \approx (6.63,\ 55.46)$$

(b) $\left(\sqrt{6.631},\ \sqrt{55.464}\right) \approx (2.58,\ 7.45)$

With 99% confidence, you can say that the population variance is between 6.63 and 55.46, and the population standard deviation is between 2.58 and 7.55 dollars per year.

15. (a) $$\left(\frac{(n-1)s^2}{\chi_R^2}, \frac{(n-1)s^2}{\chi_L^2}\right) \approx \left(\frac{9\cdot(30.244)^2}{21.666}, \frac{9\cdot(30.244)^2}{2.088}\right) \approx (380.0,\ 3942.6)$$

(b) $\left(\sqrt{380},\ \sqrt{3942.6}\right) \approx (19.5,\ 62.8)$

With 98% confidence, you can say that the population variance is between 380.0 and 3942.6, and the population standard deviation is between \$19.5 and \$62.8.

17. (a) $$\left(\frac{(n-1)s^2}{\chi_R^2}, \frac{(n-1)s^2}{\chi_L^2}\right) \approx \left(\frac{15\cdot(6.42)^2}{27.488}, \frac{15\cdot(6.42)^2}{6.262}\right) \approx (22.5,\ 98.7)$$

(b) $\left(\sqrt{22.5},\ \sqrt{98.7}\right) \approx (4.7,\ 9.9)$

With 95% confidence, you can say that the population variance is between 22.5 and 98.7, and the population standard deviation is between 4.7 and 9.9 beats per minute.

19. (a) $\left(\dfrac{(n-1)s^2}{\chi_R^2}, \dfrac{(n-1)s^2}{\chi_L^2}\right) \approx \left(\dfrac{18\cdot(15)^2}{31.526}, \dfrac{18\cdot(15)^2}{8.231}\right) \approx (128,\ 492)$

(b) $\left(\sqrt{128.465},\ \sqrt{492.042}\right) \approx (11,\ 22)$

With 95% confidence, you can say that the population variance is between 128 and 492, and the population standard deviation is between 11 and 22 grains per gallon.

21. (a) $\left(\dfrac{(n-1)s^2}{\chi_R^2}, \dfrac{(n-1)s^2}{\chi_L^2}\right) \approx \left(\dfrac{13\cdot(3725)^2}{19.812}, \dfrac{13\cdot(3725)^2}{7.042}\right) \approx (9{,}104{,}741,\ 25{,}615{,}326)$

(b) $\left(\sqrt{9{,}104{,}741},\ \sqrt{25{,}615{,}326}\right) \approx (3017,\ 5061)$

With 80% confidence, you can say that the population variance is between 9,104,741 and 25,615,326, and the population standard deviation is between \$3017 and \$5061.

23. (a) $\left(\dfrac{(n-1)s^2}{\chi_R^2}, \dfrac{(n-1)s^2}{\chi_L^2}\right) \approx \left(\dfrac{(21)(3.6)^2}{38.932}, \dfrac{(21)(3.6)^2}{8.897}\right) \approx (7.0,\ 30.6)$

(b) $\left(\sqrt{6.99},\ \sqrt{30.59}\right) \approx (2.6,\ 5.5)$

With 98% confidence, you can say that the population variance is between 7.0 and 30.6, and the population standard deviation is between 2.6 and 5.5 minutes.

25. 95% confidence interval results:

σ^2 : variance of variable

Variance	Sample Var.	DF	L. Limit	U Limit
σ^2	11.56	29	7.332092	20.891039

(2.71, 4.57)

27. 90% confidence interval results:

σ^2 : variance of variable

Variance	Sample Var.	DF	L. Limit	U Limit
σ^2	1225	17	754.8815	2401.4731

(27, 49)

29. 90% CI for σ: (0.00643, 0.0125)

Yes, because all of the values in the confidence interval are less than 0.015.

31. Answers will vary. **Sample answer:** Unlike a confidence interval for a population mean or proportion, a confidence interval for a population variance does not have a margin of error. The left and right endpoints must be calculated separately.

CHAPTER 6 REVIEW EXERCISE SOLUTIONS

1. (a) $\bar{x} \approx 103.5$
(b) $s \approx 34.663$

$$E = z_c \frac{s}{\sqrt{n}} \approx 1.645 \frac{34.663}{\sqrt{40}} \approx 9.0$$

3. $\bar{x} \pm z_c \frac{s}{\sqrt{n}} = 15.8 \pm 2.575 \frac{0.85}{\sqrt{80}} \approx 15.3 \pm 0.245 \approx (15.6,\ 16.0)$

5. $(20.75,\ 24.10) \rightarrow \bar{x} = 22.425 \rightarrow E = 24.10 - 22.425 = 1.675$

7. $s \approx 34.663$

$$n = \left(\frac{z_c \sigma}{E}\right)^2 \approx \left(\frac{1.96 \cdot 34.663}{10}\right)^2 \approx 46.158 \Rightarrow 47 \text{ people}$$

9. $n = \left(\frac{z_c \sigma}{E}\right)^2 \approx \left(\frac{(1.96)(7.098)}{2}\right)^2 \approx 48.39 \Rightarrow 49 \text{ people}$

11. $t_c = 1.383$

13. $t_c = 2.624$

15. $n = 20 \rightarrow t_c = 1{,}729$
$n = 30 \rightarrow t_c = 1.699$
$n = 20$ produces a larger critical value t_c.

17. $E = t_c \frac{s}{\sqrt{n}} = 1.753 \frac{25.6}{\sqrt{16}} \approx 11.2$

19. $E = t_c \frac{s}{\sqrt{n}} = 2.718\left(\frac{0.9}{\sqrt{12}}\right) \approx 0.7$

21. $\bar{x} \pm z_c \frac{s}{\sqrt{n}} = 72.1 \pm 1.753 \frac{25.6}{\sqrt{16}} \approx 72.1 \pm 11.219 \approx (60.9,\ 83.3)$

23. $\bar{x} \pm t_c \frac{s}{\sqrt{n}} = 6.8 \pm 2.718\left(\frac{0.9}{\sqrt{12}}\right) \approx 6.8 \pm 0.706 \approx (6.1,\ 7.5)$

25. $\bar{x} \pm t_c \frac{s}{\sqrt{n}} = 2218 \pm 1.703 \frac{523}{\sqrt{28}} \approx 2218 \pm 168.3 \approx (2050,\ 2386)$

27. $\hat{p} = \frac{x}{n} = \frac{1215}{1500} = 0.81,\ \hat{q} = 0.19$

29. $\hat{p} = \frac{x}{n} = \frac{552}{1023} \approx 0.540,\ \hat{q} \approx 0.460$

31. $\hat{p} = \frac{x}{n} = \frac{141}{1008} \approx 0.140,\ \hat{q} \approx 0.860$

33. $\hat{p} = \frac{x}{n} = \frac{346}{706} \approx 0.490,\ \hat{q} \approx 0.510$

35. $\hat{p} \pm z_c\sqrt{\dfrac{\hat{p}\hat{q}}{n}} = 0.81 \pm 1.96\sqrt{\dfrac{0.81 \cdot 0.19}{1500}} \approx 0.81 \pm 0.020 = (0.790,\ 0.830)$

With 95% confidence, you can say that the population proportion of U.S. adults who say they will participate in the 2010 Census is between 79.0% and 83.0%.

37. $\hat{p} \pm z_c\sqrt{\dfrac{\hat{p}\hat{q}}{n}} \approx 0.540 \pm 1.645\sqrt{\dfrac{0.540 \cdot 0.460}{1023}} \approx 0.540 \pm 0.026 = (0.514,\ 0.566)$

With 90% confidence, you can say that the population proportion of U.S. adults who say they have worked the night shift at some point in their lives is between 51.4% and 56.6%.

39. $\hat{p} \pm z_c\sqrt{\dfrac{\hat{p}\hat{q}}{n}} \approx 0.140 \pm 2.575\sqrt{\dfrac{0.140 \cdot 0.860}{10008}} \approx 0.140 \pm 0.028 = (0.112,\ 0.168)$

With 99% confidence, you can say that the population proportion of U.S. adults who say that the cost of healthcare is the most important financial problem facing their family today is between 11.2% and 16.8%.

41. $\hat{p} \pm z_c\sqrt{\dfrac{\hat{p}\hat{q}}{n}} \approx 0.490 \pm 1.28\sqrt{\dfrac{(0.490)(0.510)}{706}} \approx 0.49 \pm 0.024 = (0.466,\ 0.514)$

With 80% confidence, you can say that the population proportion of parents with kids 4 to 8 years old who say they know their state booster seat law is between 46.6% and 51.4%.

43. (a) $n = \hat{p}\hat{q}\left(\dfrac{z_c}{E}\right)^2 = 0.50 \cdot 0.50\left(\dfrac{1.96}{0.05}\right)^2 \approx 384.16 \to 385$ adults

(b) $n = \hat{p}\hat{q}\left(\dfrac{z_c}{E}\right)^2 = 0.63 \cdot 0.37\left(\dfrac{1.96}{0.05}\right)^2 \approx 358.19 \to 359$ adults

(c) The minimum sample size needed is smaller when a preliminary estimate is available.

45. $\chi_R^2 = 23.337,\ \chi_L^2 = 4.404$

47. $\chi_R^2 = 14.067,\ \chi_L^2 = 2.167$

49. $s \approx 7.002$

95% CI for σ^2: $\left(\dfrac{(n-1)s^2}{\chi_R^2},\ \dfrac{(n-1)s^2}{\chi_L^2}\right) \approx \left(\dfrac{16 \cdot (7.002)^2}{28.845},\ \dfrac{16 \cdot (7.002)^2}{6.908}\right) \approx (27.2,\ 113.5)$

95% CI for σ: $\left(\sqrt{27.2},\ \sqrt{113.5}\right) \approx (5.2,\ 10.7)$

51. $s \approx 1.190$

98% CI for σ^2: $\left(\dfrac{(n-1)s^2}{\chi_R^2},\ \dfrac{(n-1)s^2}{\chi_L^2}\right) \approx \left(\dfrac{(25)(1.190)^2}{44.314},\ \dfrac{(25)(1.190)}{11.524}\right) \approx (0.80,\ 3.07)$

98% CI for σ: $\left(\sqrt{0.80},\ \sqrt{3.07}\right) \approx (0.89,\ 1.75)$

CHAPTER 6 QUIZ SOLUTIONS

1. (a) $\bar{x} \approx 6.85$

(b) $s \approx 1.821$

$$E = t_c \frac{s}{\sqrt{n}} \approx 1.960 \frac{1.821}{\sqrt{30}} \approx 0.65$$

You are 95% confident that the margin of error for the population mean is about 0.65 minute.

(c) $\bar{x} \pm t_c \frac{s}{\sqrt{n}} \approx 6.848 \pm 1.960 \frac{1.821}{\sqrt{30}} = 6.848 \pm 0.652 \approx (6.196,\ 7.500)$

With 95% confidence, you can say that the population mean amount of time is between 6.20 and 7.50 minutes.

2. $n = \left(\frac{z_c \sigma}{E}\right)^2 = \left(\frac{2.575 \cdot 2.4}{1}\right)^2 \approx 38.18 \rightarrow 39$ students

3. (a) $\bar{x} \approx 33.1$

$s \approx 2.38$

(b) $\bar{x} \pm t_c \frac{s}{\sqrt{n}} \approx 33.11 \pm 1.833 \frac{2.38}{\sqrt{10}} \approx 33.11 \pm 1.38 = (31.73,\ 34.49)$

With 90% confidence you can say that the population mean time played in the season is between 31.73 and 34.49 minutes.

(c) $\bar{x} \pm z_c \frac{\sigma}{\sqrt{n}} \approx 33.11 \pm 1.645 \frac{5.25}{\sqrt{10}} \approx 33.11 \pm 2.73 = (30.38,\ 35.84)$

With 90% confidence you can say that the population mean time played in the season is between 30.38 and 35.84 minutes. This confidence interval is wider than the one found in part (b).

4. $\bar{x} \pm t_c \frac{s}{\sqrt{n}} = 6824 \pm 2.447 \frac{340}{\sqrt{7}} \approx 6824 \pm 314.46 \approx (6510,\ 7138)$

5. (a) $\hat{p} = \frac{x}{n} = \frac{1079}{1383} \approx 0.780$

(b) $\hat{p} \pm z_c \sqrt{\frac{\hat{p}\hat{q}}{n}} \approx 0.780 \pm 1.645 \sqrt{\frac{0.780 \cdot 0.220}{1383}} \approx 0.780 \pm 0.018 = (0.762,\ 0.798)$

(c) $n = \hat{p}\hat{q}\left(\frac{z_c}{E}\right)^2 \approx 0.780 \cdot 0.220 \left(\frac{2.575}{0.04}\right)^2 \approx 711.13 \rightarrow 712$ adults

6. (a) $\left(\frac{(n-1)s^2}{\chi_R^2},\ \frac{(n-1)s^2}{\chi_L^2}\right) = \left(\frac{29 \cdot (1.821)^2}{45.722},\ \frac{29 \cdot (1.821)^2}{16.047}\right) \approx (2.10,\ 5.99)$

(b) $\left(\sqrt{2.10},\ \sqrt{5.99}\right) \approx (1.45,\ 2.45)$

Hypothesis Testing with One Sample

CHAPTER 7

7.1 INTRODUCTION TO HYPOTHESIS TESTING

7.1 Try It Yourself Solutions

1a. (1) The mean is not 74 months.
$\mu \neq 74$
(2) The variance is less than or equal to 2.7.
$\sigma^2 \leq 2.7$
(3) The proportion is more than 24%
$p > 0.24$

b. (1) $\mu = 74$ (2) $\sigma^2 > 2.7$ (3) $p \leq 0.24$

c. (1) $H_0: \mu = 74$; $H_a: \mu \neq 74$ (claim)
(2) $H_0: \sigma^2 \leq 2.7$ (claim); $H_a: \sigma^2 > 2.7$
(3) $p \leq 0.24$; $H_a: p > 0.24$ (claim)

2a. $H_0: p \leq 0.01$; $H_a: p > 0.01$

b. A type I error will occur if the actual proportion is less than or equal to 0.01, but you reject H_0.
A type II error will occur if the actual proportion is greater than 0.01, but you fail to reject H_0.

c. A type II error is more serious because you would be misleading the consumer, possibly causing serious injury or death.

3a. (1) H_0: The mean life of a certain type of automobile battery is 74 months.
H_a: The mean life of a certain type of automobile battery is not 74 months.
$H_0: \mu = 74$; $H_a: \mu \neq 74$
(2) H_0: The variance of the life of the home theater system is less than or equal to 2.7.
H_a: The variance of the life of the home theater system is greater than 2.7.
$H_0: \sigma^2 \leq 2.7$; $H_a: \sigma^2 > 2.7$
(3) H_0: The proportion of homeowners who feel their house is too small for their family is less than or equal to 24%.
H_a: The proportion of homeowners who feel their house is too small for their family is greater than 24%.
$H_0: p \leq 0.24$; $H_a: p > 0.24$

b. (1) Two-tailed (2) Right-tailed (3) Right-tailed

c. (1)

$\frac{1}{2}$ *P*-value area $\frac{1}{2}$ *P*-value area
$-z$ z z

(2)

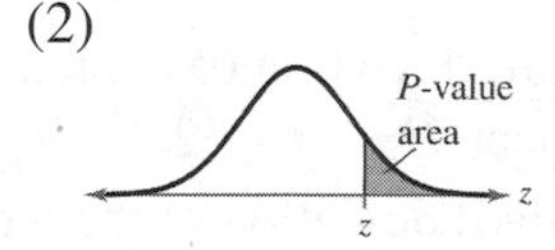

(3)

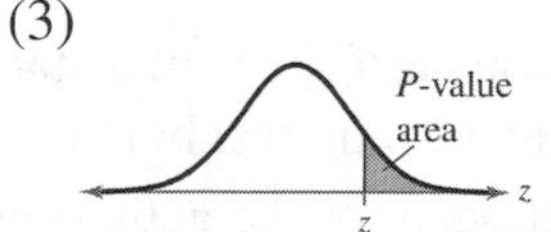

4a. There us enough evidence to support the realtor's claim that the proportion of homeowners who feel their house is too small for their family is more than 24%.

b. There is not enough evidence to support the realtor's claim that the proportion of homeowners who feel their house is too small for their family is more than 24%.

5a. (1) Support claim. (2) Reject claim.
b. (1) $H_0: \mu \geq 650$; $H_a: \mu < 650$ (claim)
(2) $H_0: \mu = 98.6$ (claim); $H_a: \mu \neq 98.6$

7.1 EXERCISE SOLUTIONS

1. The two types of hypotheses used in a hypothesis test are the null hypothesis and the alternative hypothesis.
The alternative hypothesis is the complement of the null hypothesis.

3. You can reject the null hypothesis, or you can fail to reject the null hypothesis.

5. False. In a hypothesis test, you assume the null hypothesis is true.

7. True

9. False. A small *P*-value in a test will favor a rejection of the null hypothesis.

11. $H_0: \mu \leq 645$ (claim); $H_a: \mu > 645$

13. $H_0: \sigma = 5$; $H_a: \sigma \neq 5$ (claim)

15. $H_0: p \geq 0.45$; $H_a: p < 0.45$ (claim)

17. c; $H_0: \mu \leq 3$

1 2 3 4 μ

19. b; $H_0: \mu = 3$

1 2 3 4 μ

21. Right-tailed

23. Two-tailed

25. $\mu > 750$
$H_0: \mu \leq 750$; $H_a: \mu > 750$ (claim)

27. $\sigma \leq 320$
$H_0: \sigma \leq 320$ (claim); $H_a: \sigma > 320$

29. $\mu < 45$
$H_0: \mu \geq 45$; $H_a: \mu < 45$ (claim)

31. A type I error will occur if the actual proportion of new customers who return to buy their next piece of furniture is at least 0.60, but you reject $H_0: p \geq 0.60$.
A type II error will occur if the actual proportion of new customers who return to buy their next piece of furniture is less than 0.60, but you fail to reject $H_0: p \geq 0.60$.

33. A type I error will occur if the actual standard deviation of the length of time to play a game is less than or equal to 12 minutes, but you reject $H_0: \sigma \leq 12$.
A type II error will occur of the actual standard deviation of the length of time to play a game is greater than 12 minutes, but you fail to reject $H_0: \sigma \leq 12$.

35. A type I error will occur if the actual proportion of applicants who become police officers is at most 0.20, but you reject $H_0: p \leq 0.20$.

A type II error will occur if the actual proportion of applicants who become police officers is greater than 0.20, but you fail to reject $H_0: p \leq 0.20$.

37. H_0: The proportion of homeowners who have a home security alarm is greater than or equal to 14%.

H_a: The proportion of homeowners who have a home security alarm is less than 14%.

$H_0: p \geq 0.14$; $H_a: p < 0.14$

Left-tailed because the alternative hypothesis contains <.

39. H_0: The standard deviation of the 18-hole scores for a golfer is greater than or equal to 2.1 strokes.

H_a: The standard deviation of the 18-hole scores for a golfer is less than 2.1 strokes.

$H_0: \sigma \geq 2.1$; $H_a: \sigma < 2.1$

Left-tailed because the alternative hypothesis contains <.

41. H_0: The mean length of the baseball team's games is greater than or equal to 2.5 hours.

H_a: The mean length of the baseball team's games is less than 2.5 hours.

$H_0: \mu \geq 2.5$; $H_a: \mu < 2.5$

Left-tailed because the alternative hypothesis contains <.

43. (a) There is enough evidence to support the scientist's claim that the mean incubation period for swan eggs is less than 40 days.

(b) There is not enough evidence to support the scientist's claim that the mean incubation period for swan eggs is less than 40 days.

45. (a) There is enough evidence to support U.S. Department of Labor's claim that the proportion of full-time workers earning over $450 per week is greater than 75%.

(b) There is not enough evidence to support U.S. Department of Labor's claim that the proportion of full-time workers earning over $450 per week is greater than 75%.

47. (a) There is enough evidence to support the researcher's claim that the proportion of people who have no health care visits in the past year is less than 17%.

(b) There is not enough evidence to support the researcher's claim that the proportion of people who have no health care visits in the past year is less than 17%.

49. $H_0: \mu \geq 60$; $H_a: \mu < 60$

51. (a) $H_0: \mu \geq 15$; $H_a: \mu < 15$

(b) $H_0: \mu \leq 15$; $H_a: \mu > 15$

53. If you decrease α, you are decreasing the probability that you reject H_0. Therefore, you are increasing the probability of failing to reject H_0. This could increase β, the probability of failing to reject H_0 when H_0 is false.

55. Yes; If the P-value is less than $\alpha = 0.05$, it is also less than $\alpha = 0.10$.

57. (a) Fail to reject H_0 because the CI includes values greater than 70.
(b) Reject H_0 because the CI is located below 70.
(c) Fail to reject H_0 because the CI includes values greater than 70.

59. (a) Reject H_0 because the CI is located to the right of 0.20.
(b) Fail to reject H_0 because the CI includes values less than 0.20.
(c) Fail to reject H_0 because the CI includes values less than 0.20.

7.2 HYPOTHESIS TESTING FOR THE MEAN (LARGE SAMPLES)

7.2 Try It Yourself Solutions

1a. (1) $P = 0.0347 > 0.01 = \alpha$
(2) $P = 0.0347 < 0.05 = \alpha$

b. (1) Fail to reject H_0 because $0.0347 > 0.01$.
(2) Reject H_0 because $0.0347 < 0.05$.

2a.

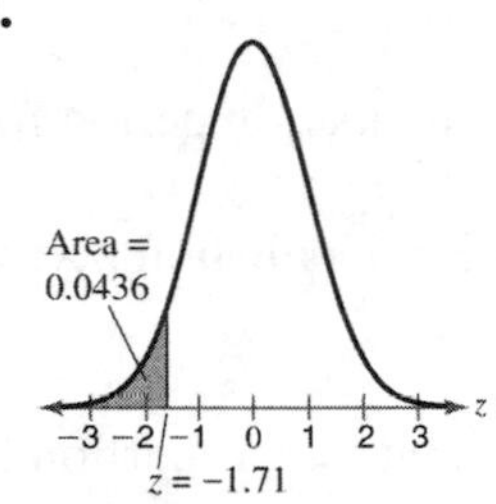

b. $P = 0.0436$

c. Reject H_0 because $P = 0.0436 < 0.05$.

3a. Area to the right of $z = 1.64$ is 0.0505.

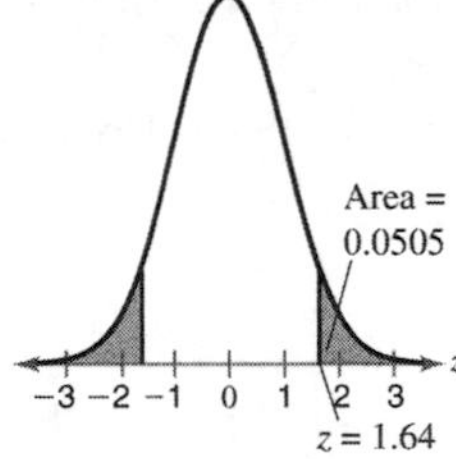

b. $p = 2(\text{area}) = 2(0.0505) = 0.1010$

c. Fail to reject H_0 because $P = 0.1010 > 0.10$.

4a. The claim is "the mean speed is greater than 35 miles per hour."
$H_0: \mu \leq 35$; $H_a: \mu > 35$ (claim)

b. $\alpha = 0.05$

c. $z = \dfrac{\bar{x}-\mu}{\frac{s}{\sqrt{n}}} = \dfrac{36-35}{\frac{4}{\sqrt{100}}} = 2.5$

d. P-value = {Area right of $z = 2.50$} = 0.0062

e. Reject H_0 because P-value = 0.0062 < 0.05.

f. There is enough evidence at the 5% level of significance to support the claim that the average speed is greater than 35 miles per hour.

5a. The claim is "one of your distributors reports an average of 150 sales per day."
$H_0: \mu = 150$ (claim); $H_\alpha: \mu \neq 150$

b. $\alpha = 0.01$

c. $z = \dfrac{\bar{x}-\mu}{\frac{s}{\sqrt{n}}} = \dfrac{143-150}{\frac{15}{\sqrt{35}}} \approx -2.76$

d. P-value = 2{Area to the left of $z = -2.76$ } = 2(0.0029) = 0.0059

e. Reject H_0 because P-value = 0.0058 < 0.01.

f. There is enough evidence at the 1% level of significance to reject the claim that the distributorship averages 150 sales per day.

6a. $P = 0.0440 > 0.01 = \alpha$

b. Fail to reject H_0.

7a.

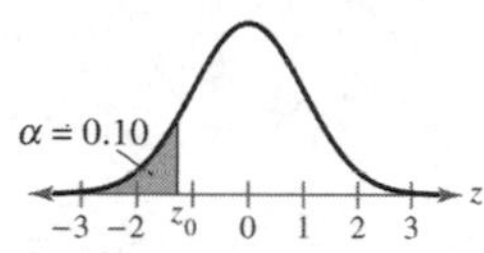

b. Area = 0.1003

c. $z_0 = -1.28$

d. $z < -1.28$

8a.

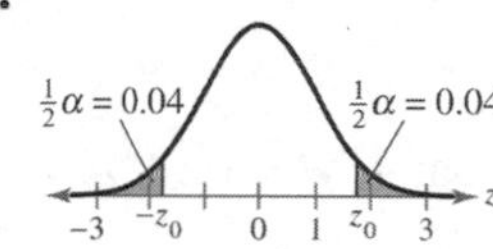

b. 0.0401 and 0.9599

c. $-z_0 = -1.75$ and $z_0 = 1.75$

d. $z < -1.75,\ z > 1.75$

9a. The claim is "the mean work day of the company's mechanical engineers is less than 8.5 hours."
$H_0: \mu \geq 8.5$; $H_a: \mu < 8.5$ (claim)

b. $\alpha = 0.01$

c. $z_0 = -2.33$; Rejection region: $z < -2.33$

d. $z = \dfrac{x-\mu}{\frac{s}{\sqrt{n}}} = \dfrac{8.2-8.5}{\frac{0.5}{\sqrt{35}}} \approx -3.55$

e.

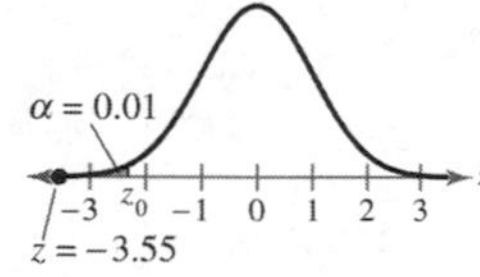

Because $-3.55 < -2.33$, reject H_0.

f. There is enough evidence at the 1% level of significance to support the claim that the mean work day is less than 8.5 hours.

10a. $\alpha = 0.01$

b. $\pm z_0 = \pm 2.575$; Rejection regions: $z < -2.575$, $z > 2.575$

c.

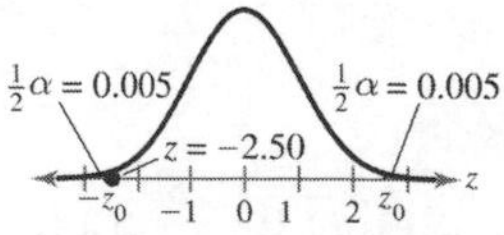

Fail to reject H_0.

d. There is not enough evidence at the 1% level of significance to reject the claim that the mean cost of raising a child from birth to age 2 by husband-wife families in the United States is $13,120.

7.2 EXERCISE SOLUTIONS

1. In the z-test using rejection region(s), the test statistic is compared with critical values. The z-test using a P-value compares the P-value with the level of significance α.

3.

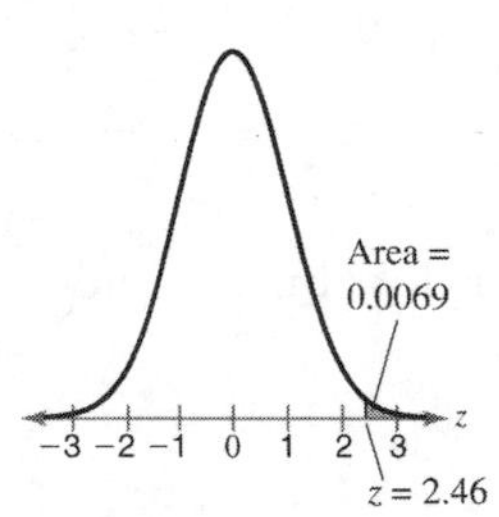

$P = 0.0934$; Reject H_0 because $P = 0.0934 < 0.10$.

5.

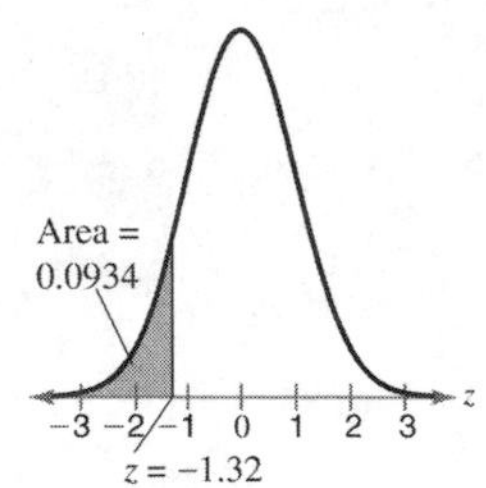

$P = 0.0069$; Reject H_0 because $P = 0.0069 < 0.01$.

7.

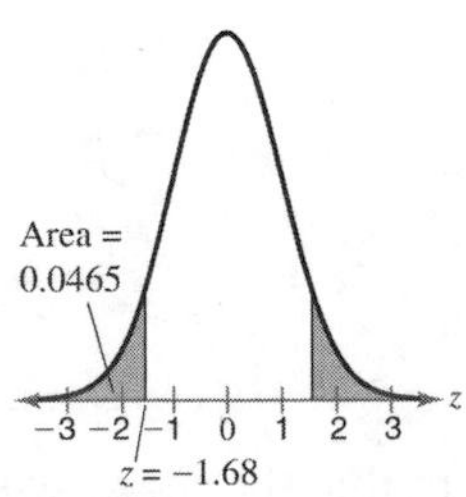

$P = 2(\text{Area}) = 2(0.0465) = 0.0930$; Fail to reject H_0 because $P = 0.0930 > 0.05$.

9. b

11. c

13. (a) Fail to reject H_0 $(P = 0.0461 > 0.01)$.
(b) Reject H_0 $(P = 0.0461 < 0.05)$.

15. Fail to reject H_0 $(P = 0.0628 > 0.05)$.

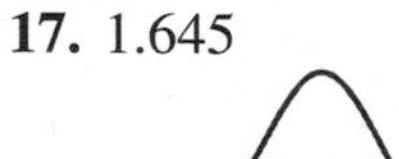

17. 1.645

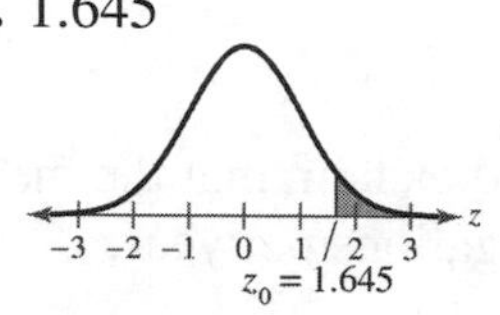

19. -1.88

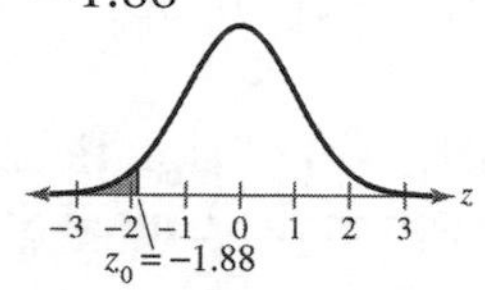

21. ± 2.33

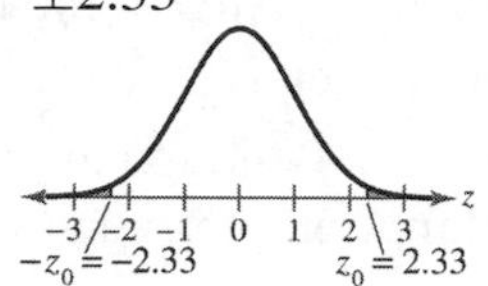

23. (a) Fail to reject H_0 because $z < 1.285$.
(b) Fail to reject H_0 because $z < 1.285$.
(c) Fail to reject H_0 because $z < 1.285$.
(d) Reject H_0 because $z > 1.285$.

25. $H_0: \mu = 40$ (claim); $H_a: \mu \neq 40$

$\alpha = 0.05 \rightarrow z_0 = \pm 1.96$

$$z = \frac{\bar{x} - \mu}{\frac{s}{\sqrt{n}}} = \frac{39.2 - 40}{\frac{3.23}{\sqrt{75}}} \approx -2.145$$

Reject H_0. There is enough evidence at the 5% level of significance to reject the claim.

27. $H_0: \mu = 8550$; $H_a: \mu \neq 8550$ (claim)

$\alpha = 0.02 \rightarrow z_0 = \pm 2.33$

$$z = \frac{\bar{x} - \mu}{\frac{s}{\sqrt{n}}} = \frac{8420 - 8550}{\frac{314}{\sqrt{38}}} \approx -2.552$$

Reject H_0. There is enough evidence at the 2% level of significance to support the claim.

29. (a) $H_0: \mu \leq 30$; $H_a: \mu > 30$ (claim)

(b) $$z = \frac{\bar{x} - \mu}{\frac{s}{\sqrt{n}}} = \frac{31 - 30}{\frac{2.5}{\sqrt{50}}} \approx 2.83$$

Area = 0.9977

(c) P-value = {Area to right of $z = 2.83$} = 0.0023

(d) Reject H_0.

(e) There is enough evidence at the 1% level of significance to support the claim that the mean raw score for the school's applicants is more than 30.

31. (a) $H_0: \mu = 28.5$ (claim); $H_a: \mu \neq 28.5$

(b) $z = \dfrac{\bar{x} - \mu}{\frac{s}{\sqrt{n}}} = \dfrac{27.8 - 28.5}{\frac{4.1}{\sqrt{100}}} \approx -1.71$

Area = 0.0436

(c) P-value = 2{Area to the left of $z = -1.71$} 2(0.0436) = 0.0872

(d) Fail to reject H_0.

(e) There is not enough evidence at the 8% level of significance to reject the claim that the mean consumption of bottled water by a person in the United States is 28.5 gallons per year.

33. (a) $H_0: \mu = 15$ (claim); $H_a: \mu \neq 15$

(b) $\bar{x} \approx 14.834$ $\quad s \approx 4.288$

$z = \dfrac{\bar{x} - \mu}{\frac{s}{\sqrt{n}}} = \dfrac{14.384 - 15}{\frac{4.288}{\sqrt{32}}} \approx -0.22$

Area = 0.4129

(c) P-value = 2{Area to left of $z = -0.22$} = 2(0.4129) = 0.8258

(d) Fail to reject H_0.

(e) There is not enough evidence at the 5% level of significance to reject the claim that the mean time it takes smokers to quit smoking permanently is 15 years.

35. (a) $H_0: \mu = 40$ (claim); $H_a: \mu \neq 40$

(b) $-z_0 = -2.575$, $z_0 = 2.575$;

Rejection regions: $z < -2.575$, $z > 2.575$

(c) $z = \dfrac{\bar{x} - \mu}{\frac{s}{\sqrt{n}}} = \dfrac{39.2 - 40}{\frac{7.5}{\sqrt{30}}} \approx -0.584$

(d) Fail to reject H_0.

(e) There is not enough evidence at the 1% level of significance to reject the claim that the mean caffeine content per 12-ounce bottle of cola is 40 milligrams.

37. (a) $H_0: \mu \geq 750$ (claim); $H_a: \mu < 750$

(b) $z_0 = -2.05$; Rejection region: $z < -2.05$

(c) $z = \dfrac{\bar{x} - \mu}{\frac{s}{\sqrt{n}}} = \dfrac{745 - 750}{\frac{60}{\sqrt{36}}} = -0.5$

(d) Fail to reject H_0.

(e) There is not enough evidence at the 2% level of significance to reject the claim that the mean life of the bulb is at least 750 hours.

39. (a) $H_0: \mu \le 32$; $H_a: \mu > 32$ (claim)

(b) $z_0 = 1.555$; Rejection region: $z > 1.555$

(c) $\bar{x} \approx 29.676$ $\quad s \approx 9.164$

$$z = \frac{\bar{x} - \mu}{\frac{s}{\sqrt{n}}} \approx \frac{29.676 - 32}{\frac{9.164}{\sqrt{34}}} \approx -1.478$$

(d) Fail to reject H_0.

(e) There is not enough evidence at the 6% level of significance to support the claim that the mean nitrogen dioxide level in Calgary is greater than 32 parts per billion.

41. (a) $H_0: \mu \ge 10$ (claim); $H_a: \mu < 10$

(b) $z_0 = -1.88$; Rejection region: $z < -1.88$

(c) $\bar{x} \approx 9.780$ $\quad s \approx 2.362$

$$z = \frac{\bar{x} - \mu}{\frac{s}{\sqrt{n}}} \approx \frac{9.780 - 10}{\frac{2.362}{\sqrt{30}}} \approx -0.51$$

(d) Fail to reject H_0.

(e) There is not enough evidence at the 3% level of significance to reject the claim that the mean weight loss after 1 month is at least 10 pounds.

43. Hypothesis test results:

μ: population mean

$H_0: \mu = 58$

$H_a: \mu \ne 58$

Standard deviation = 2.35

Mean	n	Sample Mean	Std. Err	z-Stat	P-value
μ	80	57.6	0.262738	−1.5224292	0.1279

Fail to reject H_0. There is not enough evidence at the 10% level of significance to reject the claim.

45. Hypothesis test results:

μ: population mean

$H_0: \mu = 1210$

$H_a: \mu > 1210$

Standard deviation = 205.87

Mean	n	Sample Mean	Std. Err	z-Stat	P-value
μ	250	1234.21	13.020362	1.8593953	0.0315

Reject H_0. There is enough evidence at the 8% level of significance to reject the claim.

47. $z_0 = -2.33$; Rejection region: $z < -2.33$

$$z = \frac{\bar{x} - \mu}{\frac{s}{\sqrt{n}}} = \frac{125,270 - 127,400}{\frac{6275}{\sqrt{30}}} \approx -1.86$$

Fail to reject H_0 because the standardized test statistic $z = -1.86$ is not in the rejection region $(z < -2.33)$.

49. (a) $\alpha = 0.02 \rightarrow z_0 = -2.05$; Rejection region: $z < -2.05$

Stays as fail to reject H_0 because the standardized test statistic $z = -1.86$ is not in the rejection region $(z < -2.05)$.

(b) $\alpha = 0.05 \rightarrow z_0 = -1.645$; Rejection region: $z < -1.645$

Changes to reject H_0 because the standardized test statistic $z = -1.86$ is in the rejection region $(z < -1.645)$.

(c) Same rejection region as Exercise 47: $z < -2.33$

$$z = \frac{\bar{x} - \mu}{\frac{s}{\sqrt{n}}} = \frac{125,270 - 127,400}{\frac{6275}{\sqrt{40}}} \approx -2.15$$

Stays as fail to reject H_0 because the standardized test statistic $z = -2.15$ is not in the rejection region $(z < -2.33)$.

(d) Same rejection region as Exercise 47: $z < -2.33$

$$z = \frac{\bar{x} - \mu}{\frac{s}{\sqrt{n}}} = \frac{125,270 - 127,400}{\frac{6275}{\sqrt{50}}} \approx -2.40$$

Changes to reject H_0 because the standardized test statistic $z = -2.40$ is in the rejection region $(z < -2.33)$.

7.3 HYPOTHESIS TESTING FOR THE MEAN (SMALL SAMPLES)

7.3 Try It Yourself Solutions

1a. 13 **b.** $t_0 = -2.650$

2a. 8 **b.** $t_0 = 1.397$

3a. 15 **b.** $-t_0 = -2.131$, $t_0 = 2.131$

4a. The claim is "the mean cost of insuring a 2008 Honda CR-V is less than \$1200."

H_0: $\mu \geq \$1200$; H_a: $\mu < \$1200$ (claim)

b. $\alpha = 0.10$ and d.f. = $n - 1 = 6$

c. $t_0 = -1.440$; Rejection region: $t < -1.440$

d. $z=\dfrac{\bar{x}-\mu}{\dfrac{s}{\sqrt{n}}}=\dfrac{1125-1200}{\dfrac{55}{\sqrt{7}}}\approx -3.61$

e. Reject H_0.

f. There is enough evidence at the 10% level of significance to support the claim that the mean cost of insuring a 2008 Honda CR-V is less than \$1200.

5a. The claim is "the mean conductivity of the river is 1890 milligrams per liter."
$H_0:\ \mu=1890$ (claim); $H_a:\ \mu\neq 1890$

b. $\alpha=0.01$ and d.f. = $n-1=18$

c. $-t_0=-2.878,\ t_0=2.878$; Rejection regions: $t<-2.878,\ t>2.878$

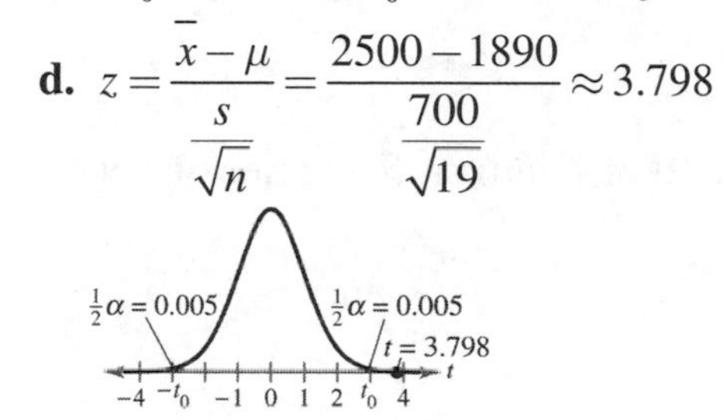

d. $z=\dfrac{\bar{x}-\mu}{\dfrac{s}{\sqrt{n}}}=\dfrac{2500-1890}{\dfrac{700}{\sqrt{19}}}\approx 3.798$

e. Reject H_0.

f. There is enough evidence at the 1% level of significance to reject the claim that the mean conductivity of the river is 1890 milligrams per liter.

6a. The claim is "the mean wait time is at most 18 minutes."
$H_0:\ \mu\leq 18$ minutes (claim); $H_a:\ \mu>18$ minutes

b. P-value = 0.9997

c. P-value = 0.9997 > 0.05 = α
Fail to reject H_0.

d. There is not enough evidence at the 5% level of significance to reject the claim that the mean wait time is at most 18 minutes.

7.3 EXERCISE SOLUTIONS

1. Identify the level of significance α and the degrees of freedom, d.f. = $n-1$. Find the critical value(s) using the t-distribution table in the row with $n-1$ d.f. If the hypothesis test is:
(1) left-tailed, use the "One Tail, α" column with a negative sign.
(2) right-tailed, use the "One Tail, α" column with a positive sign.
(3) two-tailed, use the "Two Tail, α," column with a negative and a positive sign.

3. $t_0=1.717$ **5.** $t_0=-1.328$ **7.** $t_0=\pm 2.056$

9. (a) Fail to reject H_0 because $t > -2.086$.

(b) Fail to reject H_0 because $t > -2.086$.

(c) Fail to reject H_0 because $t > -2.086$.

(d) Reject H_0 because $t < -2.086$.

11. (a) Fail to reject H_0 because $-2.602 < t < 2.602$.

(b) Fail to reject H_0 because $-2.602 < t < 2.602$.

(c) Reject H_0 because $t > 2.602$.

(d) Reject H_0 because $t < -2.602$.

13. $H_0: \mu = 15$ (claim); $H_a: \mu \neq 15$

$\alpha = 0.01$ and d.f. = $n - 1 = 5$

$t_0 = \pm 4.032$

$$t = \frac{\bar{x} - \mu}{\frac{s}{\sqrt{n}}} = \frac{13.9 - 15}{\frac{3.23}{\sqrt{6}}} \approx -0.834$$

Fail to reject H_0. There is not enough evidence at the 1% level of significance to reject the claim.

15. $H_0: \mu \geq 8000$ (claim); $H_a: \mu < 8000$

$\alpha = 0.01$ and d.f. = $n - 1 = 24$

$t_0 = -2.492$

$$t = \frac{\bar{x} - \mu}{\frac{s}{\sqrt{n}}} = \frac{7700 - 8000}{\frac{450}{\sqrt{25}}} \approx -3.333$$

Reject H_0. There is enough evidence at the 1% level of significance to reject the claim.

17. (a) $H_0: \mu = 18{,}000$ (claim); $H_a: \mu \neq 18{,}000$

(b) $-t_0 = -2.145,\ t_0 = 2.145$; Rejection region: $t < -2.145,\ t > 2.145$

(c) $t = \dfrac{\bar{x} - \mu}{\frac{s}{\sqrt{n}}} = \dfrac{18{,}550 - 18{,}000}{\frac{1767}{\sqrt{15}}} \approx 1.21$

(d) Fail to reject H_0.

(e) There is not enough evidence at the 5% level of significance to reject the claim that the mean price of a 2008 Subaru Forester is $18,000.

19. (a) $H_0: \mu \leq 60$; $H_a: \mu > 60$ (claim)

(b) $t_0 = 1.943$; Rejection region: $t > 1.943$

(c) $t = \dfrac{\bar{x} - \mu}{\frac{s}{\sqrt{n}}} = \dfrac{70 - 60}{\frac{12.5}{\sqrt{7}}} \approx 2.12$

(d) Reject H_0.

(e) There is enough evidence at the 5% level of significance to support the claim that the mean number of hours worked per week by surgical faculty who teach at an academic institution is more than 60 hours.

21. (a) $H_0: \mu \leq 1$; $H_a: \mu > 1$ (claim)

(b) $t_0 = 1.356$; Rejection region: $t > 1.356$

(c) $t = \dfrac{\bar{x} - \mu}{\frac{s}{\sqrt{n}}} = \dfrac{1.50 - 1}{\frac{0.28}{\sqrt{13}}} \approx 6.44$

(d) Reject H_0.

(e) There is enough evidence at the 10% level of significance to support the claim that the mean amount of waste recycled by adults in the United States is more than 1 pound per person per day.

23. (a) $H_0: \mu = \$26,000$ (claim); $H_a: \mu \neq \$26,000$

(b) $-t_0 = -2.262$, $t_0 = 2.262$; Rejection region: $t < -2.262$, $t > 2.262$

(c) $\bar{x} \approx 25,852.2$, $s \approx \$3197.1$

$$t = \frac{\bar{x} - \mu}{\frac{s}{\sqrt{n}}} \approx \frac{25,852.2 - 26,000}{\frac{3197.1}{\sqrt{10}}} \approx -0.15$$

(d) Fail to reject H_0.

(e) There is not enough evidence at the 5% level of significance to reject the claim that the mean salary for full-time male workers over age 25 without a high school diploma is \$26,000.

25. (a) $H_0: \mu \leq 45$; $H_a: \mu > 45$ (claim)

(b) $t = \dfrac{\bar{x} - \mu}{\frac{s}{\sqrt{n}}} = \dfrac{48 - 45}{\frac{5.4}{\sqrt{25}}} \approx 2.78$

P-value = {Area to right of t = 2.78 } = 0.0052

(c) Reject H_0.

(d) There is enough evidence at the 10% level of significance to support the claim that the mean speed of the vehicles is greater than 45 miles per hour.

27. (a) $H_0: \mu = \$105$ (claim); $H_a: \mu \neq \$105$

(b) $t = \dfrac{\bar{x} - \mu}{\frac{s}{\sqrt{n}}} = \dfrac{110 - 105}{\frac{8.50}{\sqrt{20}}} \approx 2.63$

P-value = 2{Area to right of t = 2.63} = 2(0.00825) = 0.0165

(c) Fail to reject H_0.

(d) There is not enough evidence at the 1% level of significance to reject the claim that the mean daily meal cost for two adults traveling together on vacation in San Francisco is \$105.

29. (a) $H_0: \mu \geq 32$; $H_a: \mu < 32$ (claim)

(b) $\bar{x} \approx 30.167$ $\quad s \approx 4.004$

$$t = \frac{\bar{x} - \mu}{\frac{s}{\sqrt{n}}} = \frac{30.167 - 32}{\frac{4.004}{\sqrt{18}}} \approx -1.942$$

P-value = {Area to left of $t = -1.942$} = 0.0344

(c) Reject H_0.

(d) There is enough evidence at the 5% level of significance to support the claim that the mean class size for full-time faculty is fewer than 32 students.

31. Hypothesis test results:

μ: population mean

$H_0: \mu = 75$

$H_A: \mu > 75$

Mean	Sample Mean	Std. Err	*DF*	*T*-Stat	*P*-value
μ	73.6	0.62757164	25	−2.2308211	0.9825

Fail to reject H_0.

There is not enough evidence at the 5% level of significance to reject the claim.

33. Hypothesis test results:

μ: population mean

$H_0: \mu = 188$

$H_A: \mu < 188$

Mean	Sample Mean	Std. Err	*DF*	*T*-Stat	*P*-value
μ	186	4	8	−0.5	0.3153

Fail to reject H_0.

There is not enough evidence at the 5% level of significance to support the claim.

35. $H_0: \mu \leq 5000$; $H_a: \mu > 5000$ (claim)

$$t = \frac{\bar{x} - \mu}{\frac{s}{\sqrt{n}}} \approx \frac{5434 - 5000}{\frac{625}{\sqrt{6}}} \approx 1.70$$

P-value = {Area right of $t = 1.70$} ≈ 0.0748

Because 0.0748 > 0.05, fail to reject H_0.

37. Because σ is unknown, $n < 30$, and the gas mileage is normally distributed, use the t-distribution.
$H_0: \mu \geq 23$ (claim); $H_a: \mu < 23$

$$t = \frac{\bar{x} - \mu}{\frac{s}{\sqrt{n}}} \approx \frac{22 - 23}{\frac{4}{\sqrt{5}}} \approx -0.559$$

P-value = {Area left of $t = -0.559$} = 0.303

Fail to reject H_0. There is not enough evidence at the 5% level of significance to reject the claim that the mean gas mileage for the luxury sedan is at least 23 miles per gallon.

39. More likely; For degrees of freedom less than 30, the tails of the t-distribution curve are thicker than those of a standard normal distribution curve. So, if you incorrectly used a standard normal sampling distribution instead of a t-sampling distribution, the area under the curve at the tails will be smaller than what it would be for the t-test, meaning the critical value(s) will lie closer to the mean. This makes it more likely for the test statistic to be in the rejection region(s). This result is the same regardless of whether the test is left-tailed, right-tailed, or two-tailed; in each case, the tail thickness affects the location of the critical value(s).

7.4 HYPOTHESIS TESTING FOR PROPORTIONS

7.4 Try It Yourself Solutions

1a. $np = (125)(0.25) = 31.25 > 5,\ nq = (125)(0.75) = 93.75 > 5$

b. The claim is "more than 25% of U.S. adults have used a cellular phone to access the internet."
$H_0: p \leq 0.25$; $H_a: p > 0.25$ (claim)

c. $\alpha = 0.05$

d. $z_0 = 1.645$; Rejection region: $z > 1.645$

e. $$z = \frac{\hat{p} - p}{\sqrt{\frac{pq}{n}}} = \frac{0.32 - 0.25}{\sqrt{\frac{(0.25)(0.75)}{125}}} \approx 1.81$$

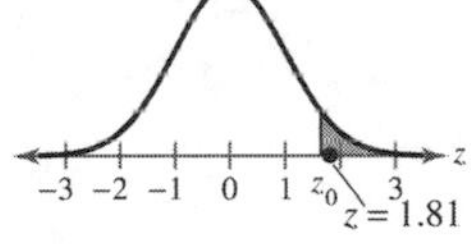

f. Reject H_0.

g. There is enough evidence at the 5% level of significance to support the claim that more than 25% of U.S. adults have used a cellular phone to access the Internet.

2a. $np = (250)(0.30) = 75 > 5,\ nq = (250)(0.70) = 175 > 5$

b. The claim is "30% of U.S. adults have not purchased a certain brand because they found the advertisements distasteful."
$H_0: p = 0.30$ (claim); $H_a: p \neq 0.30$

c. $\alpha = 0.10$

d. $-z_0 = -1.645,\ z_0 = 1.645$; Rejection region: $z < -1.645,\ z > 1.645$

e. $$z=\frac{\hat{p}-p}{\sqrt{\frac{pq}{n}}}=\frac{0.36-0.30}{\sqrt{\frac{(0.30)(0.70)}{250}}}\approx 2.07$$

f. Reject H_0.

g. There is enough evidence at the 10% level of significance to reject the claim that 30% of U.S. adults have not purchased a certain brand because they found the advertisements distasteful.

7.4 EXERCISE SOLUTIONS

1. If $np \geq 5$ and $nq \geq 5$, the normal distribution can be used.

3. $np=(40)(0.12)=4.8<5$

   $nq=(40)(0.88)=35.2>5$

   Cannot use normal distribution because $np<5$.

5. $np=(500)(0.15)=75>5$

   $nq=(500)(0.85)=425>5 \rightarrow$ use normal distribution

   $H_0: p=0.15$; $H_a: p\neq 0.15$ (claim)

   $-z_0=-1.96$, $z_0=1.96$; Rejection region: $z<-1.96$, $z>1.96$

   $$z=\frac{\hat{p}-p}{\sqrt{\frac{pq}{n}}}=\frac{0.12-0.15}{\sqrt{\frac{(0.15)(0.85)}{500}}}\approx -1.88$$

   Fail to reject H_0. There is not enough evidence at the 5% level of significance to support the claim.

7. $np=(100)(0.45)=45>5$

   $nq=(100)(0.55)=55>5 \rightarrow$ use normal distribution

   $H_0: p\leq 0.45$ (claim); $H_a: p>0.45$

   $z_0=1.645$; Rejection region: $z>1.645$

   $$z=\frac{\hat{p}-p}{\sqrt{\frac{pq}{n}}}=\frac{0.52-0.45}{\sqrt{\frac{(0.45)(0.55)}{100}}}\approx 1.41$$

   Fail to reject H_0. There is not enough evidence at the 5% level of significance to reject the claim.

9. (a) $H_0: p\geq 0.25$; $H_a: p<0.25$ (claim)

   (b) $z_0=-1.645$; Rejection region: $z<-1.645$

(c) $z = \dfrac{\hat{p} - p}{\sqrt{\dfrac{pq}{n}}} = \dfrac{0.185 - 0.25}{\sqrt{\dfrac{(0.25)(0.75)}{200}}} \approx -2.12$

(d) Reject H_0.

(e) There is enough evidence at the 5% level of significance to support the claim that less than 25% of U.S. adults are smokers.

11. (a) $H_0: p \leq 0.50$ (claim); $H_a: p > 0.50$

(b) $z_0 = 2.33$; Rejection region: $z > 2.33$

(c) $z = \dfrac{\hat{p} - p}{\sqrt{\dfrac{pq}{n}}} = \dfrac{0.58 - 0.50}{\sqrt{\dfrac{(0.50)(0.50)}{150}}} \approx 1.96$

(d) Fail to reject H_0.

(e) There is not enough evidence at the 1% level of significance to reject the claim that at most 50% of people believe that drivers should be allowed to use cellular phone with hands-free devices while driving.

13. (a) $H_0: p \leq 0.75$; $H_a: p > 0.75$ (claim)

(b) $z_0 = 1.28$; Rejection region: $z > 1.28$

(c) $z = \dfrac{\hat{p} - p}{\sqrt{\dfrac{pq}{n}}} = \dfrac{0.82 - 0.75}{\sqrt{\dfrac{(0.75)(0.25)}{150}}} \approx 1.98$

(d) Reject H_0.

(e) There is enough evidence at the 10% level of significance to support the claim that more than 75% of females ages 20-29 are taller than 62 inches.

15. (a) $H_0: p \geq 0.35$; $H_a: p < 0.35$ (claim)

(b) $z_0 = -1.28$; Rejection region: $z < -1.28$

(c) $z = \dfrac{\hat{p} - p}{\sqrt{\dfrac{pq}{n}}} = \dfrac{0.39 - 0.35}{\sqrt{\dfrac{(0.35)(0.65)}{400}}} \approx 1.68$

(d) Fail to reject H_0.

(e) There is not enough evidence at the 10% level of significance to support the claim that less than 35% of U.S. households own a dog.

17. $H_0: p \geq 0.52$ (claim); $H_a: p < 0.52$

$z_0 = -1.645$: Rejection region: $z < -1.645$

$$z = \frac{\hat{p} - p}{\sqrt{\dfrac{pq}{n}}} = \frac{0.48 - 0.52}{\sqrt{\dfrac{(0.52)(0.48)}{50}}} \approx -0.566$$

Fail to reject H_0. There is not enough evidence at the 5% level of significance to reject the claim that at least 52% of adults are more likely to buy a product when there are free samples.

19. $H_0: p \geq 0.35$; $H_a: p < 0.35$ (claim)

$z_0 = -1.28$: Rejection region: $z < -1.28$

$$z = \frac{x - np}{\sqrt{npq}} = \frac{156 - (400)(0.35)}{\sqrt{(400)(0.35)(0.65)}} \approx 1.68$$

Reject H_0. The results are the same.

7.5 HYPOTHESIS TESTING FOR VARIANCE AND STANDARD DEVIATION

7.5 Try It Yourself Solutions

1a. $df = 17$, $\alpha = 0.01$

b. $\chi_0^2 = 33.409$

2a. $df = 29$, $\alpha = 0.05$

b. $\chi_0^2 = 17.708$

3a. $df = 50$, $\alpha = 0.01$

b. $\chi_R^2 = 79.490$

c. $\chi_L^2 = 27.991$

4a. The claim is "the variance of the amount of sports drink in a 12-ounce bottle is no more than 0.40."

$H_0: \sigma^2 \leq 0.40$ (claim); $H_a: \sigma^2 > 0.40$

b. $\alpha = 0.01$ and d.f. = $n - 1 = 30$

c. $\chi_0^2 = 50.892$; Rejection region: $\chi^2 > 50.892$

d. $\chi^2 = \dfrac{(n-1)s^2}{\sigma^2} = \dfrac{(30)(0.75)}{0.40} = 56.250$

e. Reject H_0.

f. There is enough evidence at the 1% level of significance to reject the claim that the variance of the amount of sports drink in a 12-ounce bottle is no more than 0.40.

5a. The claim is "the standard deviation of the lengths of response times is less than 3.7 minutes."

$H_0: \sigma \geq 3.7$; $H_a: \sigma < 3.7$ (claim)

b. $\alpha = 0.05$ and d.f. = $n - 1 = 8$

c. $\chi_0^2 = 2.733$; Rejection region: $\chi^2 < 2.733$

d. $\chi^2 = \dfrac{(n-1)s^2}{\sigma^2} = \dfrac{(8)(3.0)}{(3.7)^2} \approx 5.259$

e. Fail to reject H_0.

f. There is not enough evidence at the 5% level of significance to support the claim that the standard deviation of the lengths of response times is less than 3.7 minutes.

6a. The claim is "the variance of the weight losses is 25.5."
$H_0: \sigma^2 = 25.5$ (claim); $H_a: \sigma^2 \neq 25.5$

b. $\alpha = 0.10$ and d.f. $= n-1 = 12$

c. $\chi_L^2 = 5.226$ and $\chi_R^2 = 21.026$; Rejection region: $\chi^2 > 21.026$, $\chi^2 < 5.226$

d. $\chi^2 = \dfrac{(n-1)s^2}{\sigma^2} = \dfrac{(12)(10.8)}{25.5} \approx 5.082$

e. Reject H_0.

f. There is enough evidence at the 10% level of significance to reject the claim that the variance of the weight losses of users is 25.5.

7.5 EXERCISE SOLUTIONS

1. Specify the level of significance α. Determine the degrees of freedom. Determine the critical values using the χ^2-distribution. For a right-tailed test, use the value that corresponds to d.f. and α. For a left-tailed test, use the value that corresponds to d.f. and $1-\alpha$. For a two-tailed test, use the value that corresponds to d.f. and $\frac{1}{2}\alpha$, and d.f. and $1-\frac{1}{2}\alpha$.

3. The requirement of a normal distribution is more important when testing a standard deviation than when testing a mean. If the population is not normal, the results of the χ^2-test can be misleading because the χ^2-test is not as robust as the tests for the population mean.

5. $\chi_0^2 = 38.885$

7. $\chi_0^2 = 0.872$

9. $\chi_L^2 = 60.391$, $\chi_R^2 = 101.879$

11. (a) Fail to reject H_0 because $\chi^2 < 6.251$.
(b) Fail to reject H_0 because $\chi^2 < 6.251$.
(c) Fail to reject H_0 because $\chi^2 < 6.251$.
(d) Reject H_0 because $\chi^2 > 6.251$.

13. (a) Fail to reject H_0 because $8.547 < \chi^2 < 22.307$.
(b) Reject H_0 because $\chi^2 > 22.307$.
(c) Reject H_0 because $\chi^2 < 8.547$.
(d) Fail to reject H_0 because $8.547 < \chi^2 < 22.307$.

15. $H_0: \sigma^2 = 0.52$ (claim); $H_a: \sigma^2 \neq 0.52$
$\chi_L^2 = 7.564$, $\chi_R^2 = 30.191$; Rejection regions: $\chi^2 < 7.564$, $\chi^2 > 30.191$

$$\chi^2 = \frac{(n-1)s^2}{\sigma^2} = \frac{(17)(0.508)^2}{(0.52)} \approx 16.608$$

Fail to reject H_0. There is not enough evidence at the 5% level of significance to reject the claim.

17. $H_0: \sigma = 24.9$ (claim); $H_a: \sigma \neq 24.9$

$\chi_L^2 = 34.764$, $\chi_R^2 = 67.505$; Rejection regions: $\chi^2 < 34.764$, $\chi^2 > 67.505$

$$\chi^2 = \frac{(n-1)s^2}{\sigma^2} = \frac{(50)(29.1)^2}{(24.9)^2} \approx 68.29$$

Reject H_0. There is enough evidence at the 10% level of significance to reject the claim.

19. (a) $H_0: \sigma^2 = 1.25$ (claim); $H_a: \sigma^2 \neq 1.25$

(b) $\chi_L^2 = 10.283$, $\chi_R^2 = 35.479$; Rejection regions: $\chi^2 < 10.283$, $\chi^2 > 35.479$

(c) $\chi^2 = \frac{(n-1)s^2}{\sigma^2} = \frac{(21)(1.35)}{1.25} = 22.68$

(d) Fail to reject H_0.

(e) There is not enough evidence at the 5% level of significance to reject the claim that the variance of the number of grams of carbohydrates in servings of its tortilla chips is 1.25.

21. (a) $H_0: \sigma \geq 36$; $H_a: \sigma < 36$ (claim)

(b) $\chi_0^2 = 13.240$; Rejection region: $\chi^2 < 13.240$

(c) $\chi^2 = \frac{(n-1)s^2}{\sigma^2} = \frac{(21)(33.4)^2}{(36)^2} \approx 18.076$

(d) Fail to reject H_0.

(e) There is not enough evidence at the 10% level of significance to support the claim that the standard deviation for eighth graders on the examination is less than 36 points.

23. (a) $H_0: \sigma \leq 25$ (claim); $H_a: \sigma > 25$

(b) $\chi_0^2 = 36.741$; Rejection region: $\chi^2 > 36.741$

(c) $\chi^2 = \frac{(n-1)s^2}{\sigma^2} = \frac{(27)(31)^2}{(25)^2} \approx 41.515$

(d) Reject H_0.

(e) There is enough evidence at the 10% level of significance to reject the claim that the standard deviation of the number of fatalities per year from tornadoes is no more than 25.

25. (a) $H_0: \sigma \geq \$3500$; $H_a: \sigma < \$3500$ (claim)

(b) $\chi_0^2 = 18.114$; Rejection region: $\chi^2 < 18.114$

(c) $\chi^2 = \frac{(n-1)s^2}{\sigma^2} = \frac{(27)(4100)^2}{(3500)^2} \approx 37.051$

(d) Fail to reject H_0.

(e) There is not enough evidence at the 10% level of significance to support the claim that the standard deviation of the total charges for patients involved in a crash in which the vehicle struck a construction baracade is less than $3500.

27. (a) $H_0: \sigma \le \$6100$; $H_a: \sigma > \$6100$ (claim)

(b) $\chi_0^2 = 27.587$; Rejection region: $\chi^2 > 27.587$

(c) $s \approx 7814.23$

$$\chi^2 = \frac{(n-1)s^2}{\sigma^2} = \frac{(17)(7814.23)^2}{(6100)^2} \approx 27.897$$

(d) Reject H_0.

(e) There is enough evidence at the 5% level of significance to support the claim that the standard deviation of the annual salaries for environmental engineers is greater than \$6100.

29. Hypothesis test results:

σ^2 : variance of variable

$H_0: \sigma^2 = 9$

$H_A: \sigma^2 < 9$

Variance	Sample Var.	*DF*	Chi-Squared Stat	*P*-value
σ^2	2.03	9	2.03	0.009

Reject H_0. There is enough evidence at the 1% level of significance to reject the claim.

31. $\sigma^2 = (4.5)^2 = 20.25$

$s^2 = (5.8)^2 = 33.64$

Hypothesis test results:

σ^2 : variance of variable

$H_0: \sigma^2 = 20.25$

$H_A: \sigma^2 > 20.25$

Variance	Sample Var.	*DF*	Chi-Squared Stat	*P*-value
σ^2	33.64	14	23.257284	0.0562

Fail to reject H_0. There is not enough evidence at the 5% level of significance to support the claim.

33. $\chi^2 = 37.051$

P-value = {Area left of $\chi^2 = 37.051$} = 0.9059

Fail to reject H_0 because *P*-value = 0.9059 > 0.10.

35. $\chi^2 = 27.897$

P-value = {Area right of $\chi^2 = 27.897$} = 0.0462

Reject H_0 because *P*-value = 0.0462 < 0.05.

CHAPTER 7 REVIEW EXERCISE SOLUTIONS

1. $H_0: \mu \leq 375$ (claim); $H_a: \mu > 375$

3. $H_0: p \geq 0.205$; $H_a: p < 0.205$ (claim)

5. $H_0: \sigma \leq 1.9$; $H_a: \sigma > 1.9$ (claim)

7a. H_0: p $= 0.71$ (claim); $H_a: p \neq 0.71$

b. A type I error will occur if the actual proportion of Americans who support plans to order deep cuts in executive compensation at companies that have received federal bailout funds is 71%, but you reject $H_0: p = 0.71$.

A type II error will occur if the actual proportion is not 71%, but you fail to reject $H_0: p = 0.71$.

c. Two-tailed because the alternative hypothesis contains ≠.

d. There is enough evidence to reject the news outlet's claim that the proportion of Americans who support plans to order deep cuts in executive compensation at companies that have received federal bailout funds is 71%.

e. There is not enough evidence to reject the news outlet's claim that the proportion of Americans who support plans to order deep cuts in executive compensation at companies that have received federal bailout funds is 71%.

9a. $H_0: \sigma \leq 50$ (claim); $H_a: \sigma > 50$

b. A type I error will occur if the actual standard deviation of the sodium content in one serving of a certain soup is no more than 50 milligrams, but you reject $H_0: \sigma \leq 50$.

A type II error will occur if the actual standard deviation of the sodium content in one serving of a certain soup is more than 50 milligrams, but you fail to reject $H_0: \sigma \leq 50$.

c. Right-tailed because the alternative hypothesis contains >.

d. There is enough evidence to reject the soup maker's claim that the standard deviation of the sodium content in one serving of a certain soup is no more than 50 milligrams.

e. There is not enough evidence to reject the soup maker's claim that the standard deviation of the sodium content in one serving of a certain soup is no more than 50 milligrams.

11. P-value = {Area to left of $z = -0.94$} = 0.1736

Fail to reject H_0.

13. $H_0: \mu \leq 0.05$ (claim); $H_a: \mu > 0.05$

$$z = \frac{\bar{x} - \mu}{\frac{s}{\sqrt{n}}} = \frac{0.057 - 0.05}{\frac{0.018}{\sqrt{32}}} \approx 2.20$$

P-value = {Area to right of $z = 2.20$} = 0.0139

$\alpha = 0.10 \Rightarrow$ Reject H_0.

$\alpha = 0.05 \Rightarrow$ Reject H_0.

$\alpha = 0.01 \Rightarrow$ Fail to reject H_0.

15. $z_0 \approx -2.05$

$z_0 = -2.05$

17. $z_0 = 1.96$

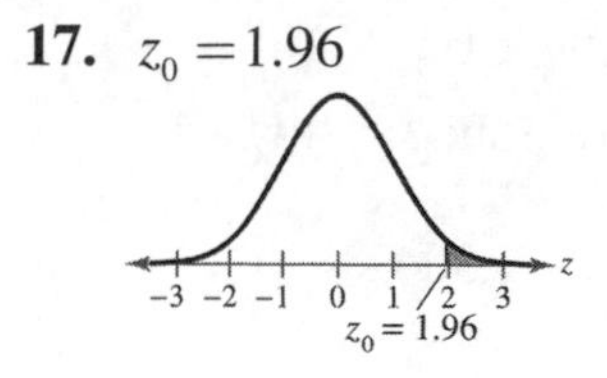

19. Fail to reject H_0 because $-1.645 < z < 1.645$.

21. Fail to reject H_0 because $-1.645 < z < 1.645$.

23. $H_0: \mu \leq 45$ (claim); $H_a: \mu > 45$

$z_0 = 1.645$; Rejection region: $z > 1.645$

$$z = \frac{\bar{x} - \mu}{\frac{s}{\sqrt{n}}} = \frac{47.2 - 45}{\frac{6.7}{\sqrt{42}}} \approx 2.128$$

Reject H_0. There is enough evidence at the 5% level of significance to reject the claim.

25. $H_0: \mu \geq 5.500$; $H_a: \mu < 5.500$ (claim)

$z_0 = -2.33$; Rejection region: $z < -2.33$

$$z = \frac{\bar{x} - \mu}{\frac{s}{\sqrt{n}}} = \frac{5.497 - 5.500}{\frac{0.011}{\sqrt{36}}} \approx -1.636$$

Fail to reject H_0. There is not enough evidence at the 1% level of significance to support the claim.

27. $H_0: \mu = \$10,380$ (claim); $H_a: \mu \neq \$10,380$

$$z = \frac{\bar{x} - \mu}{\frac{s}{\sqrt{n}}} = \frac{10,240 - 10,380}{\frac{1561}{\sqrt{800}}} \approx -2.54$$

P-value = 2{Area left of $z = -2.54$} = 2(0.0055) = 0.0110

Fail to reject H_0. There is not enough evidence at the 1% level of significance to reject the claim that the mean cost of raising a child from birth to age 2 by husband-wife families in rural areas is $10,380.

29. $-t_0 = -2.093$, $t_0 = 2.093$

31. $t_0 = -2.977$

33. $H_0: \mu = 95$; $H_a: \mu \neq 95$ (claim)

$-t_0 = -2.201$, $t_0 = 2.201$; Rejection regions: $t < -2.201$, $t > 2.201$

$$t = \frac{\bar{x} - \mu}{\frac{s}{\sqrt{n}}} = \frac{94.1 - 95}{\frac{1.53}{\sqrt{12}}} \approx -2.038$$

Fail to reject H_0. There is not enough evidence at the 5% level of significance to support the claim.

35. $H_0: \mu \geq 0$ (claim); $H_a: \mu < 0$

$t_0 = -1.341$; Rejection region: $t < -1.341$

$$t = \frac{\bar{x} - \mu}{\frac{s}{\sqrt{n}}} = \frac{-0.45 - 0}{\frac{1.38}{\sqrt{16}}} \approx -1.304$$

Fail to reject H_0. There is not enough evidence at the 10% level of significance to reject the claim.

37. $H_0: \mu \leq 48$ (claim); $H_a: \mu > 48$

$t_0 = 3.143$; Rejection region: $t > 3.143$

$$t = \frac{\bar{x} - \mu}{\frac{s}{\sqrt{n}}} = \frac{52 - 48}{\frac{2.5}{\sqrt{7}}} \approx 4.233$$

Reject H_0. There is enough evidence at the 1% level of significance to reject the claim.

39. $H_0: \mu = \$25$ (claim); $H_a: \mu \neq \$25$

$-t_0 = -1.740$, $t_0 = 1.740$; Rejection regions: $t < -1.740$, $t > 1.740$

$$t = \frac{\bar{x} - \mu}{\frac{s}{\sqrt{n}}} = \frac{26.25 - 25}{\frac{3.23}{\sqrt{18}}} \approx 1.642$$

Fail to reject H_0. There is not enough evidence at the 10% level of significance to reject the claim that the mean monthly cost of joining a health club is \$25.

41. $H_0: \mu \geq \$10{,}200$ (claim); $H_a: \mu < \$10{,}200$

$\bar{x} \approx 9895.8$ $\quad\quad$ $s \approx 490.88$

$$t = \frac{\bar{x} - \mu}{\frac{s}{\sqrt{n}}} = \frac{9895.8 - 10{,}200}{\frac{490.88}{\sqrt{16}}} \approx -2.48$$

P-value = {Area to left of $t = -2.48$ } ≈ 0.0128

Fail to reject H_0. There is not enough evidence at the 1% level of significance to reject the claim that the mean expenditure per student in public elementary and secondary schools is at least \$10,200.

43. $np = (40)(0.15) = 6 > 5$

$nq = (40)(0.85) = 34 > 5 \rightarrow$ can use normal distribution

$H_0: p = 0.15$ (claim); $H_a: p \neq 0.15$

$-z_0 = -1.96$, $z_0 = 1.96$; Rejection regions: $z < -1.96$, $z > 1.96$

$$z = \frac{\hat{p} - p}{\sqrt{\frac{pq}{n}}} = \frac{0.09 - 0.15}{\sqrt{\frac{(0.15)(0.85)}{40}}} \approx -1.063$$

Fail to reject H_0. There is not enough evidence at the 5% level of significance to reject the claim.

45. $np = (75)(0.09) = 6.75 > 5$

$nq = (75)(0.91) = 68.25 > 5 \rightarrow$ can use normal distribution

$H_0: p \geq 0.09$; $H_a: p < 0.09$ (claim)

$z_0 = -1.405$; Rejection region: $z < -1.405$

$$z = \frac{\hat{p} - p}{\sqrt{\frac{pq}{n}}} = \frac{0.07 - 0.09}{\sqrt{\frac{(0.09)(0.91)}{75}}} \approx -0.605$$

Fail to reject H_0. There is not enough evidence at the 8% level of significance to support the claim.

47. Because $np = 1.2 < 5$, the normal distribution cannot be used to approximate the binomial distribution.

49. $np = (50)(0.24) = 12 > 5$

$nq = (50)(0.76) = 38 > 5 \rightarrow$ can use normal distribution

$H_0: p = 0.24$; $H_a: p \neq 0.24$ (claim)

$-z_0 = -2.33$, $z_0 = 2.33$; Rejection regions: $z < -2.33$, $z > 2.33$

$$z = \frac{\hat{p} - p}{\sqrt{\frac{pq}{n}}} = \frac{0.32 - 0.24}{\sqrt{\frac{(0.24)(0.76)}{50}}} \approx 1.32$$

Fail to reject H_0. There is not enough evidence at the 2% level of significance to support the claim.

51. $H_0: p \leq 0.16$; $H_a: p > 0.16$ (claim)

$z_0 = 2.055$; Rejection region: $z > 2.055$

$$\hat{p} = \frac{x}{n} = \frac{256}{1420} \approx 0.180$$

$$z = \frac{\hat{p} - p}{\sqrt{\frac{pq}{n}}} \approx \frac{0.180 - 0.16}{\sqrt{\frac{(0.16)(0.84)}{1420}}} \approx 2.056$$

Reject H_0. There is enough evidence at the 2% level of significance to support the claim that over 16% of U.S. adults are without health care coverage.

53. $\chi_0^2 = 30.144$

55. $\chi_0^2 = 63.167$

57. $H_0: \sigma^2 \leq 2$; $H_a: \sigma^2 > 2$ (claim)

$\chi_0^2 = 24.769$; Rejection region: $\chi^2 > 24.769$

$$\chi^2 = \frac{(n-1)s^2}{\sigma^2} = \frac{(17)(2.95)}{(2)} = 25.075$$

Reject H_0. There is enough evidence at the 10% level of significance to support the claim.

59. $H_0: \sigma = 1.25$ (claim); $H_a: \sigma \neq 1.25$

$\chi_L^2 = 0.831$, $\chi_R^2 = 12.833$; Rejection regions: $\chi^2 < 0.831$, $\chi^2 > 12.833$

$$\chi^2 = \frac{(n-1)s^2}{\sigma^2} = \frac{(5)(1.03)^2}{(1.25)^2} \approx 3.395$$

Fail to reject H_0. There is not enough evidence at the 5% level of significance to reject the claim.

61. $H_0: \sigma^2 \leq 0.01$ (claim); $H_a: \sigma^2 > 0.01$

$\chi_0^2 = 49.645$; Rejection region: $\chi^2 > 49.645$

$$\chi^2 = \frac{(n-1)s^2}{\sigma^2} = \frac{(27)(0.064)}{(0.01)} = 172.8$$

Reject H_0. There is enough evidence at the 0.5% level of significance to reject the claim that the variance is at most 0.01.

63. $\chi_L^2 = 13.844$, $\chi_R^2 = 41.923$; Rejection regions: $\chi^2 < 13.844$, $\chi^2 > 41.923$

From Exercise 62, $\chi^2 = 43.94$.

You can reject H_0 at the 5% level of significance because $\chi^2 = 43.94 > 41.923$.

CHAPTER 7 QUIZ SOLUTIONS

1. (a) $H_0: \mu \geq 170$ (claim); $H_a: \mu < 170$

(b) One-tailed because the alternative hypothesis contains <; *z*-test because $n > 30$.

(c) $z_0 = -1.88$; Rejection region: $z < -1.88$

(d) $$z = \frac{\bar{x} - \mu}{\frac{s}{\sqrt{n}}} = \frac{168.5 - 170}{\frac{11}{\sqrt{360}}} \approx -2.59$$

(e) Reject H_0.

(f) There is enough evidence at the 3% level of significance to reject the claim that the mean consumption of vegetables and melons by people in the United States is at least 170 pounds per person.

2. (a) $H_0: \mu \geq 7.25$ (claim); $H_a: \mu < 7.25$

(b) One-tailed because the alternative hypothesis contains <; *t*-test because $n < 30$, σ is unknown, and the population is normally distributed.

(c) $t_0 = -1.796$; Rejection region: $t < -1.796$

(d) $$t = \frac{\bar{x} - \mu}{\frac{s}{\sqrt{n}}} = \frac{7.15 - 7.25}{\frac{0.27}{\sqrt{12}}} \approx -1.283$$

(e) Fail to reject H_0.

(f) There is not enough evidence at the 5% level of significance to reject the claim that the mean hat size for a male is at least 7.25 inches.

3. (a) $H_0: p \leq 0.10$ (claim); $H_a: p > 0.10$

(b) One-tailed because the alternative hypothesis contains >; z-test because $np > 5$ and $nq > 5$.

(c) $z_0 = 1.75$; Rejection region: $z > 1.75$

(d) $$z = \frac{\hat{p} - p}{\sqrt{\frac{pq}{n}}} = \frac{0.13 - 0.10}{\sqrt{\frac{(0.10)(0.90)}{57}}} \approx 0.75$$

(e) Fail to reject H_0.

(f) There is not enough evidence at the 4% level of significance to reject the claim that no more than 10% of microwaves need repair during the first 5 years of use.

4. (a) $H_0: \sigma = 112$ (claim); $H_a: \sigma \neq 112$

(b) Two-tailed because the alternative hypothesis contains $\neq$; χ^2-test because the test is for a standard deviation and the population is normally distributed.

(c) $\chi_L^2 = 9.390$, $\chi_R^2 = 28.869$; Rejection regions: $\chi^2 < 9.390$, $\chi^2 > 28.869$

(d) $$\chi^2 = \frac{(n-1)s^2}{\sigma^2} = \frac{(18)(143)^2}{(112)^2} \approx 29.343$$

(e) Reject H_0.

(f) There is enough evidence at the 10% level of significance to reject the claim that the standard deviation of SAT critical reading test scores is 112.

5. (a) $H_0: \mu = \$62,569$ (claim); $H_a: \mu \neq \$62,569$

(b) Two-tailed because the alternative hypothesis contains $\neq$; t-test because $n < 30$, σ is unknown, and the population is normally distributed.

(c) Not necessary

(d) $$t = \frac{\bar{x} - \mu}{\frac{s}{\sqrt{n}}} = \frac{59,231 - 62,569}{\frac{5945}{\sqrt{15}}} \approx -2.175$$

P-value = 2{Area to left of $t = -2.175$} = 2(0.02365) = 0.0473

(e) Reject H_0.

(f) There is enough evidence at the 5% level of significance to reject the claim that the mean income for full-time workers ages 25 to 34 with a master's degree is \$62,569.

6. (a) $H_0: \mu = \$201$ (claim); $H_a: \mu \neq \$201$

(b) Two-tailed because the alternative hypothesis contains $\neq$; z-test because $n \geq 30$.

(c) Not necessary

(d) $$z = \frac{\bar{x} - \mu}{\frac{s}{\sqrt{n}}} = \frac{216 - 201}{\frac{30}{\sqrt{35}}} \approx 2.958$$

P-value = 2{Area right of $z = 2.958$} = 2(0.0015) = 0.0030

(e) Reject H_0.

(f) There is enough evidence at the 5% level of significance to reject the claim that the mean daily cost of meals and lodging for a family of 4 traveling in Kansas is \$201.

CHAPTER 8

Hypothesis Testing with Two Samples

8.1 TESTING THE DIFFERENCE BETWEEN MEANS (LARGE INDEPENDENT SAMPLES)

8.1 Try It Yourself Solutions

Note: Answers may differ due to rounding.

1a. (1) Independent
(2) Dependent

b. (1) Because each sample represents blood pressures of different individuals, and it is not possible to form a pairing between the members of the samples.
(2) Because the samples represent exam scores of the same students, the samples can be paired with respect to each student.

2. The claim is "there is a difference in the mean annual wages for forensic science technicians working for local and state governments."

a. $H_0: \mu_1 = \mu_2$; $H_a: \mu_1 \neq \mu_2$(claim)

b. $\alpha = 0.10$

c. $z_0 = 1.645$; Rejection regions: $z > 1.645$, $z < -1.645$

d. $$z = \frac{(\bar{x}_1 - \bar{x}_2) - (\mu_1 - \mu_2)}{\sqrt{\frac{s_1^2}{n_1} + \frac{s_2^2}{n_2}}} = \frac{(53{,}000 - 51{,}910) - (0)}{\sqrt{\frac{(6200)^2}{100} + \frac{(5575)^2}{100}}} \approx 1.667$$

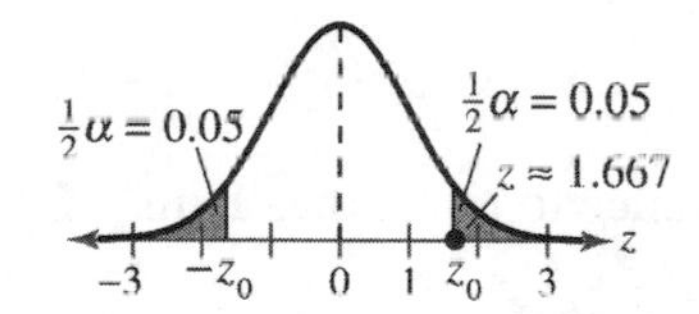

e. Reject H_0.

f. There is enough evidence at the 10% level of significance to support the claim that there is a difference in the mean annual wages for forensic science technicians working for local and state governments.

3a. $$z = \frac{(\bar{x}_1 - \bar{x}_2) - (\mu_1 - \mu_2)}{\sqrt{\frac{s_1^2}{n_1} + \frac{s_2^2}{n_2}}} = \frac{(274 - 271) - (0)}{\sqrt{\frac{(22)^2}{150} + \frac{(18)^2}{200}}} \approx 1.36$$

$\rightarrow$ P-value = {area right of $z = 1.36$} = 0.0865

b. Fail to reject H_0.

c. There is not enough evidence at the 5% level of significance to support the travel agency's claim that the average daily cost of meals and lodging for vacationing in Alaska is greater than the same average cost for vacationing in Colorado.

8.1 EXERCISE SOLUTIONS

1. Two samples are dependent if each member of one sample corresponds to a member of the other sample. Example: The weights of 22 people before starting an exercise program and the weights of the same 22 people 6 weeks after starting the exercise program.

Two samples are independent if the sample selected from one population is not related to the sample from the other population. Example: The weights of 25 cats and the weights of 20 dogs.

3. Use *P*-values.

5. Independent because different students were samples.

7. Dependent because the same football players were sampled.

9. Independent because different boats were sampled.

11. Dependent because the same tire sets were sampled.

13. $H_0: \mu_1 = \mu_2$ (claim); $H_a: \mu_1 \neq \mu_2$
Rejection region: $z_0 < 2.575$, $z_0 > 2.575$
(a) $\bar{x}_1 - \bar{x}_2 = 16 - 14 = 2$

(b) $$z = \frac{(\bar{x}_1 - \bar{x}_2) - (\mu_1 - \mu_2)}{\sqrt{\frac{s_1^2}{n_1} + \frac{s_2^2}{n_2}}} = \frac{(16-14)-(0)}{\sqrt{\frac{(3.4)^2}{30} + \frac{(1.5)^2}{30}}} \approx 2.95$$

(c) z is in the rejection region because $2.95 > 2.575$.
(d) Reject H_0. There is enough evidence at the 1% level of significance to reject the claim.

15. $H_0: \mu_1 \geq \mu_2$; $H_a: \mu_1 < \mu_2$ (claim)
Rejection region: $z_0 < -1.645$
(a) $\bar{x}_1 - \bar{x}_2 = 2435 - 2432 = 3$

(b) $$z = \frac{(\bar{x}_1 - \bar{x}_2) - (\mu_1 - \mu_2)}{\sqrt{\frac{s_1^2}{n_1} + \frac{s_2^2}{n_2}}} = \frac{(2435-2432)-(0)}{\sqrt{\frac{(75)^2}{35} + \frac{(105)^2}{90}}} \approx 0.18$$

(c) z is not in the rejection region because $0.18 > -1.645$.
(d) Fail to reject H_0. There is not enough evidence at the 5% level of significance to support the claim.

17. $H_0: \mu_1 \le \mu_2$; $H_a: \mu_1 > \mu_2$ (claim)

$z_0 = 2.33$; Rejection region: $z > 2.33$

$$z = \frac{(\bar{x}_1 - \bar{x}_2) - (\mu_1 - \mu_2)}{\sqrt{\frac{s_1^2}{n_1} + \frac{s_2^2}{n_2}}} = \frac{(5.2 - 5.5) - (0)}{\sqrt{\frac{(0.2)^2}{45} + \frac{(0.3)^2}{37}}} \approx -5.207$$

Fail to reject H_0. There is not enough evidence at the 1% level of significance to support the claim.

19. Because $z \approx 2.96 > 1.96$ and $P \approx 0.0031 < 0.05$, reject H_0.

21. The claim is "the mean braking distances are different for the two types of tires."

(a) $H_0: \mu_1 = \mu_2$; $H_a: \mu_1 \ne \mu_2$ (claim)

(b) $z_0 = \pm 1.645$; Rejection regions: $z < -1.645$, $z > 1.645$

(c) $$z = \frac{(\bar{x}_1 - \bar{x}_2) - (\mu_1 - \mu_2)}{\sqrt{\frac{s_1^2}{n_1} + \frac{s_2^2}{n_2}}} = \frac{(42 - 45) - (0)}{\sqrt{\frac{(4.7)^2}{35} + \frac{(4.3)^2}{35}}} \approx -2.786$$

(d) Reject H_0.

(e) There is enough evidence at the 10% level of significance to support the claim that the mean braking distances for the two types of tires.

23. The claim is "Region A's average wind speed is greater than Region B's."

(a) $H_0: \mu_1 \le \mu_2$; $H_a: \mu_1 > \mu_2$ (claim)

(b) $z_0 > 1.645$; Rejection region: $z > 1.645$

(c) $$z = \frac{(\bar{x}_1 - \bar{x}_2) - (\mu_1 - \mu_2)}{\sqrt{\frac{s_1^2}{n_1} + \frac{s_2^2}{n_2}}} = \frac{(13.2 - 12.5) - (0)}{\sqrt{\frac{(2.3)^2}{60} + \frac{(2.7)^2}{60}}} \approx 1.53$$

(d) Fail to reject H_0.

(e) There is not enough evidence at the 5% level of significance to conclude that Region A's average wind speed is greater than Region B's.

25. The claim is "male and female high school students have equal ACT scores."

(a) $H_0: \mu_1 = \mu_2$ (claim); $H_a: \mu_1 \ne \mu_2$

(b) $z_0 = \pm 2.575$; Rejection regions: $z < -2.575$, $z > 2.575$

(c) $$z = \frac{(\bar{x}_1 - \bar{x}_2) - (\mu_1 - \mu_2)}{\sqrt{\frac{s_1^2}{n_1} + \frac{s_2^2}{n_2}}} = \frac{(21.0 - 20.8) - (0)}{\sqrt{\frac{(5.0)^2}{43} + \frac{(4.7)^2}{56}}} \approx 0.202$$

(d) Fail to reject H_0.

(e) There is not enough evidence at the 1% level of significance to reject the claim that the male and female high school students have equal ACT scores.

27. The claim is "the average home sales price in Dallas, Texas is the same as in Austin, Texas."

(a) $H_0: \mu_1 = \mu_2$ (claim); $H_a: \mu_1 \neq \mu_2$

(b) $z_0 = \pm 1.645$; Rejection regions: $z < -1.645$, $z > 1.645$

(c) $$z = \frac{(\bar{x}_1 - \bar{x}_2) - (\mu_1 - \mu_2)}{\sqrt{\frac{s_1^2}{n_1} + \frac{s_2^2}{n_2}}} = \frac{(240{,}993 - 249{,}237) - (0)}{\sqrt{\frac{(25{,}875)^2}{35} + \frac{(27{,}110)^2}{35}}} \approx -1.30$$

(d) Fail to reject H_0.

(e) There is not enough evidence at the 10% level of significance to reject the claim that the average home sales price in Dallas, Texas is the same as in Austin, Texas.

29. The claim is "the average home sales price in Dallas, Texas is the same as in Austin, Texas."

(a) $H_0: \mu_1 = \mu_2$ (claim); $H_a: \mu_1 \neq \mu_2$

(b) $z_0 = \pm 1.645$; Rejection regions: $z < -1.645$, $z > 1.645$

(c) $$z = \frac{(\bar{x}_1 - \bar{x}_2) - (\mu_1 - \mu_2)}{\sqrt{\frac{s_1^2}{n_1} + \frac{s_2^2}{n_2}}} = \frac{(247{,}245 - 239{,}150) - (0)}{\sqrt{\frac{(23{,}740)^2}{50} + \frac{(20{,}690)^2}{50}}} \approx 1.86$$

(d) Reject H_0.

(e) There is enough evidence at the 10% level of significance to reject the claim that the average home sales price in Dallas, Texas is the same as in Austin, Texas. The new samples do lead to a different conclusion.

31. The claim is "children ages 6–17 spent more time watching television in 1981 than children ages 6–17 do today."

(a) $H_0: \mu_1 \leq \mu_2$; $H_a: \mu_1 > \mu_2$ (claim)

(b) $z_0 = 1.96$; Rejection region: $z > 1.96$

(c) $\bar{x}_1 \approx 2.130$, $s_1 \approx 0.490$, $n_1 = 30$

$\bar{x}_2 \approx 1.757$, $s_2 \approx 0.470$, $n_2 = 30$

$$z = \frac{(\bar{x}_1 - \bar{x}_2) - (\mu_1 - \mu_2)}{\sqrt{\frac{s_1^2}{n_1} + \frac{s_2^2}{n_2}}} = \frac{(2.130 - 1.757) - (0)}{\sqrt{\frac{(0.490)^2}{30} + \frac{(0.470)^2}{30}}} \approx 3.01$$

(d) Reject H_0.

(e) There is enough evidence at the 2.5% level of significance to support the claim that children ages 6–17 spent more time watching television in 1981 than children ages 6–17 do today.

33. The claim is "there is no difference in the mean washer diameter manufactured by two different methods."

(a) $H_0: \mu_1 = \mu_2$; (claim) $H_a: \mu_1 \neq \mu_2$

(b) $z_0 = \pm 2.575$; Rejection region: $z < -2.575$, $z > 2.575$

(c) $\bar{x}_1 \approx 0.875,\ s_1 \approx 0.011,\ n_1 = 35$

$\bar{x}_2 \approx 0.701,\ s_2 \approx 0.011,\ n_2 = 35$

$$z = \frac{(\bar{x}_1 - \bar{x}_2) - (\mu_1 - \mu_2)}{\sqrt{\frac{s_1^2}{n_1} + \frac{s_2^2}{n_2}}} = \frac{(0.875 - 0.701) - (0)}{\sqrt{\frac{(0.011)^2}{35} + \frac{(0.011)^2}{35}}} \approx 66.172$$

(d) Reject H_0.

(e) There is enough evidence at the 1% level of significance to reject the claim that there is no difference in the mean washer diameter manufactured by two different methods.

35. They are equivalent through algebraic manipulation of the equation.

$\mu_1 = \mu_2 \rightarrow \mu_1 - \mu_2 = 0$

37. Hypothesis test results:

μ_1 : mean of population 1 (Std. Dev. = 5.4)

μ_2 : mean of population 2 (Std. Dev. = 7.5)

$\mu_1 - \mu_2$: mean difference

$H_0 : \mu_1 - \mu_2 = 0$

$H_a : \mu_1 - \mu_2 \neq 0$

Difference	n_1	n_2	Sample Mean
$\mu_1 - \mu_2$	50	45	4

Std. Err.	*z*-stat.	*P*-value
1.3539572	2.9543033	0.0031

$P = 0.0031 < 0.01$, so reject H_0.

There is enough evidence at the 1% level of significance to support the claim.

39. Hypothesis test results:

μ_1 : mean of population 1 (Std. Dev. = 0.92)

μ_2 : mean of population 2 (Std. Dev. = 0.73)

$\mu_1 - \mu_2$: mean difference

$H_0 : \mu_1 - \mu_2 = 0$

$H_a : \mu_1 - \mu_2 < 0$

Difference	n_1	n_2	Sample Mean
$\mu_1 - \mu_2$	35	40	−0.32

Std. Err.	*z*-stat.	*P*-value
0.193663	−1.6523548	0.0492

P-value = $0.0492 < 0.05$, so reject H_0.

There is enough evidence at the 5% level of significance to reject the claim.

41. $H_0: \mu_1 - \mu_2 = -9$ (claim); $H_a: \mu_1 - \mu_2 \neq -9$

$z_0 = \pm 2.575$; Rejection region: $z < -2.575, z > 2.575$

$$z = \frac{(\bar{x}_1 - \bar{x}_2) - (\mu_1 - \mu_2)}{\sqrt{\frac{s_1^2}{n_1} + \frac{s_2^2}{n_2}}} = \frac{(11.5 - 20) - (-9)}{\sqrt{\frac{(3.8)^2}{70} + \frac{(6.7)^2}{65}}} \approx 0.528$$

Fail to reject H_0. There is not enough evidence at the 1% level of significance to reject the claim that children spend 9 hours a week more in day care or preschool today than in 1981.

43. $H_0: \mu_1 - \mu_2 \leq 10{,}000$; $H_a: \mu_1 - \mu_2 > 10{,}000$ (claim)

$z_0 = 1.645$; Rejection region: $z > 1.645$

$$z = \frac{(\bar{x}_1 - \bar{x}_2) - (\mu_1 - \mu_2)}{\sqrt{\frac{s_1^2}{n_1} + \frac{s_2^2}{n_2}}} = \frac{(94{,}980 - 80{,}830) - (10{,}000)}{\sqrt{\frac{(8795)^2}{42} + \frac{(9250)^2}{38}}} \approx 2.05$$

Reject H_0. There is enough evidence at the 5% level of significance to support the claim that the mean annual salaries of microbiologists in California and Maryland is more than $10,000.

45.

$$(\bar{x}_1 - \bar{x}_2) - z_c\sqrt{\frac{s_1^2}{n_1} + \frac{s_2^2}{n_2}} < \mu_1 - \mu_2 < (\bar{x}_1 - \bar{x}_2) + z_c\sqrt{\frac{s_1^2}{n_1} + \frac{s_2^2}{n_2}}$$

$$(123.1 - 125) - 1.96\sqrt{\frac{(9.9)^2}{269} + \frac{(10.1)^2}{268}} < \mu_1 - \mu_2 < (123.1 - 125) + 1.96\sqrt{\frac{(9.9)^2}{269} + \frac{(10.1)^2}{268}}$$

$$-1.9 - 1.96\sqrt{0.745} < \mu_1 - \mu_2 < -1.9 + 1.96\sqrt{0.745}$$

$$-3.6 < \mu_1 - \mu_2 < -0.2$$

47. $H_0: \mu_1 - \mu_2 \geq 0$; $H_a: \mu_1 - \mu_2 < 0$ (claim)

$z_0 = -1.645$; Rejection region: $z < -1.645$

$$z = \frac{(\bar{x}_1 - \bar{x}_2) - (\mu_1 - \mu_2)}{\sqrt{\frac{s_1^2}{n_1} + \frac{s_2^2}{n_2}}} = \frac{(123.1 - 125) - (0)}{\sqrt{\frac{(9.9)^2}{269} + \frac{(10.1)^2}{268}}} \approx -2.20$$

Reject H_0. There is enough evidence at the 5% level of significance to support the claim. You should recommend using the DASH diet and exercise program over the traditional diet and exercise program, because the mean systolic blood pressure was significantly lower in the DASH program.

49. $H_0: \mu_1 - \mu_2 \geq 0$; $H_a: \mu_1 - \mu_2 < 0$ (claim)

The 95% CI for $\mu_1 - \mu_2$ in Exercise 45 contained only values less than zero and, as found in Exercise 47, there was enough evidence at the 5% level of significance to support the claim. If the CI for $\mu_1 - \mu_2$ contains only negative numbers you reject H_0 because the null hypothesis states that $\mu_1 - \mu_2$ is greater than or equal to zero.

8.2 TESTING THE DIFFERENCE BETWEEN MEANS (SMALL INDEPENDENT SAMPLES)

8.2 Try It Yourself Solutions

1a. The claim is "there is a difference in the mean annual earnings based on level of education."
$H_0: \mu_1 = \mu_2$; $H_a: \mu_1 \neq \mu_2$ (claim)

b. $\alpha = 0.01$; d.f. $= \min\{n_1 - 1, n_2 - 1\} = \min\{15-1, 12-1\} = 11$

c. $t_0 = \pm 3.106$; Rejection regions: $t < -3.106$, $t > 3.106$

d. $$t = \frac{(\bar{x}_1 - \bar{x}_2) - (\mu_1 - \mu_2)}{\sqrt{\frac{s_1^2}{n_1} + \frac{s_2^2}{n_2}}} = \frac{(27{,}136 - 34{,}329) - (0)}{\sqrt{\frac{(2318)^2}{15} + \frac{(4962)^2}{12}}} \approx -4.63$$

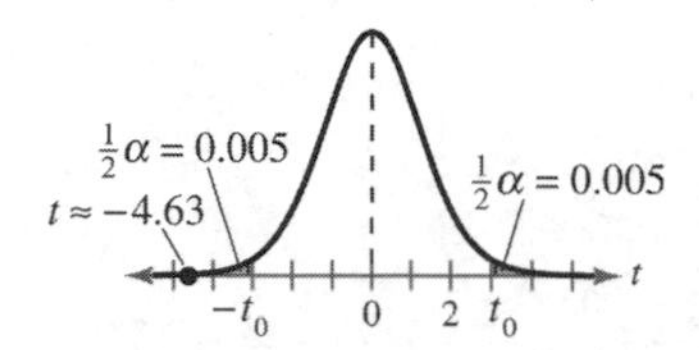

e. Reject H_0.

f. There is evidence that "there is a difference in the mean annual earnings based on level of education."

2a. The claim is "the watts usage of a manufacturer's 17-inch flat panel monitors is less than that of its leading competitor."
$H_0: \mu_1 \geq \mu_2$; $H_a: \mu_1 < \mu_2$ (claim)

b. $\alpha = 0.10$; d.f. $= n_1 + n_2 - 2 = 12 + 15 - 2 = 25$

c. $t_0 = -1.316$; Rejection regions: $t < -1.316$

d. $$t = \frac{(\bar{x}_1 - \bar{x}_2) - (\mu_1 - \mu_2)}{\sqrt{\frac{(n-1)s_1^2 + (n_2-1)s_2^2}{n_1 + n_2 - 2}}\sqrt{\frac{1}{n_1} + \frac{1}{n_2}}} = \frac{(32 - 35) - (0)}{\sqrt{\frac{(12-1)(2.1)^2 + (15-1)(1.8)^2}{12+15-2}}\sqrt{\frac{1}{12} + \frac{1}{15}}} \approx -3.997$$

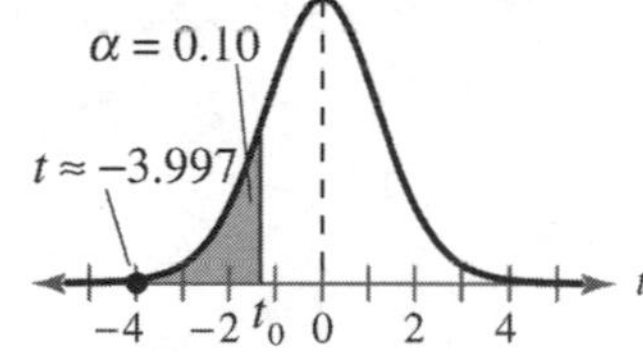

e. Reject H_0.

f. There is enough evidence to support the claim.

8.2 EXERCISE SOLUTIONS

1. (1) The samples must be randomly selected.
(2) The samples must be independent
(3) Each population must have a normal distribution.

3. (a) d.f. $= n_1 + n_2 - 2 = 23$

$t_0 = \pm 1.714$

(b) d.f. $= \min\{n_1 - 1,\ n_2 - 1\} = 10$

$t_0 = \pm 1.812$

5. (a) d.f. $= n_1 + n_2 - 2 = 16$

$t_0 = -1.746$

(b) d.f. $= \min\{n_1 - 1,\ n_2 - 1\} = 6$

$t_0 = -1.943$

7. (a) d.f. $= n_1 + n_2 - 2 = 19$

$t_0 = 1.729$

(b) d.f. $= \min\{n_1 - 1,\ n_2 - 1\} = 7$

$t_0 = 1.895$

9. $H_0: \mu_1 = \mu_2$ (claim); $H_a: \mu_1 \neq \mu_2$

d.f. $= n_1 + n_2 - 2 = 27$

Rejection regions: $t < -2.771,\ t > 2.771$

(a) $\bar{x}_1 - \bar{x}_2 = 33.7 - 33.5 = -1.8$

(b) $$t = \frac{(\bar{x}_1 - \bar{x}_2) - (\mu_1 - \mu_2)}{\sqrt{\frac{(n-1)s_1^2 + (n_2-1)s_2^2}{n_1 + n_2 - 2}}\sqrt{\frac{1}{n_1} + \frac{1}{n_2}}} = \frac{(33.7 - 35.5) - (0)}{\sqrt{\frac{(12-1)(3.5)^2 + (17-1)(2.2)^2}{12 + 17 - 2}}\sqrt{\frac{1}{12} + \frac{1}{17}}} \approx -1.70$$

(c) t is not in the rejection region.

(d) Fail to reject H_0.

11. $H_0: \mu_1 \leq \mu_2$ (claim); $H_a: \mu_1 > \mu_2$

d.f. $= \min\{n_1 - 1,\ n_2 - 1\} = 9$

Rejection region: $t > 1.833$

(a) $\bar{x}_1 - \bar{x}_2 = 2410 - 2305 = 105$

(b) $$t = \frac{(\bar{x}_1 - \bar{x}_2) - (\mu_1 - \mu_2)}{\sqrt{\frac{s_1^2}{n_1} + \frac{s_2^2}{n_2}}} = \frac{(2410 - 2305) - (0)}{\sqrt{\frac{(175)^2}{13} + \frac{(52)^2}{10}}} \approx 2.05$$

(c) t is in the rejection region.

(d) Reject H_0.

13. (a) The claim is "the mean annual costs of routine veterinarian visits for dogs and cats are the same."

$H_0: \mu_1 = \mu_2$ (claim); $H_a: \mu_1 \neq \mu_2$

(b) d.f. $= \min\{n_1 - 1,\ n_2 - 1\} = 6$

$t_0 = \pm 1.943$; Rejection regions: $t < -1.943,\ t > 1.943$

(c) $t=\dfrac{(\bar{x}_1-\bar{x}_2)-(\mu_1-\mu_2)}{\sqrt{\dfrac{s_1^2}{n_1}+\dfrac{s_2^2}{n_2}}}=\dfrac{(225-203)-(0)}{\sqrt{\dfrac{(28)^2}{7}+\dfrac{(15)^2}{10}}}\approx 1.90$

(d) Fail to reject H_0.

(e) There is not enough evidence at the 10% level of significance to reject the claim that the mean annual costs of routine veterinarians visits for dogs and cats are the same.

15. (a) The claim is "the mean bumper repair cost is less for mini cars than for midsize cars."
$H_0: \mu_1 \geq \mu_2;\ H_a: \mu_1 < \mu_2$ (claim)

(b) d.f. $= n_1 + n_2 - 2 = 20$
$t_0 = -1.325$; Rejection region: $t < -1.325$

(c) $t=\dfrac{(\bar{x}_1-\bar{x}_2)-(\mu_1-\mu_2)}{\sqrt{\dfrac{(n-1)s_1^2+(n_2-1)s_2^2}{n_1+n_2-2}}\sqrt{\dfrac{1}{n_1}+\dfrac{1}{n_2}}}=\dfrac{(1621-1895)-(0)}{\sqrt{\dfrac{(6-1)(493)^2+(16-1)(648)^2}{6+16-2}}\sqrt{\dfrac{1}{6}+\dfrac{1}{16}}}\approx -0.9338$

(d) Fail to reject H_0.

(e) There is not enough evidence at the 10% level of significance to support the claim that the mean bumper repair cost is less for mini cars than for midsize cars.

17. (a) The claim is "the mean household income is greater in Allegheny County than it is in Erie County."
$H_0: \mu_1 \leq \mu_2;\ H_a: \mu_1 > \mu_2$ (claim)

(b) d.f. $= \min\{n_1-1,\ n_2-1\} = 14$
$t_0 = 1.761$; Rejection region: $t > 1.761$

(c) $t=\dfrac{(\bar{x}_1-\bar{x}_2)-(0)}{\sqrt{\dfrac{s_1^2}{n_1}+\dfrac{s_2^2}{n_2}}}=\dfrac{(48{,}800-44{,}400)-(0)}{\sqrt{\dfrac{(8800)^2}{19}+\dfrac{(5100)^2}{15}}}\approx 1.99$

(d) Reject H_0.

(e) There is enough evidence at the 5% level of significance to support the claim that the mean household income is greater in Allegheny County than it is in Erie County.

19. (a) The claim is "the new treatment makes a difference in the tensile strength of steel bars."
$H_0: \mu_1 = \mu_2;\ H_a: \mu_1 \neq \mu_2$ (claim)

(b) d.f. $= n_1 + n_2 - 2 = 21$
$t_0 = \pm 2.831$; Rejection regions: $t < -2.831,\ t > 2.831$

(c) $\bar{x}_1 = 368.3,\ s_1 \approx 22.301,\ n_1 = 10$
$\bar{x}_2 \approx 389.538,\ s_2 \approx 14.512,\ n_2 = 13$

$t=\dfrac{(\bar{x}_1-\bar{x}_2)-(\mu_1-\mu_2)}{\sqrt{\dfrac{(n-1)s_1^2+(n_2-1)s_2^2}{n_1+n_2-2}}\sqrt{\dfrac{1}{n_1}+\dfrac{1}{n_2}}}\approx\dfrac{(368.3-389.538)-(0)}{\sqrt{\dfrac{(10-1)(22.301)^2+(13-1)(14.512)^2}{10+13-2}}\sqrt{\dfrac{1}{10}+\dfrac{1}{13}}}\approx -2.76$

(d) Fail to reject H_0.

(e) There is not enough evidence at the 1% level of significance to support the claim that the new treatment makes a difference in the tensile strength of steel bars.

21. (a) The claim is "the new method of teaching reading produces higher reading test scores than the old method"

$H_0: \mu_1 \geq \mu_2$; $H_a: \mu_1 < \mu_2$ (claim)

(b) d.f. $= n_1 + n_2 - 2 = 42$

$t_0 = -1.282$; Rejection region: $t < -1.282$

(c) $\bar{x}_1 \approx 56.684$, $s_1 \approx 6.961$, $n_1 = 19$

$\bar{x}_2 \approx 67.4$, $s_2 \approx 9.014$, $n_2 = 25$

$$t = \frac{(\bar{x}_1 - \bar{x}_2) - (\mu_1 - \mu_2)}{\sqrt{\frac{(n-1)s_1^2 + (n_2-1)s_2^2}{n_1 + n_2 - 2}}\sqrt{\frac{1}{n_1} + \frac{1}{n_2}}} \approx \frac{(56.684 - 67.4) - (0)}{\sqrt{\frac{(19-1)(6.961)^2 + (25-1)(9.014)^2}{19+25-2}}\sqrt{\frac{1}{19} + \frac{1}{25}}} \approx -4.295$$

(d) Reject H_0.

(e) There is enough evidence at the 10% level of significance to support the claim that the new method of teaching reading produces higher reading test scores than the old method.

23. Hypothesis test results:

μ_1 : mean of population 1

μ_2 : mean of population 2

$\mu_1 - \mu_2$: mean difference

$H_0: \mu_1 - \mu_2 = 0$

$H_a: \mu_1 - \mu_2 > 0$

Difference	Sample Mean	Std. Err.
$\mu_1 - \mu_2$	−8	16.985794

DF	*T*-stat.	*P*-value
22	−0.47098184	0.6789

$P = 0.6789 > 0.10$, so fail to reject H_0.

There is not enough evidence at the 10% level of significance to support the claim.

25. Hypothesis test results:

μ_1 : mean of population 1

μ_2 : mean of population 2

$\mu_1 - \mu_2$: mean difference

$H_0: \mu_1 - \mu_2 = 0$

$H_a: \mu_1 - \mu_2 < 0$

(without pooled variances)

Difference	Sample Mean	Std. Err.
$\mu_1 - \mu_2$	−43	28.12301

DF	*T*-stat.	*P*-value
18.990595	−1.5289971	0.0714

$P = 0.0714 > 0.05$, so fail to reject H_0.

There is enough evidence at the 5% level of significance to support the claim.

27. $\sigma = \sqrt{\dfrac{(n-1)s_1^2 + (n_2-1)s_2^2}{n_1+n_2-2}} = \sqrt{\dfrac{(21-1)(166)^2+(11-1)(204)^2}{21+11-2}} \approx 179.56$

$(\bar{x}_1 - \bar{x}_2) \pm t_c\hat{\sigma}\sqrt{\dfrac{1}{n_1}+\dfrac{1}{n_2}} \to (1805-1629) \pm 1.96 \cdot 179.56\sqrt{\dfrac{1}{21}+\dfrac{1}{11}}$

$\to 176 \pm 130.99 \to 45.01 < \mu_1 - \mu_2 < 306.99 \to 45 < \mu_1 - \mu_2 < 307$

29. $(\bar{x}_1 - \bar{x}_2) \pm t_c\sqrt{\dfrac{s_1^2}{n_1}+\dfrac{s_2^2}{n_2}} \to (267-244) \pm 2.132\sqrt{\dfrac{(6)^2}{9}+\dfrac{(12)^2}{5}}$

$\to 23 \pm 12.21 \to 10.78 < \mu_1 - \mu_2 < 35.21 \to 11 < \mu_1 - \mu_2 < 35$

8.3 TESTING THE DIFFERENCE BETWEEN MEANS (DEPENDENT SAMPLES)

8.3 Try It Yourself Solutions

1a. The claim is "athletes can decrease their times in the 40-yard dash."
$H_0: \mu_d \le 0$; $H_a: \mu_d > 0$ (claim)

b. $\alpha = 0.05$ d.f. $= n - 1 = 11$

c. $t_0 = 1.796$; Rejection region: $t > 1.796$

d.

Before	After	d	d^2
4.85	4.78	0.07	0.0049
4.90	4.90	0.00	0.0000
5.08	5.05	0.03	0.0009
4.72	4.65	0.07	0.0049
4.62	4.64	−0.02	0.0004
4.54	4.50	0.04	0.0016
5.25	5.24	0.01	0.0001
5.18	5.27	−0.09	0.0081
4.81	4.75	0.06	0.0036
4.57	4.43	0.14	0.0196
4.63	4.61	0.02	0.0004
4.77	4.82	−0.05	0.0025
		$\sum d = 0.28$	$\sum d^2 = 0.047$

$$\bar{d} = \frac{\sum d}{n} = \frac{0.28}{12} \approx 0.0233$$

$$s_d = \sqrt{\frac{\left(\sum d^2\right) - \left[\dfrac{\left(\sum d\right)^2}{n}\right]}{n-1}} = \sqrt{\frac{0.047 - \dfrac{(0.28)^2}{12}}{11}} \approx 0.0607$$

e. $t = \dfrac{\bar{d} - \mu_d}{\dfrac{s_d}{\sqrt{n}}} \approx \dfrac{0.0233 - 0}{\dfrac{0.0607}{\sqrt{12}}} \approx 1.330$

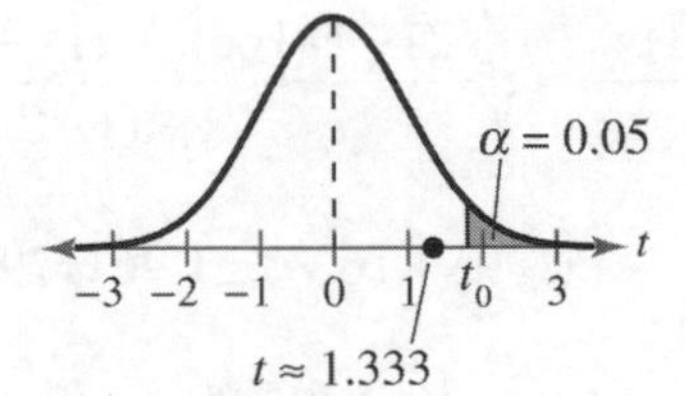

f. Fail to reject H_0.

g. There is not enough evidence at the 5% level of significance to support the claim that athletes can decrease their times in the 40-yard dash using new strength shoes.

2a. The claim is "the drug changes the body's temperature."

$H_0: \mu_d = 0$; $H_a: \mu_d \neq 0$ (claim)

b. $\alpha = 0.05$ d.f. $= n - 1 = 6$

c. $t_0 = \pm 2.447$; Rejection region: $t < -2.447$, $t > 2.447$

d.

Before	After	d	d^2
101.8	99.2	2.6	6.76
98.5	98.4	0.1	0.01
98.1	98.2	−0.1	0.01
99.4	99	0.4	0.16
98.9	98.6	0.3	0.09
100.2	99.7	0.5	0.25
97.9	97.8	0.1	0.01
		$\sum d = 3.9$	$\sum d^2 = 7.29$

$$\bar{d} = \frac{\sum d}{n} = \frac{3.9}{7} \approx 0.5771$$

$$s_d = \sqrt{\frac{\sum d^2 - \left[\frac{\left(\sum d\right)^2}{n}\right]}{n-1}} = \sqrt{\frac{7.29 - \frac{(3.9)^2}{7}}{6}} \approx 0.9235$$

e. $t = \dfrac{\bar{d} - \mu_d}{\dfrac{s_d}{\sqrt{n}}} \approx \dfrac{0.5571 - 0}{\dfrac{0.9235}{\sqrt{7}}} \approx 1.596$

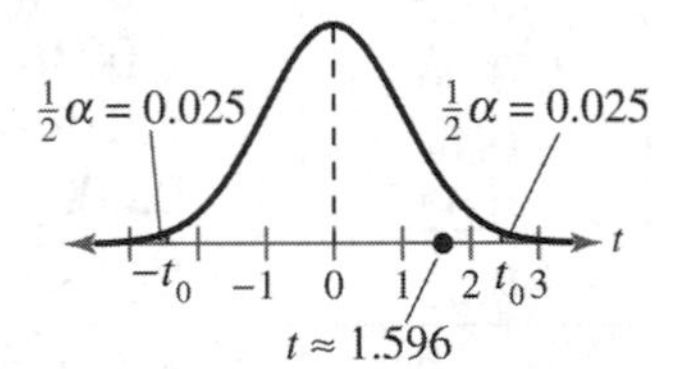

f. Fail to reject H_0.

g. There is not enough evidence at the 5% level of significance to conclude that the drug changes the body's temperature.

8.3 EXERCISE SOLUTIONS

1. (1) Each sample must be randomly selected.
(2) Each member of the first sample must be paired with a member of the second sample.
(3) Each population should be normally distributed.

3. $H_0: \mu_d \geq 0$; $H_a: \mu_d < 0$ (claim)

$\alpha = 0.05$ and d.f. $= n - 1 = 13$

$t_0 = -1.771$ (left-tailed); Rejection region: $t < -1.771$

$$t = \frac{\bar{d} - \mu_d}{\frac{s_d}{\sqrt{n}}} = \frac{1.5 - 0}{\frac{3.2}{\sqrt{14}}} \approx 1.754$$

Fail to reject H_0.

5. $H_0: \mu_d \leq 0$ (claim); $H_a: \mu_d > 0$

$\alpha = 0.10$ and d.f. $= n - 1 = 15$

$t_0 = 1.341$ (Right-tailed); Rejection region: $t > 1.341$

$$t = \frac{\bar{d} - \mu_d}{\frac{s_d}{\sqrt{n}}} = \frac{6.5 - 0}{\frac{9.54}{\sqrt{16}}} \approx 2.725$$

Reject H_0.

7. $H_0: \mu_d \geq 0$ (claim); $H_a: \mu_d < 0$

$\alpha = 0.01$ and d.f. $= n - 1 = 14$

$t_0 = -2.624$ (Left-tailed); Rejection region: $t < -2.624$

$$t = \frac{\bar{d} - \mu_d}{\frac{s_d}{\sqrt{n}}} = \frac{-2.3 - 0}{\frac{1.2}{\sqrt{15}}} \approx -7.423$$

Reject H_0.

9. (a) The claim is a "grammar seminar will help students reduce the number of grammatical errors."

$H_0: \mu_d \leq 0$; $H_a: \mu_d > 0$ (claim)

(b) $t_0 = 3.143$; Rejection region: $t > 3.143$

(c) $\bar{d} = 3.143$ and $s_d \approx 2.035$

(d) $$t = \frac{\bar{d} - \mu_d}{\frac{s_d}{\sqrt{n}}} = \frac{3.143 - 0}{\frac{2.035}{\sqrt{7}}} \approx 4.086$$

(e) Reject H_0.

(f) There is enough evidence at the 1% level of significance to support the claim that grammar seminar will help students reduce the number of grammatical errors."

11. (a) The claim is "a particular exercise program will help participants lose weight after one month."

$H_0: \mu_d \le 0$; $H_a: \mu_d > 0$ (claim)

(b) $t_0 = 1.363$; Rejection region: $t > 1.363$

(c) $\bar{d} = 3.75$ and $s_d \approx 7.841$

(d) $$t = \frac{\bar{d} - \mu_d}{\frac{s_d}{\sqrt{n}}} = \frac{3.75 - 0}{\frac{7.841}{\sqrt{12}}} \approx 1.657$$

(e) Reject H_0.

(f) There is enough evidence at the 10% level of significance to support the claim that the exercise program helps participants lose weight after one month.

13. (a) The claim is "soft tissue therapy and spinal manipulation help to reduce the length of time patients suffer from headaches."

$H_0: \mu_d \le 0$; $H_a: \mu_d > 0$ (claim)

(b) $t_0 = 2.764$; Rejection region; $t > 2.764$

(c) $\bar{d} = 1.225$ and $s_d \approx 0.441$

(d) $$t = \frac{\bar{d} - \mu_d}{\frac{s_d}{\sqrt{n}}} = \frac{1.225 - 0}{\frac{0.441}{\sqrt{11}}} \approx 9.438$$

(e) Reject H_0.

(f) There is enough evidence at the 1% level of significance to support the claim that soft tissue therapy and spinal manipulation help reduce the length of time patients suffer from headaches.

15. (a) The claim is "the new drug reduces systolic blood pressure."

$H_0: \mu_d \le 0$; $H_a: \mu_d > 0$ (claim)

(b) $t_0 = 1.895$; Rejection region: $t > 1.895$

(c) $\bar{d} = 14.75$ and $s_d \approx 6.861$

(d) $$t = \frac{\bar{d} - \mu_d}{\frac{s_d}{\sqrt{n}}} = \frac{14.75 - 0}{\frac{6.861}{\sqrt{8}}} \approx 6.081$$

(e) Reject H_0.

(f) There is enough evidence at the 5% level of significance to support the claim that the new drug reduces systolic blood pressure

17. (a) The claim is "the product ratings have changed from last year to this year."

$H_0: \mu_d = 0$; $H_a: \mu_d \ne 0$ (claim)

(b) $t_0 = \pm 2.365$; Rejection regions: $t < -2.365$, $t > 2.365$

(c) $\bar{d} = -1$ and $s_d \approx 1.309$

(d) $t = \dfrac{\bar{d} - \mu_d}{\frac{s_d}{\sqrt{n}}} = \dfrac{-1-0}{\frac{1.309}{\sqrt{8}}} \approx -2.161$

(e) Fail to reject H_0.

(f) There is not enough evidence at the 5% level of significance to support the claim that the product ratings have changed from last year to this year.

19. Hypothesis test results:

$\mu_1 - \mu_2$: mean of the paired difference between cholesterol (before) and cholesterol (after)

$H_0 : \mu_1 - \mu_2 = 0$

$H_a : \mu_1 - \mu_2 > 0$

Difference	Sample Diff.
Cholesterol (before) – Cholesterol (after)	2.857143

Std. Err.	DF	T-stat.	*P*-value
1.6822401	6	1.6984155	0.0702

$P = 0.0702 > 0.05$, so fail to reject H_0. There is not enough evidence at the 5% level of significance to support the claim that the new cereal lowers total blood cholesterol levels.

21. Yes; $P \approx 0.0003 < 0.05$, so you reject H_0.

23. $\bar{d} \approx -1.525$ and $s_d \approx 0.542$

$$\bar{d} - t_{\alpha/2}\frac{s_d}{\sqrt{n}} < \mu_d < \bar{d} - t_{\alpha/2}\frac{s_d}{\sqrt{n}}$$

$$-1.525 - 1.753\left(\frac{0.542}{\sqrt{16}}\right) < \mu_d < -1.525 + 1.753\left(\frac{0.542}{\sqrt{16}}\right)$$

$$-1.525 - 0.238 < \mu_d < -1.525 + 0.238$$

$$-1.76 < \mu_d < -1.29$$

8.4 TESTING THE DIFFERENCE BETWEEN PROPORTIONS

8.4 Try It Yourself Solutions

1a. The claim is "there is a difference between the proportion of male high school students who smoke cigarettes and the proportion of female high school students who smoke cigarettes.

$H_0 : p_1 = p_2$; $H_a : p_1 \neq p_2$ (claim)

b. $\alpha = 0.05$

c. $z_0 = \pm 1.96$; Rejection region: $z < -1.96$, $z > 1.96$

d. $\bar{p} = \dfrac{x_1 + x_2}{n_1 + n_2} == \dfrac{1484 + 1378}{7000 + 7489} \approx 0.1975$

$\bar{q} \approx 0.8025$

e. $n_1\overline{p} \approx 1382.5 > 5$, $n_1\overline{q} \approx 5617.5 > 5$, $n_2\overline{p} \approx 1479.0775 > 5$, and $n_2\overline{q} \approx 6009.9225 > 5$.

f. $z = \dfrac{(\hat{p}_1 - \hat{p}_2) - (p_1 - p_2)}{\sqrt{\overline{p}\overline{q}\left(\dfrac{1}{n_1} + \dfrac{1}{n_2}\right)}} = \dfrac{(0.212 - 0.184) - (0)}{\sqrt{0.1975 \cdot 0.8025\left(\dfrac{1}{7000} + \dfrac{1}{7489}\right)}} \approx 4.23$

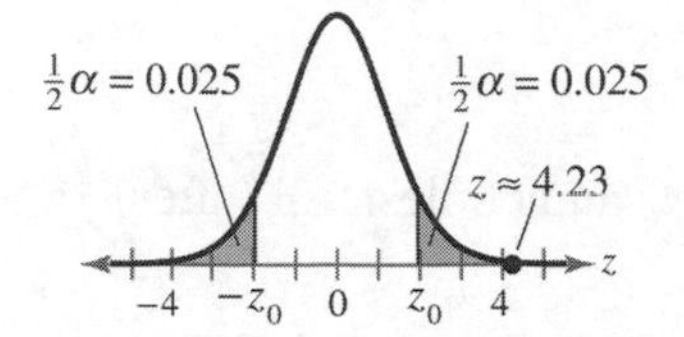

g. Reject H_0.

h. There is enough evidence at the 5% level of significance to support the claim that there is a difference between the proportion of male high school students who smoke cigarettes and the proportion of female high school students who smoke cigarettes.

2a. The claim is "the proportion of male high school students who smoke cigars is greater than the proportion of female high school students who smoke cigars."

$H_0: p_1 \le p_2$; $H_a: p_1 > p_2$ (claim)

b. $\alpha = 0.05$

c. $z_0 = 1.645$; Rejection region: $z > 1.645$

d. $\overline{p} = \dfrac{x_1 + x_2}{n_1 + n_2} == \dfrac{1162 + 539}{7000 + 7489} \approx 0.1174$

$\overline{q} \approx 0.8826$

e. $n_1\overline{p} \approx 821.8 > 5$, $n_1\overline{q} \approx 6178.2 > 5$, $n_2\overline{p} \approx 879.2086 > 5$, and $n_2\overline{q} \approx 6609.7914 > 5$.

f. $z = \dfrac{(\hat{p}_1 - \hat{p}_2) - (p_1 - p_2)}{\sqrt{\overline{p}\overline{q}\left(\dfrac{1}{n_1} + \dfrac{1}{n_2}\right)}} = \dfrac{(0.166 - 0.072) - (0)}{\sqrt{0.1174 \cdot 0.8826\left(\dfrac{1}{7000} + \dfrac{1}{7489}\right)}} \approx 17.565$

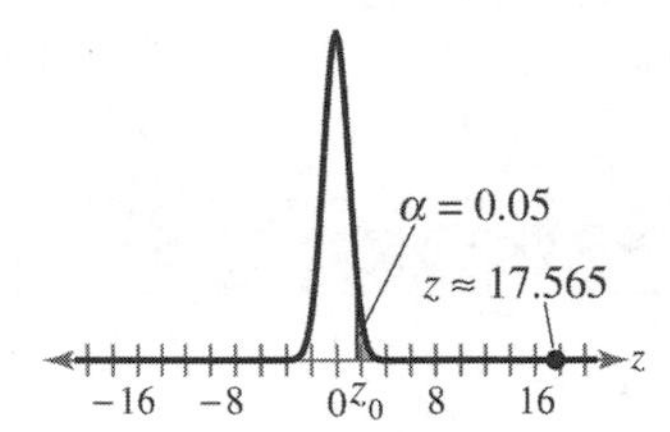

g. Reject H_0.

h. There is enough evidence at the 5% level of significance to support the claim that the proportion of male high school students who smoke cigars is greater than the proportion of female high school students who smoke cigars.

8.4 EXERCISE SOLUTIONS

1. (1) The samples must be randomly selected.
(2) The samples must be independent.
(3) $n_1\overline{p} \ge 5$, $n_1\overline{q} \ge 5$, $n_2\overline{p} \ge 5$, $n_2\overline{q} \ge 5$

3. $H_0: p_1 = p_2$; $H_a: p_1 \neq p_2$ (claim)

$z_0 = \pm 2.575$; Rejection regions: $z < -2.575$, $z > 2.575$

$$\bar{p} = \frac{x_1 + x_2}{n_1 + n_2} = \frac{35 + 36}{70 + 60} \approx 0.546$$

$\bar{q} \approx 0.454$

$n_1\bar{p} \approx 38.22 > 5$, $n_1\bar{q} \approx 31.78 > 5$, $n_2\bar{p} \approx 32.76 > 5$, and $n_2\bar{q} \approx 27.24 > 5$.

Can use normal sampling distribution.

$$z = \frac{(\hat{p}_1 - \hat{p}_2) - (p_1 - p_2)}{\sqrt{\bar{p}\bar{q}\left(\frac{1}{n_1} + \frac{1}{n_2}\right)}} = \frac{(0.5 - 0.6) - (0)}{\sqrt{0.546 \cdot 0.454\left(\frac{1}{70} + \frac{1}{60}\right)}} \approx -1.142$$

Fail to reject H_0.

5. $H_0: p_1 = p_2$ (claim); $H_a: p_1 \neq p_2$

$z_0 = \pm 1.645$; Rejection regions: $z < -1.645$, $z > 1.645$

$$\bar{p} = \frac{x_1 + x_2}{n_1 + n_2} = \frac{42 + 76}{150 + 200} \approx 0.337$$

$\bar{q} \approx 0.663$

$n_1\bar{p} \approx 50.55 > 5$, $n_1\bar{q} \approx 99.45 > 5$, $n_2\bar{p} \approx 67.4 > 5$, and $n_2\bar{q} \approx 132.6 > 5$.

Can use normal sampling distribution.

$$z = \frac{(\hat{p}_1 - \hat{p}_2) - (p_1 - p_2)}{\sqrt{\bar{p}\bar{q}\left(\frac{1}{n_1} + \frac{1}{n_2}\right)}} = \frac{(0.28 - 0.38) - (0)}{\sqrt{(0.337)(0.663)\left(\frac{1}{150} + \frac{1}{200}\right)}} \approx -1.96$$

Reject H_0.

7. $H_0: p_1 \leq p_2$ (claim); $H_a: p_1 > p_2$

$z_0 = 1.28$; Rejection region: $z > 1.28$

$$\bar{p} = \frac{x_1 + x_2}{n_1 + n_2} = \frac{344 + 304}{860 + 800} \approx 0.390$$

$\bar{q} \approx 0.610$

$n_1\bar{p} \approx 335.4 > 5$, $n_1\bar{q} \approx 524.6 > 5$, $n_2\bar{p} \approx 312 > 5$, and $n_2\bar{q} \approx 488 > 5$.

Can use normal sampling distribution.

$$z=\frac{(\hat{p}_1-\hat{p}_2)-(p_1-p_2)}{\sqrt{\bar{p}\bar{q}\left(\frac{1}{n_1}+\frac{1}{n_2}\right)}}=\frac{(0.400-0.380)-(0)}{\sqrt{(0.390)(0.610)\left(\frac{1}{860}+\frac{1}{800}\right)}}\approx 0.835$$

Fail to reject H_0.

9a. The claim is "there is a difference in the proportion of subjects who feel all or mostly better after 4 weeks between subjects who used magnetic insoles and subjects who used nonmagnetic insoles."

$H_0: p_1 = p_2$ (claim); $H_a: p_1 \neq p_2$

b. $z_0 = 2.575$; Rejection regions: $z < -2.575$; $z > 2.575$

c. $\bar{p}=\frac{x_1+x_2}{n_1+n_2}=\frac{17+18}{54+41}\approx 0.368$

$\bar{q}\approx 0.632$

$$z=\frac{(\hat{p}_1-\hat{p}_2)-(p_1-p_2)}{\sqrt{\bar{p}\bar{q}\left(\frac{1}{n_1}+\frac{1}{n_2}\right)}}=\frac{(0.315-0.439)-(0)}{\sqrt{(0.368)(0.632)\left(\frac{1}{54}+\frac{1}{41}\right)}}\approx -1.24$$

d. Fail to reject H_0.

e. There is not enough evidence at the 1% level of significance to support the claim that there is a difference in the proportion of subjects who feel all or mostly better after 4 weeks between subjects who used magnetic insoles and subjects who used nonmagnetic insoles.

11a. The claim is "the proportion of males who enrolled in college is less than the proportion of females who enrolled in college."

$H_0: p_1 \geq p_2$; $H_a: p_1 < p_2$ (claim)

b. $z_0 = -1.645$; Rejection region: $z < -1.645$

c. $\bar{p}=\frac{x_1+x_2}{n_1+n_2}=\frac{575{,}750+595{,}561}{875{,}000+901{,}000}\approx 0.660$

$\bar{q}\approx 0.340$

$$z=\frac{(\hat{p}_1-\hat{p}_2)-(p_1-p_2)}{\sqrt{\bar{p}\bar{q}\left(\frac{1}{n_1}+\frac{1}{n_2}\right)}}=\frac{(0.658-0.661)-(0)}{\sqrt{(0.660)(0.340)\left(\frac{1}{875{,}000}+\frac{1}{901{,}000}\right)}}\approx -4.22$$

d. Reject H_0.

e. There is enough evidence at the 5% level of significance to support the claim that the proportion of males who enrolled in college is less than the proportion of females who enrolled in college.

13a. The claim is "the proportion of subjects who are pain-free is the same for the two groups."
$H_0: p_1 = p_2$ (claim); $H_a: p_1 \neq p_2$

b. $z_0 = \pm 1.96$; Rejection region: $z < -1.96$, $z > 1.96$

c. $\bar{p} = \dfrac{x_1 + x_2}{n_1 + n_2} = \dfrac{100 + 41}{400 + 407} \approx 0.175$

$\bar{q} \approx 0.825$

$$z = \frac{(\hat{p}_1 - \hat{p}_2) - (p_1 - p_2)}{\sqrt{\bar{p}\bar{q}\left(\frac{1}{n_1} + \frac{1}{n_2}\right)}} = \frac{(0.25 - 0.10) - (0)}{\sqrt{(0.175)(0.825)\left(\frac{1}{400} + \frac{1}{407}\right)}} \approx 5.61$$

d. Reject H_0.

e. There is enough evidence at the 5% level of significance to reject the claim that the proportion of subject who are pain-free is the same for the two groups.

15a. The claim is "the proportion of motorcyclists who wear a helmet is now greater."
$H_0: p_1 \leq p_2$; $H_a: p_1 > p_2$ (claim)

b. $z_0 = 1.645$; Rejection region: $z > 1.645$

c. $\bar{p} = \dfrac{x_1 + x_2}{n_1 + n_2} = \dfrac{404 + 317}{600 + 500} \approx 0.655$

$\bar{q} \approx 0.345$

$$z = \frac{(\hat{p}_1 - \hat{p}_2) - (p_1 - p_2)}{\sqrt{\bar{p}\bar{q}\left(\frac{1}{n_1} + \frac{1}{n_2}\right)}} = \frac{(0.673 - 0.634) - (0)}{\sqrt{(0.655)(0.345)\left(\frac{1}{600} + \frac{1}{500}\right)}} \approx 1.37$$

d. Fail to reject H_0.

e. There is not enough evidence at the 5% level of significance to support the claim that the proportion of motorcyclists who wear a helmet is now greater.

17a. The claim is "the proportion of Internet users is the same for the two age groups."
$H_0: p_1 = p_2$ (claim); $H_a: p_1 \neq p_2$

b. $z_0 = \pm 2.575$; Rejection region: $z < -2.575$, $z > 2.575$

c. $\bar{p} = \dfrac{x_1 + x_2}{n_1 + n_2} = \dfrac{419 + 324}{450 + 400} \approx 0.874$

$\bar{q} \approx 0.126$

$$z = \frac{(\hat{p}_1 - \hat{p}_2) - (p_1 - p_2)}{\sqrt{\bar{p}\bar{q}\left(\frac{1}{n_1} + \frac{1}{n_2}\right)}} = \frac{(0.931 - 0.81) - (0)}{\sqrt{(0.874)(0.126)\left(\frac{1}{450} + \frac{1}{400}\right)}} \approx 5.31$$

d. Reject H_0.

e. There is enough evidence at the 1% level of significance to reject the claim that the proportion of Internet users is the same for the two age groups.

19. $H_0: p_1 = p_2$ (claim); $H_a: p_1 \neq p_2$

$z_0 = 1.96$; Rejection regions: $z < -1.96$, $z > 1.96$

$$\bar{p} = \frac{x_1 + x_2}{n_1 + n_2} = \frac{240 + 204}{400 + 400} \approx 0.555$$

$\bar{q} \approx 0.445$

$$z = \frac{(\hat{p}_1 - \hat{p}_2) - (p_1 - p_2)}{\sqrt{\bar{p}\bar{q}\left(\frac{1}{n_1} + \frac{1}{n_2}\right)}} = \frac{(0.60 - 0.51) - (0)}{\sqrt{(0.555)(0.445)\left(\frac{1}{400} + \frac{1}{400}\right)}} \approx 2.56$$

Reject H_0.
There is enough evidence at the 5% level of significance to reject the claim that the proportion of customers who wait 20 minutes or less is the same at the Fairfax North and Fairfax South offices.

21. $H_0: p_1 \geq p_2$; $H_a: p_1 < p_2$ (claim)

$z_0 = -1.28$; Rejection region: $z < -1.28$

$$\bar{p} = \frac{x_1 + x_2}{n_1 + n_2} = \frac{224 + 252}{400 + 400} \approx 0.595$$

$\bar{q} \approx 0.405$

$$z = \frac{(\hat{p}_1 - \hat{p}_2) - (p_1 - p_2)}{\sqrt{\bar{p}\bar{q}\left(\frac{1}{n_1} + \frac{1}{n_2}\right)}} = \frac{(0.56 - 0.63) - (0)}{\sqrt{(0.595)(0.405)\left(\frac{1}{400} + \frac{1}{400}\right)}} \approx -2.02$$

Reject H_0.
There is enough evidence at the 10% level of significance to support the claim that the proportion of customers who wait 20 minutes or less at the Roanoke office is less than the proportion of customers who wait 20 minutes or less at the Staunton office.

23. No; When $\alpha = 0.01$, the rejection region becomes $z < -2.33$. Because $-2.02 > -2.33$, you fail to reject H_0.

25. Hypothesis test results:

p_1 : proportion of successes for population 1

p_2 : proportion of successes for population 2

$p_1 - p_2$: difference in proportions

$H_0 : p_1 - p_2 = 0$

$H_A : p_1 - p_2 > 0$

Difference	Count 1	Total 1	Count 2	Total 2
$p_1 - p_2$	7501	13300	8120	14500

Sample Diff.	Std. Err.	Z-Stat	*P*-value
0.0039849626	0.0059570055	0.66895396	0.2518

$P = 0.2518 > 0.05$, so fail to reject H_0.

There is not enough evidence at the 5% level of significance to support the claim that the proportion of men ages 18–24 living in their parents' homes was greater in 2000 than in 2009.

27. Hypothesis test results:

p_1 : proportion of successes for population 1

p_2 : proportion of successes for population 2

$p_1 - p_2$: difference in proportions

$H_0 : p_1 - p_2 = 0$

$H_A : p_1 - p_2 \neq 0$

Difference	Count1	Total1	Count2	Total2
$p_1 - p_2$	7501	13300	5610	13200

Sample Diff.	Std. Err.	Z-Stat	*P*-value
0.13898496	0.006142657	22.626196	<0.0001

$P < 0.0001 < 0.01$, so reject H_0.

There is enough evidence at the 1% level of significance to reject the claim that the proportion of 18 to 24-year olds living in their parents' homes in 2000 was the same for men and women.

29.

$$(\hat{p}_1 - \hat{p}_2) - z_c\sqrt{\frac{\hat{p}_1\hat{q}_1}{n_1} + \frac{\hat{p}_2\hat{q}_2}{n_2}} < p_1 - p_2 < (\hat{p}_1 - \hat{p}_2) + z_c\sqrt{\frac{\hat{p}_1\hat{q}_1}{n_1} + \frac{\hat{p}_2\hat{q}_2}{n_2}}$$

$$(0.07 - 0.09) - 1.96\sqrt{\frac{(0.07)(0.93)}{10{,}000} + \frac{(0.09)(0.91)}{8000}} < p_1 - p_2 < (0.07 - 0.09) + 1.96\sqrt{\frac{(0.07)(0.93)}{10{,}000} + \frac{(0.09)(0.91)}{8000}}$$

$$-0.02 - 0.008 < p_1 - p_2 < -0.02 + 0.008$$

$$-0.028 < p_1 - p_2 < -0.012$$

CHAPTER 8 REVIEW EXERCISE SOLUTIONS

1. Dependent because the same cities were sampled.

3. $H_0: \mu_1 \geq \mu_2$ (claim); $H_a: \mu_1 < \mu_2$

$z_0 = -1.645$; Rejection region: $z < -1.645$

$$z = \frac{(\bar{x}_1 - \bar{x}_2) - (\mu_1 - \mu_2)}{\sqrt{\frac{s_1^2}{n_1} + \frac{s_2^2}{n_2}}} = \frac{(1.28 - 1.34) - (0)}{\sqrt{\frac{(0.30)^2}{96} + \frac{(0.23)^2}{85}}} \approx -1.519$$

Fail to reject H_0. There is not enough evidence at the 5% level of significance to support the claim.

5. $H_0: \mu_1 \geq \mu_2$; $H_a: \mu_1 < \mu_2$ (claim)

$z_0 = -1.28$; Rejection regions: $z < -1.28$

$$z = \frac{(\bar{x}_1 - \bar{x}_2) - (\mu_1 - \mu_2)}{\sqrt{\frac{s_1^2}{n_1} + \frac{s_2^2}{n_2}}} = \frac{(0.28 - 0.33) - (0)}{\sqrt{\frac{(0.11)^2}{41} + \frac{(0.10)^2}{34}}} \approx -2.060$$

Reject H_0. There is enough evidence at the 10% level of significance to support the claim.

7. (a) $H_0: \mu_1 \geq \mu_2$; $H_a: \mu_1 < \mu_2$ (claim)

(b) $z_0 = -1.645$; Rejection region: $z < -1.645$

(c) $$z = \frac{(\bar{x}_1 - \bar{x}_2) - (\mu_1 - \mu_2)}{\sqrt{\frac{s_1^2}{n_1} + \frac{s_2^2}{n_2}}} = \frac{(1010 - 1180) - (0)}{\sqrt{\frac{(75)^2}{42} + \frac{(90)^2}{39}}} \approx -9.20$$

(d) Reject H_0.

(e) There is enough evidence at the 5% level of significance to support the claim that the Wendy's fish sandwich has less sodium than the Long John Silver's fish sandwich.

9. Yes; The new rejection region is $z < -2.33$, which contains $z = -9.20$, so you still reject H_0.

11. $H_0: \mu_1 = \mu_2$ (claim); $H_a: \mu_1 \neq \mu_2$

d.f. = $n_1 + n_2 - 2 = 31$

$t_0 = \pm 1.96$; Rejection regions: $t < -1.96$, $t > 1.96$

$$t = \frac{(\bar{x}_1 - \bar{x}_2) - (\mu_1 - \mu_2)}{\sqrt{\frac{(n_1 - 1)s_1^2 + (n_2 - 1)s_2^2}{n_1 + n_2 - 2}}\sqrt{\frac{1}{n_1} + \frac{1}{n_2}}} = \frac{(228 - 207) - (0)}{\sqrt{\frac{(20-1)(27)^2 + (13-1)(25)}{20 + 13 - 2}}\sqrt{\frac{1}{20} + \frac{1}{13}}} \approx 2.25$$

Reject H_0. There is not enough evidence at the 5% level of significance to support the claim.

13. $H_0: \mu_1 \le \mu_2$ (claim); $H_a: \mu_1 > \mu_2$

d.f. = $\min\{n_1 - 1, n_2 - 1\} = 24$

$t_0 = 1.711$; Rejection regions: $t > 1.711$

$$t = \frac{(\bar{x}_1 - \bar{x}_2) - (\mu_1 - \mu_2)}{\sqrt{\frac{s_1^2}{n_1} + \frac{s_2^2}{n_2}}} = \frac{(183.5 - 184.7) - (0)}{\sqrt{\frac{(1.3)^2}{25} + \frac{(3.9)^2}{25}}} \approx -1.460$$

Fail to reject H_0. There is not enough evidence at the 5% level of significance to reject the claim.

15. $H_0: \mu_1 = \mu_2$; $H_a: \mu_1 \ne \mu_2$ (claim)

d.f. = $n_1 + n_2 - 2 = 10$

$t_0 = \pm 3.169$; Rejection regions: $t < -3.169$, $t > 3.169$

$$t = \frac{(\bar{x}_1 - \bar{x}_2) - (\mu_1 - \mu_2)}{\sqrt{\frac{(n-1)s_1^2 + (n_2 - 1)s_2^2}{n_1 + n_2 - 2}} \cdot \sqrt{\frac{1}{n_1} + \frac{1}{n_2}}} = \frac{(61 - 55) - (0)}{\sqrt{\frac{(5-1)(3.3)^2 + (7-1)(1.2)^2}{5 + 7 - 2}} \cdot \sqrt{\frac{1}{5} + \frac{1}{7}}} \approx 4.484$$

Reject H_0. There is enough evidence at the 1% level of significance to support the claim.

17. (a) The claim is "third graders taught with the directed reading activities scored higher than those taught without the activities."

$H_0: \mu_1 \le \mu_2$; $H_a: \mu_1 > \mu_2$ (claim)

(b) d.f. = $n_1 + n_2 - 2 = 42$

$t_0 = 1.645$; Rejection region: $t > 1.645$

(c) $\bar{x}_1 \approx 51.476$, $s_1 \approx 11.007$, $n_1 = 21$

$\bar{x}_2 \approx 41.522$, $s_2 \approx 17.149$, $n_2 = 23$

$$t = \frac{(\bar{x}_1 - \bar{x}_2) - (\mu_1 - \mu_2)}{\sqrt{\frac{(n-1)s_1^2 + (n_2 - 1)s_2^2}{n_1 + n_2 - 2}}} = \frac{(51.476 - 41.522) - (0)}{\sqrt{\frac{(21-1)(11.007)^2 + (23-1)(17.149)^2}{21 + 23 - 2}} \sqrt{\frac{1}{21} + \frac{1}{23}}} \approx 2.266$$

(d) Reject H_0.

(e) There is enough evidence at the 5% level of significance to support the claim that third graders taught with the directed reading activities scored higher than those taught without the activities.

19. $H_0: \mu_d = 0$ (claim); $H_a: \mu_d \ne 0$

$\alpha = 0.01$ and d.f. $= n - 1 = 15$

$t_0 = \pm 2.947$ (Two-tailed test); Rejection regions: $t < -2.947$, $t > 2.947$

$$t = \frac{\bar{d} - \mu_d}{\frac{s_d}{\sqrt{n}}} = \frac{85 - 0}{\frac{10.7}{\sqrt{16}}} \approx 3.178$$

Reject H_0.

21. $H_0 : \mu_d \le 0$ (claim); $H_a : \mu_d > 0$

$\alpha = 0.10$ and d.f. $= n - 1 = 32$

$t_0 = 1.282$ (Right-tailed test); Rejection region: $t > 1.282$

$$t = \frac{\bar{d} - \mu_d}{\frac{s_d}{\sqrt{n}}} = \frac{10.3 - 0}{\frac{18.19}{\sqrt{33}}} \approx 3.253$$

Reject H_0.

23. (a) The claim is "the men's systolic blood pressure decreased."
$H_0 : \mu_d \le 0$; $H_a : \mu_d > 0$ (claim)

(b) $t_0 = 1.383$; Rejection region: $t > 1.383$

(c) $\bar{d} = 5$ and $s_d \approx 8.743$

(d) $$t = \frac{\bar{d} - \mu_d}{\frac{s_d}{\sqrt{n}}} = \frac{5 - 0}{\frac{8.743}{\sqrt{10}}} \approx 1.808$$

(e) Reject H_0.

(f) There is enough evidence at the 10% level of significance to support the claim that the man's systolic blood pressure decreased.

25. $H_0 : p_1 = p_2$ (claim); $H_a : p_1 \ne p_2$

$z_0 = \pm 1.96$; Rejection regions: $z < -1.96$, $z > 1.96$

$$\bar{p} = \frac{x_1 + x_2}{n_1 + n_2} = \frac{425 + 410}{840 + 760} \approx 0.522$$

$\bar{q} \approx 0.478$

$n_1\bar{p} \approx 438.48 > 5$, $n_1\bar{q} \approx 401.52 > 5$, $n_2\bar{p} \approx 396.72 > 5$, and $n_2\bar{q} \approx 363.28 > 5$.

Can use normal sampling distribution.

$$z = \frac{(\hat{p}_1 - \hat{p}_2) - (p_1 - p_2)}{\sqrt{\bar{p}\bar{q}\left(\frac{1}{n_1} + \frac{1}{n_2}\right)}} = \frac{(0.506 - 0.539) - (0)}{\sqrt{0.522 \cdot 0.478\left(\frac{1}{840} + \frac{1}{760}\right)}} \approx -1.320$$

Fail to reject H_0.

27. $H_0: p_1 \le p_2$; $H_a: p_1 > p_2$ (claim)

$z_0 = 1.28$; Rejection region: $z > 1.28$

$$\bar{p} = \frac{x_1 + x_2}{n_1 + n_2} = \frac{261 + 207}{556 + 483} \approx 0.450$$

$\bar{q} \approx 0.550$

$n_1\bar{p} \approx 250.2 > 5$, $n_1\bar{q} \approx 305.8 > 5$, $n_2\bar{p} \approx 217.35 > 5$, and $n_2\bar{q} \approx 265.65 > 5$.

Can use normal sampling distribution.

$$z = \frac{(\hat{p}_1 - \hat{p}_2) - (p_1 - p_2)}{\sqrt{\bar{p}\bar{q}\left(\frac{1}{n_1} + \frac{1}{n_2}\right)}} = \frac{(0.469 - 0.429) - (0)}{\sqrt{0.450 \cdot 0.550\left(\frac{1}{556} + \frac{1}{483}\right)}} \approx 1.293$$

Reject H_0.

29. (a) The claim is "the proportions of U.S. adults who considered the amount of federal income tax they had to pay to be too high were the same for the two years."
$H_0: p_1 = p_2$ (claim); $H_a: p_1 \ne p_2$

(b) $z_0 = \pm 2.575$; Rejection regions: $z < -2.575$, $z > 2.575$

(c) $\bar{p} = \frac{x_1 + x_2}{n_1 + n_2} = \frac{468 + 472}{900 + 1027} \approx 0.488$

$\bar{q} \approx 0.512$

$$z = \frac{(\hat{p}_1 - \hat{p}_2) - (p_1 - p_2)}{\sqrt{\bar{p}\bar{q}\left(\frac{1}{n_1} + \frac{1}{n_2}\right)}} = \frac{(0.520 - 0.460) - (0)}{\sqrt{0.488 \cdot 0.512\left(\frac{1}{900} + \frac{1}{1027}\right)}} \approx 2.63$$

(d) Reject H_0.

(e) There is enough evidence at the 1% level of significance to reject the claim that the proportions of U.S. adults who considered the amount of federal income tax they had to pay to be too high were the same for the two years.

31. Yes; When $\alpha = 0.05$, the rejection regions become $z < -1.96$ and $z > 1.96$. Because $2.65 > 1.96$, you still reject H_0.

CHAPTER 8 QUIZ SOLUTIONS

1. (a) $H_0: \mu_1 \le \mu_2$; $H_a: \mu_1 > \mu_2$ (claim)

(b) One-tailed because H_a contains >; z-test because n_1 and n_2 are each greater than 30.

(c) $z_0 = 1.645$; Rejection region: $z > 1.645$

(d) $$z = \frac{(\bar{x}_1 - \bar{x}_2) - (\mu_1 - \mu_2)}{\sqrt{\frac{s_1^2}{n_1} + \frac{s_2^2}{n_2}}} = \frac{(149 - 145) - (0)}{\sqrt{\frac{(35)^2}{49} + \frac{(33)^2}{50}}} \approx 0.585$$

(e) Fail to reject H_0.

(f) There is not enough evidence at the 5% level of significance to support the claim that the mean score on the science assessment for male high school students was higher than for the female high school students.

2. (a) $H_0: \mu_1 = \mu_2$ (claim); $H_a: \mu_1 \ne \mu_2$

(b) Two-tailed because H_a contains $\ne$; t-test because n_1 and n_2 are less than 30, the samples are independent, and the population are normally distributed.

(c) d.f. = $n_1 + n_2 - 2 = 26$

$t_0 = \pm 2.779$; Rejection regions: $t < -2.779$, $t > 2.779$

(d) $$t = \frac{(\bar{x}_1 - \bar{x}_2) - (\mu_1 - \mu_2)}{\sqrt{\frac{(n-1)s_1^2 + (n_2 - 1)s_2^2}{n_1 + n_2 - 2}} \cdot \sqrt{\frac{1}{n_1} + \frac{1}{n_2}}} = \frac{(153 - 149) - (0)}{\sqrt{\frac{(13-1)(32)^2 + (15-1)(30)^2}{13 + 15 - 2}} \cdot \sqrt{\frac{1}{13} + \frac{1}{15}}} \approx 0.341$$

(e) Fail to reject H_0.

(f) There is not enough evidence at the 1% level of significance to reject the claim that the mean score on the science assessment test are the same for fourth grade boys and girls.

3. (a) $H_0: p_1 = p_2$ (claim); $H_a: p_1 \ne p_2$

(b) Two-tailed because H_a contains $\ne$; z-test because you are testing proportions and $n_1\bar{p}$, $n_1\bar{q}$, $n_2\bar{p}$, and $n_2\bar{q} \ge 5$.

(c) $z_0 = 1.645$; Rejection region: $z < -1.645$, $z > 1.645$

(d) $$\bar{p} = \frac{x_1 + x_2}{n_1 + n_2} = \frac{336 + 429}{800 + 1100} \approx 0.403$$

$\bar{q} \approx 0.597$

$$z = \frac{(\hat{p}_1 - \hat{p}_2) - (p_1 - p_2)}{\sqrt{\bar{p}\bar{q}\left(\frac{1}{n_1} + \frac{1}{n_2}\right)}} = \frac{(0.42 - 0.39) - (0)}{\sqrt{0.403 \cdot 0.597\left(\frac{1}{800} + \frac{1}{1100}\right)}} \approx 1.32$$

(e) Fail to reject H_0.

(f) There is not enough evidence at the 10% level of significance to reject the claim that the proportion of U.S. adults who are worried that their family will become a victim of terrorism has not changed.

4. (a) $H_0: \mu_d \geq 0$; $H_a: \mu_d < 0$ (claim)

(b) One-tailed because H_a contains <; t-test because both populations are normally distributed and the samples are dependent.

(c) $t_0 = -2.718$; Rejection region: $t < -2.718$

(d) $\bar{d} \approx -51.17$ and $s_d \approx 34.94$

$$t = \frac{\bar{d} - \mu_d}{\frac{s_d}{\sqrt{n}}} = \frac{-51.17 - 0}{\frac{34.94}{\sqrt{12}}} \approx -5.07$$

(e) Reject H_0.

(f) There is enough evidence at the 1% level of significance to support the claim that the seminar helps adults increase their credit scores.

CUMULATIVE REVIEW, CHAPTERS 6–8

1. (a) $\hat{p} = 0.13$, $\hat{q} = 0.87$

$$\hat{p} \pm z_c\sqrt{\frac{\hat{p}\hat{q}}{n}} = 0.13 \pm 1.96\sqrt{\frac{(0.13)(0.87)}{1000}} \approx 0.13 \pm 0.021 = (0.109,\ 0.151)$$

(b) $H_0: p \leq 0.10$; $H_a: p > 0.10$ (claim)

$z_0 = 1.645$; Rejection region: $z > 1.645$

$$z = \frac{\hat{p} - p}{\sqrt{\frac{pq}{n}}} = \frac{0.13 - 0.10}{\sqrt{\frac{(0.10)(0.90)}{1000}}} \approx 3.16$$

Reject H_0.

There is enough evidence at the 5% level of significance to support the claim that more than 10% of people who attend community college are age 40 or older.

2. $H_0: \mu_d \geq 0$; $H_a: \mu_d < 0$ (claim)

$\alpha = 0.10$ d.f. $= n - 1 = 7$

$t_0 = -1.415$; Rejection region: $t < -1.415$

$\bar{d} = -1.575$ and $s_d \approx 0.803$

$$t = \frac{\bar{d} - \mu_d}{\frac{s_d}{\sqrt{n}}} \approx \frac{-1.575 - 0}{\frac{0.803}{\sqrt{8}}} \approx -5.548$$

Reject H_0.

There is enough evidence at the 10% level of significance to support the claim that the fuel additive improved gas mileage.

3. $\bar{x} \pm z_c \frac{s}{\sqrt{n}} = 26.97 \pm 1.96\frac{3.4}{\sqrt{42}} \approx 26.97 \pm 1.03 = (25.94,\ 28.00)$; z-distribution

4. $\bar{x} \pm t_c \frac{s}{\sqrt{n}} = 3.46 \pm 1.753 \frac{1.63}{\sqrt{16}} \approx 3.46 \pm 0.71 = (2.75,\ 4.17)$; t-distribution

5. $\bar{x} \pm t_c \frac{s}{\sqrt{n}} = 12.1 \pm 2.787 \frac{2.64}{\sqrt{26}} \approx 12.1 \pm 1.4 = (10.7,\ 13.5)$; t-distribution

6. $\bar{x} \pm t_c \frac{s}{\sqrt{n}} = 8.21 \pm 2.365 \frac{0.62}{\sqrt{8}} \approx 8.21 \pm 0.52 = (7.69,\ 8.73)$; t-distribution

7. $H_0: \mu_1 \le \mu_2$; $H_a: \mu_1 > \mu_2$ (claim)

$z_0 = 1.28$; Rejection region: $z > 1.28$

$$z = \frac{(\bar{x}_1 - \bar{x}_2) - 0}{\sqrt{\frac{s_1^2}{n_1} + \frac{s_2^2}{n_2}}} = \frac{(3086 - 2263) - (0)}{\sqrt{\frac{(563)^2}{85} + \frac{(624)^2}{68}}} \approx 8.464$$

Reject H_0. There is enough evidence at the 10% level of significance to support the claim that the mean birth weight of a single-birth baby is greater than the mean birth weight of a baby that has a twin.

8. $H_0: \mu \ge 33$
$H_a: \mu < 33$ (claim)

9. $H_0: p \ge 0.19$ (claim)
$H_a: p < 0.19$

10. $H_0: \sigma = 0.63$ (claim)
$H_a: \sigma \ne 0.63$

11. $H_0: \mu = 2.28$
$H_a: \mu \ne 2.28$ (claim)

12. (a) $\left(\frac{(n-1)s^2}{\chi_R^2}, \frac{(n-1)s^2}{\chi_L^2}\right) = \left(\frac{(26-1)(3.1)^2}{46.928}, \frac{(26-1)(3.1)^2}{10.520}\right) \approx (5.1,\ 22.8)$

(b) $\left(\sqrt{5.1}, \sqrt{22.8}\right) \approx (2.3,\ 4.8)$

(c) $H_0:\ \sigma \ge 2.5$; $H_a:\ \sigma < 2.5$ (claim)

$\chi_0^2 = 11.524$; Rejection region: $\chi^2 < 11.524$

$$\chi^2 = \frac{(n-1)s^2}{\sigma^2} = \frac{(25)(3.1)}{(2.5)^2} = 38.44$$

Fail to reject H_0.

There is not enough evidence at the 1% level of significance to support the claim that the standard deviation of the mean number of chronic medications taken by elderly adults in the community is less than 2.5 medications.

13. $H_0: \mu_1 = \mu_2$; $H_a: \mu_1 \neq \mu_2$ (claim)

d.f. $= n_1 + n_2 - 2 = 26 + 18 - 2 = 42$

$t_0 = \pm 1.96$; Rejection regions: $t < -1.960$, $t > 1.960$

$$t = \frac{(\bar{x}_1 - \bar{x}_2) - (0)}{\sqrt{\frac{(n_1 - 1)s_1^2 + (n_2 - 1)s_2^2}{n_1 + n_2 - 2}}\sqrt{\frac{1}{n_1} + \frac{1}{n_2}}} = \frac{1783 - 2064}{\sqrt{\frac{(26-1)(218)^2 + (18-1)(186)^2}{26 + 18 - 2}}\sqrt{\frac{1}{20} + \frac{1}{18}}} \approx -4.456$$

Reject H_0. There is enough evidence at the 5% level of significance to support the claim that the mean SAT scores for male athletes and male non-athletes at a college are different.

14. (a) $\bar{x} \approx 38{,}896.46$

$s \approx 2881.83$

$$\bar{x} \pm t_c \frac{s}{\sqrt{n}} \approx 38{,}896.46 \pm 2.060 \frac{2881.83}{\sqrt{26}}$$

$$\approx 38{,}896.46 \pm 1164.26$$

$$\approx (37{,}732.2,\ 40{,}060.7)$$

(b) $H_0: \mu = 40{,}000$ (claim); $H_a: \mu \neq 40{,}000$

$t_0 = \pm 2.060$; Rejection regions: $t < -2.060$, $t > 2.060$

$$t = \frac{\bar{x} - \mu}{\frac{s}{\sqrt{n}}} \approx \frac{38{,}896.46 - 40{,}000}{\frac{2881.83}{\sqrt{26}}} = -1.95$$

Fail to reject H_0. There is not enough evidence at the 5% level of significance to reject the claim that the mean annual earnings for translators is $40,000.

15. $H_0: p_1 = p_2$ (claim); $H_a: p_1 \neq p_2$

$$\bar{p} = \frac{x_1 + x_2}{n_1 + n_2} = \frac{195 + 204}{319 + 323} \approx 0.621$$

$\bar{q} \approx 0.379$

$z_0 = \pm 1.645$; Rejection regions: $z < -1.645$, $z > 1.645$

$$z_0 = \frac{(\hat{p}_1 - \hat{p}_2) - 0}{\sqrt{\bar{p}\bar{q}\left(\frac{1}{n_1} + \frac{1}{n_2}\right)}} = \frac{(0.611 - 0.632)}{\sqrt{(0.621)(0.379)\left(\frac{1}{319} + \frac{1}{323}\right)}} \approx -0.548$$

Fail to reject H_0. There is not enough evidence at the 10% level of significance to reject the claim that the proportions of players sustaining head and neck injuries are the same for the two groups.

16. (a) $\bar{x} \pm z\frac{s}{\sqrt{n}} = 42 \pm 1.96\frac{1.6}{\sqrt{40}} \approx 42 \pm 0.496 \approx (41.5,\ 42.5)$

(b) $H_0: \mu \geq 45$ (claim); $H_a: \mu < 45$

$z_0 = -1.96$; Rejection region: $z < -1.96$

$$z = \frac{\bar{x} - \mu}{\frac{s}{\sqrt{n}}} = \frac{42 - 45}{\frac{1.6}{\sqrt{40}}} \approx -11.86$$

Reject H_0. There is enough evidence at the 5% level of significance to reject the claim that the mean incubation period for ostriches is at least 45 days.

Correlation and Regression

9.1 CORRELATION

9.1 Try It Yourself Solutions

1ab.

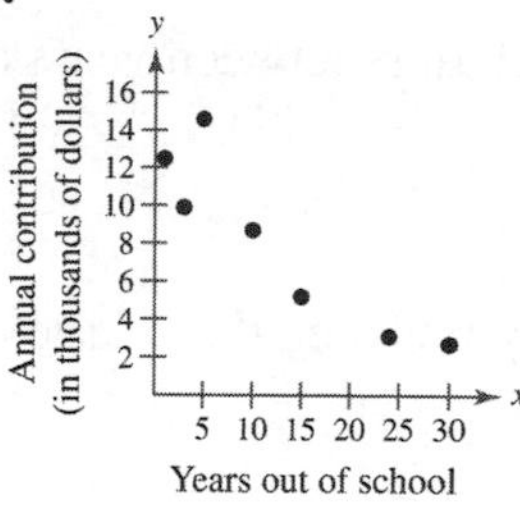

c. Yes, it appears that there is a negative linear correlation. As the number of years out of school increases, the annual contribution decreases.

2ab.

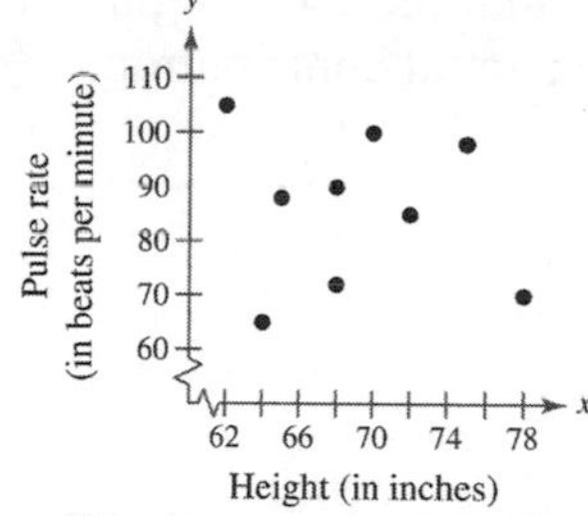

c. No, it appears that there is no linear correlation between height and pulse rate.

3ab.

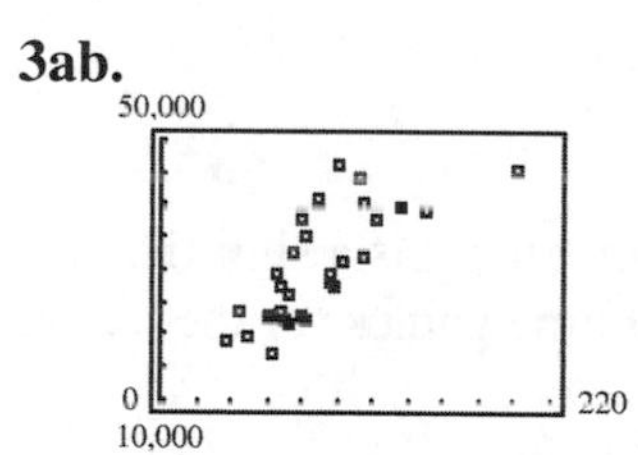

c. Yes, it appears that there is a positive linear correlation. As the team salary increases, the average attendance per home game increases.

4. (a)

x	y	xy	x^2	y^2
1	12.5	12.5	1	156.25
10	8.7	87.0	100	75.69
5	14.6	73.0	25	213.16
15	5.2	78.0	225	27.04
3	9.9	29.7	9	98.01
24	3.1	74.4	576	9.61
30	2.7	81.0	900	7.29
$\sum x = 88$	$\sum y = 56.7$	$\sum xy = 435.6$	$\sum x^2 = 1836$	$\sum y^2 = 587.05$

(b) $$r=\frac{n\sum xy-(\sum x)(\sum y)}{\sqrt{n\sum x^2-(\sum x)^2}\sqrt{n\sum y^2-(\sum y)^2}}$$
$$=\frac{(7)(435.6)-(88)(56.7)}{\sqrt{(7)(1836)-(88)^2}\sqrt{(7)(587.05)-(56.7)^2}}$$
$$=\frac{-1940.4}{\sqrt{5108}\sqrt{894.46}}\approx -0.908$$

(c) Because r is close to -1, this suggests a strong negative linear correlation between years out of school and annual contribution.

5ab. $r\approx 0.750$

c. Because r is close to 1, this suggests a strong positive linear correlation between the salaries and the average attendances at home games.

6a. $n=7$

b. $\alpha=0.01$

c. 0.875

d. $|r|\approx|0.908|>0.875$; The correlation is significant.

e. There is enough evidence at the 1% level of significance to conclude that there is a significant linear correlation between the number of years out of school and the annual contribution.

7a. $H_0: \rho=0;\ H_a: \rho\neq 0$

b. $\alpha=0.01$

c. d.f. = $n-2=28$

d. $t_0=\pm 2.763$; Rejection regions: $t<-2.763$ or $t>2.763$

e. $$t=\frac{r}{\sqrt{\frac{1-r^2}{n-2}}}=\frac{0.74972}{\sqrt{\frac{1-(0.74972)^2}{30-2}}}\approx 5.995$$

f. Reject H_0.

g. There is enough evidence at the 1% level of significance to conclude that there is a significant linear correlation between the salaries and the average attendances at home games for the teams in Major League Baseball.

9.1 EXERCISE SOLUTIONS

1. Increase

3. The range of values for the correlation coefficient is -1 to 1.

5. Answers will vary. Sample answer:
Perfect positive linear correlation: price per gallon of gasoline and total cost of gasoline
Perfect negative linear correlation: distance from door and height of wheelchair ramp

7. r is the sample correlation coefficient, while ρ is the population correlation coefficient.

9. Negative linear correlation

11. Perfect negative linear correlation

13. Positive linear correlation

15. (c), You would expect a positive linear correlation between age and income.

17. (b), You would expect a negative linear correlation between age and balance on student loans.

19. Explanatory variable: Amount of water consumed
Response variable: Weight loss

21. (a)

(b)

x	y	xy	x^2	y^2
16	109	1744	256	11,881
25	122	3050	625	14,884
39	143	5577	1521	20,449
45	132	5940	2025	17,424
49	199	9751	2401	39,601
64	185	11,840	4096	34,225
70	199	13,930	4900	39,601
29	130	3770	841	16,900
57	175	9975	3249	30,625
22	118	2596	484	13,924
$\sum x = 416$	$\sum y = 1512$	$\sum xy = 68{,}173$	$\sum x^2 = 20{,}398$	$\sum y^2 = 239{,}514$

$$r = \frac{n\sum xy - (\sum x)(\sum y)}{\sqrt{n\sum x^2 - (\sum x)^2}\sqrt{n\sum y^2 - (\sum y)^2}}$$

$$= \frac{10(68{,}173) - (416)(1512)}{\sqrt{10(20{,}398) - (416)^2}\sqrt{10(239{,}514) - (1512)^2}}$$

$$= \frac{52{,}738}{\sqrt{30{,}924}\sqrt{108{,}996}} \approx 0.908$$

(c) Strong positive linear correlation

23. (a)

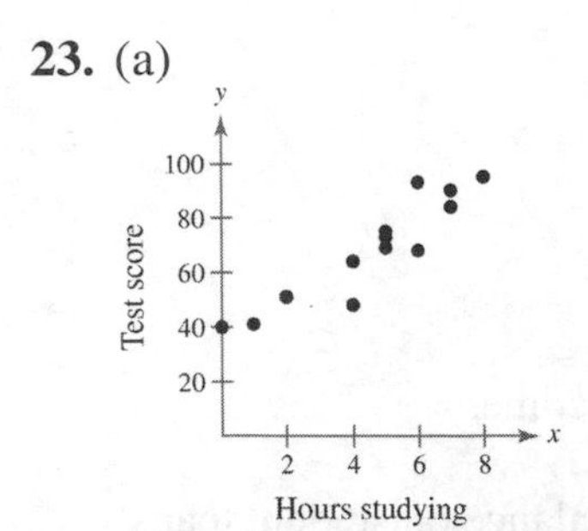

(b)

x	y	xy	x^2	y^2
0	40	0	0	1600
1	41	41	1	1681
2	51	102	4	2601
4	48	192	16	2304
4	64	256	16	4096
5	69	345	25	4761
5	73	365	25	5329
5	75	375	25	5625
6	68	408	36	4624
6	93	558	36	8649
7	84	588	49	7056
7	90	630	49	8100
8	95	760	64	9025
$\sum x=60$	$\sum y=891$	$\sum xy=4620$	$\sum x^2=346$	$\sum y^2=65{,}451$

$$r=\frac{n\sum xy-(\sum x)(\sum y)}{\sqrt{n\sum x^2-(\sum x)^2}\sqrt{n\sum y^2-(\sum y)^2}}$$

$$=\frac{13(4620)-(60)(891)}{\sqrt{13(346)-(60)^2}\sqrt{13(65{,}451)-(891)^2}}$$

$$=\frac{6600}{\sqrt{898}\sqrt{56{,}982}}\approx 0.923$$

(c) Strong positive linear correlation

25. (a)

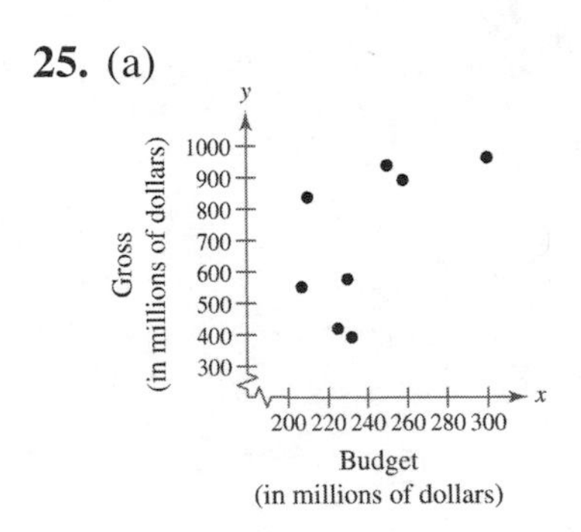

(b)

x	y	xy	x^2	y^2
300	961	288,300	90,000	923,521
258	891	229,878	66,564	793,881
250	937	234,250	62,500	877,969
210	836	175,560	44,100	698,896
232	391	90,712	53,824	152,881
230	576	132,480	52,900	331,776
225	419	94,275	50,625	175,561
207	551	114,057	42,849	303,601
$\sum x = 1912$	$\sum y = 5562$	$\sum xy = 1,359,512$	$\sum x^2 = 463,362$	$\sum y^2 = 4,258,086$

$$r = \frac{n\sum xy - (\sum x)(\sum y)}{\sqrt{n\sum x^2 - (\sum x)^2}\sqrt{n\sum y^2 - (\sum y)^2}}$$

$$= \frac{8(1,359,512) - (1912)(5562)}{\sqrt{8(463,362) - (1912)^2}\sqrt{8(4,258,086) - (5562)^2}}$$

$$= \frac{241,552}{\sqrt{51,152}\sqrt{3,128,844}} \approx 0.604$$

(c) Weak positive linear correlation

27. (a)

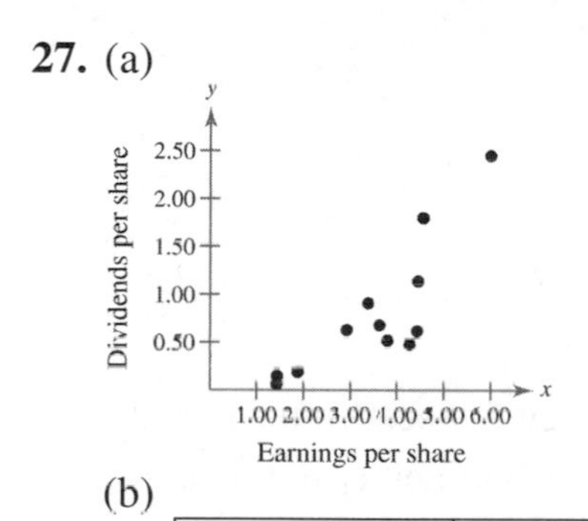

(b)

x	y	xy	x^2	y^2
6.00	2.45	14.7000	36.0000	6.0025
1.44	0.15	0.2160	2.0736	0.0225
4.44	0.62	2.7528	19.7136	0.3844
3.38	0.91	3.0758	11.4244	0.8281
3.63	0.68	2.4684	13.1769	0.4624
4.46	1.14	5.0844	19.8916	1.2996
3.80	0.52	1.9760	14.4400	0.2704
1.43	0.06	0.0858	2.0449	0.0036
1.88	0.19	0.3572	3.5344	0.0361
4.57	1.80	8.2260	20.8849	3.2400
4.28	0.48	2.0544	18.3184	0.2304
2.92	0.63	1.8396	8.5264	0.3969
$\sum x = 42.23$	$\sum y = 9.63$	$\sum xy = 42.8364$	$\sum x^2 = 170.0291$	$\sum y^2 = 13.1769$

$$r = \frac{n\sum xy - (\sum x)(\sum y)}{\sqrt{n\sum x^2 - (\sum x)^2}\sqrt{n\sum y^2 - (\sum y)^2}}$$

$$= \frac{12(42.8364) - (42.23)(9.63)}{\sqrt{12(170.0291) - (42.23)^2}\sqrt{12(13.1769) - (9.63)^2}}$$

$$= \frac{107.3619}{\sqrt{256.9763}\sqrt{65.3859}} \approx 0.828$$

(c) Strong positive linear correlation

29. The correlation coefficient becomes $r \approx 0.621$. The new data entry is an outlier, so the linear correlation is weaker.

31. $r \approx 0.623$
$n = 8$ and $\alpha = 0.01$

critical value $= 0.834$

$|r| \approx 0.623 < 0.834 \Rightarrow$ The correlation is not significant.

or

$H_0: \rho = 0;\ H_a: \rho \neq 0$

$\alpha = 0.01$

$\text{d.f.} = n - 2 = 6$

$t_0 = \pm 3.703$; Rejection regions: $t < -3.707$ or $t > 3.707$.

$$t = \frac{r}{\sqrt{\frac{1-r^2}{n-2}}} = \frac{0.623}{\sqrt{\frac{1-(0.623)^2}{8-2}}} \approx 1.951$$

Fail to reject H_0. There is not enough evidence at the 1% level of significance to conclude that there is a significant linear correlation between vehicle weight and the variability in braking distance.

33. $r \approx 0.923$
$n = 13$ and $\alpha = 0.01$

critical value $= 0.684$

$|r| = 0.923 > 0.684 \Rightarrow$ The correlation is significant.

or

$H_0: \rho = 0;\ H_a: \rho \neq 0$

$\alpha = 0.01$

$\text{d.f.} = n - 2 = 11$

$t_0 = \pm 3.106$; Rejection regions: $t < -3.106$ or $t > 3.106$.

$r \approx 0.923$

$$t = \frac{r}{\sqrt{\frac{1-r^2}{n-2}}} = \frac{0.923}{\sqrt{\frac{1-(0.923)^2}{13-2}}} \approx 7.955$$

Reject H_0. There is enough evidence at the 1% level of significance to conclude that a significant linear correlation exists.

35. $r \approx 0.828$

$n = 12$ and $\alpha = 0.01$

critical value $= 0.708$

$|r| \approx 0.828 > 0.708 \Rightarrow$ The correlation is significant.

or

$H_0: \rho = 0;\ H_a: \rho \neq 0$

$\alpha = 0.01$

d.f. $= n - 2 = 10$

$t_0 = \pm 3.169$; Rejection regions: $t < -3.169$ or $t > 3.169$.

$r \approx 0.828$

$$t = \frac{r}{\sqrt{\frac{1-r^2}{n-2}}} = \frac{0.828}{\sqrt{\frac{1-(0.828)^2}{12-2}}} \approx 4.670$$

Reject H_0. There is enough evidence at the 1% level of significance to conclude that there is a significant linear correlation between earnings per share and dividends per share.

37. (a)

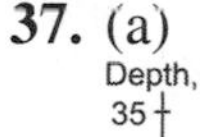

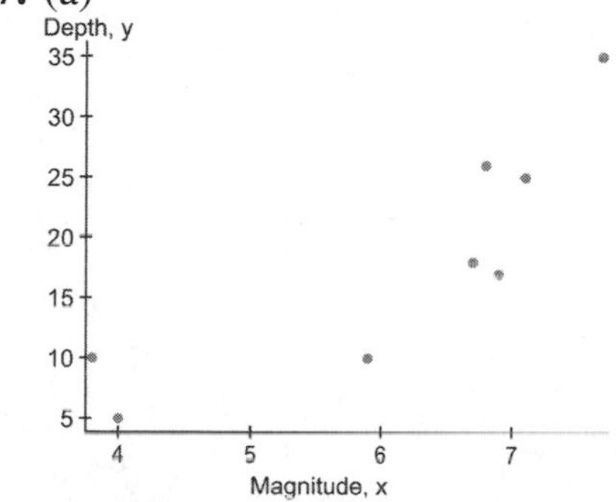

(b) 0.848

(c) Reject H_0. There is enough evidence at the 1% level of significance to conclude that there is a significant linear correlation between the magnitudes of earthquakes and their depths below the surface at the epicenter.

39. The correlation coefficient becomes $r \approx 0.085$. The new rejection regions are $t < -3.499$ and $t > 3.499$ and the new standardized test statistic is $t \approx 0.227$. So, now you fail to reject H_0.

41. 0.883; 0.883; The correlation coefficient remains unchanged when the x-values and y-values are switched.

43. Answers will vary.

9.2 LINEAR REGRESSION

9.2 Try It Yourself Solutions

1a. $n=7$

$$\sum x=88$$
$$\sum y=56.7$$
$$\sum xy=435.6$$
$$\sum x^2=1836$$

b. $$m=\frac{n\sum xy-(\sum x)(\sum y)}{n\sum x^2-(\sum x)^2}=\frac{(7)(435.6)-(88)(56.7)}{(7)(1836)-(88)^2}=\frac{-1940.4}{5108}\approx-0.379875$$

$$b=\bar{y}-m\bar{x}=\frac{\sum y}{n}-m\frac{\sum x}{n}\approx\frac{(56.7)}{7}-(-0.379875)\frac{(88)}{7}\approx 12.8756$$

c. $\hat{y}=-0.380x+12.876$

2a. Enter the data.

b. $m\approx 189.038015$

$b\approx 13,497.9583$

c. $\hat{y}=189.038x+13,497.958$

3a. (1) $\hat{y}=12.481(2)+33.683$ (2) $\hat{y}=12.481(3.32)+33.683$

b. (1) 58.645 (2) 75.120

c. (1) 58.645 minutes (2) 75.120 minutes

9.2 EXERCISE SOLUTIONS

1. A residual is the difference between the observed y-value of a data point and the predicted y-value on the regression line for the x-coordinate of the data point. A residual is positive when the data point is above the line, negative when the point is below the line, and zero when the observed y-value equals the predicted y-value.

3. Substitute a value of x into the equation of a regression line and solve for y.

5. The correlation between variables must be significant.

7. b **9.** e **11.** f **13.** c **15.** a

17.

x	y	xy	x^2
869	60	52,140	755,161
820	50	41,000	672,400
771	50	38,550	594,441
696	52	36,192	484,416
692	40	27,680	478,864
676	47	31,772	456,976
656	41	26,896	430,336
492	39	19,188	242,064
486	26	12,636	236,196
$\sum x = 6158$	$\sum y = 405$	$\sum xy = 286{,}054$	$\sum x^2 = 4{,}350{,}854$

$$m = \frac{n\sum xy - (\sum x)(\sum y)}{n\sum x^2 - (\sum x)^2}$$

$$= \frac{(9)(286{,}054) - (6158)(405)}{(9)(4{,}350{,}854) - (6158)^2}$$

$$= \frac{80{,}496}{1{,}236{,}722} \approx 0.065088$$

$$b = \bar{y} - m\bar{x} = \frac{\sum y}{n} - m\frac{\sum x}{n}$$

$$\approx \frac{405}{9} - (0.065088)\left(\frac{6158}{9}\right)$$

$$\approx 0.4653$$

$$\hat{y} = mx + b = 0.065x + 0.465$$

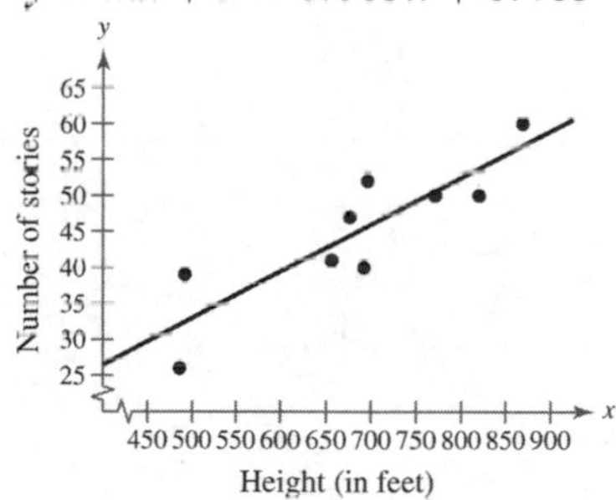

(a) $\hat{y} = 0.065(800) + 0.465 \approx 52$

(b) $\hat{y} = 0.065(750) + 0.465 \approx 49$

(c) It is not meaningful to predict the value of y for $x = 400$ because $x = 400$ is outside the range of the original data.

(d) $\hat{y} = 0.065(625) + 0.465 \approx 41$

19.

x	y	xy	x^2
0	40	0	0
1	41	41	1
2	51	102	4
4	48	192	16
4	64	256	16
5	69	345	25
5	73	365	25
5	75	375	25
6	68	408	36
6	93	558	36
7	84	588	49
7	90	630	49
8	95	760	64
$\sum x=60$	$\sum y=891$	$\sum xy=4620$	$\sum x^2=346$

$$m=\frac{n\sum xy-(\sum x)(\sum y)}{n\sum x^2-(\sum x)^2}$$

$$=\frac{(13)(4620)-(60)(891)}{(13)(346)-(60)^2}$$

$$=\frac{6600}{898}\approx 7.349666$$

$$b=\bar{y}-m\bar{x}\approx\left(\frac{891}{13}\right)-(7.349666)\left(\frac{60}{13}\right)\approx 34.6169$$

$$\hat{y}=7.350x+34.617$$

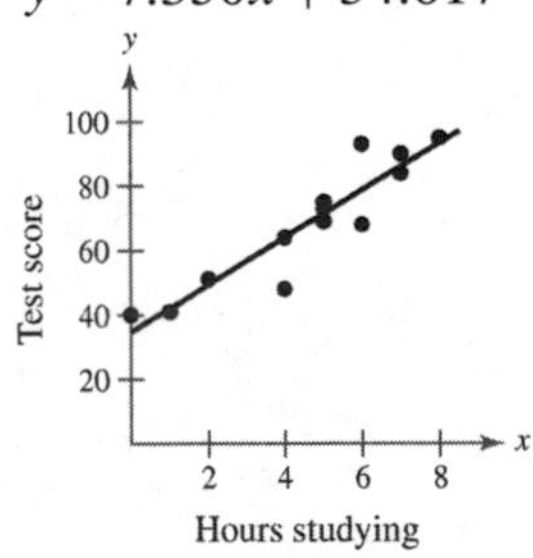

(a) $\hat{y}=7.350(3)+34.617\approx 57$

(b) $\hat{y}=7.350(6.5)+34.617\approx 82$

(c) It is not meaningful to predict the value of y for $x = 13$ because $x = 13$ is outside the range of the original data.

(d) $\hat{y}=7.350(4.5)+34.617\approx 68$

21.

x	y	xy	x^2
150	420	63,000	22,500
170	470	79,900	28,900
120	350	42,000	14,400
120	360	43,200	14,400
90	270	24,300	8,100
180	550	99,000	32,400
170	530	90,100	28,900
140	460	64,400	19,600
90	380	34,200	8,100
110	330	36,300	12,100
$\sum x = 1340$	$\sum y = 4120$	$\sum xy = 576,400$	$\sum x^2 = 189,400$

$$m = \frac{x\sum xy - (\sum x)(\sum y)}{n\sum x^2 - (\sum x)^2}$$

$$= \frac{(10)(576,400) - (1340)(4120)}{(10)(189,400) - (1340)^2}$$

$$= \frac{243,200}{98,400} \approx 2.471545$$

$$b = \bar{y} - m\bar{x} \approx \left(\frac{4120}{10}\right) - (2.471545)\left(\frac{1340}{10}\right) \approx 80.8130$$

$$\hat{y} = 2.472x + 80.813$$

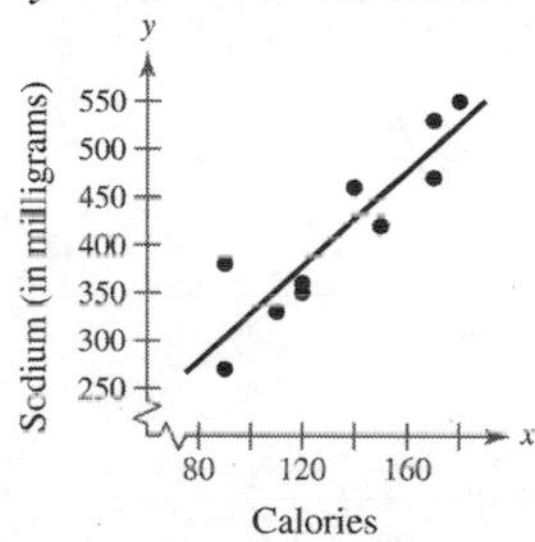

(a) $\hat{y} = 2.472(170) + 80.813 = 501.053$ milligrams

(b) $\hat{y} = 2.472(100) + 80.813 = 328.013$ milligrams

(c) $\hat{y} = 2.472(140) + 80.813 = 426.893$ milligrams

(d) It is not meaningful to predict the value of y for $x = 210$ because $x = 210$ is outside the range of the original data.

23.

x	y	xy	x^2
8.5	66.0	561.00	72.25
9.0	68.5	616.50	81.00
9.0	67.5	607.50	81.00
9.5	70.0	665.00	90.25
10.0	70.0	700.00	100.00
10.0	72.0	720.00	100.00
10.5	71.5	750.75	110.25
10.5	69.5	729.75	110.25
11.0	71.5	786.50	121.00
11.0	72.0	792.00	121.00
11.0	73.0	803.00	121.00
12.0	73.5	882.00	144.00
12.0	74.0	888.00	144.00
12.5	74.0	925.00	156.25
$\sum x = 146.5$	$\sum y = 993.0$	$\sum xy = 10{,}427.0$	$\sum x^2 = 1552.25$

$$m = \frac{x\sum xy - (\sum x)(\sum y)}{n\sum x^2 - (\sum x)^2}$$

$$= \frac{(14)(10{,}427.0) - (146.5)(993.0)}{(14)(1552.25) - (146.5)^2}$$

$$= \frac{503.5}{269.25} \approx 1.870009$$

$$b = \bar{y} - m\bar{x} \approx \left(\frac{993.0}{14}\right) - (1.870)\left(\frac{146.5}{14}\right) \approx 51.3603$$

$$\hat{y} = 1.870x + 51.360$$

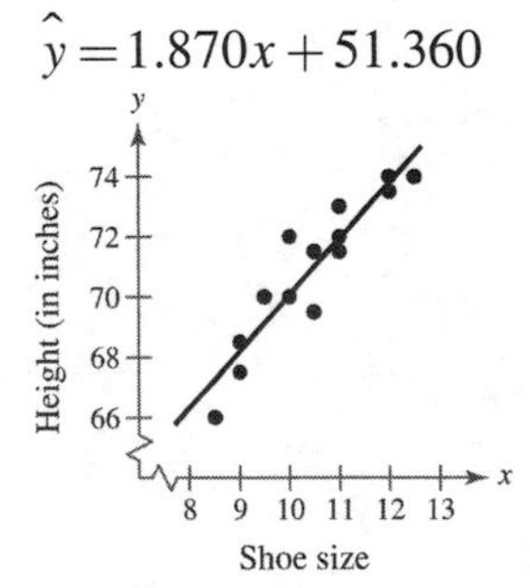

(a) $\hat{y} = 1.870(11.5) + 51.360 = 72.865$ inches

(b) $\hat{y} = 1.870(8.0) + 51.360 = 66.32$ inches

(c) It is not meaningful to predict the value of y for $x = 15.5$ because $x = 15.5$ is outside the range of the original data.

(d) $\hat{y} = 1.870(10.0) + 51.360 = 70.06$ inches

25. Strong positive linear correlation; As the years of experience of registered nurses increase. their salaries tend to increase.

27. No, it is not meaningful to predict a salary for a registered nurse with 28 years of experience because $x = 28$ is outside the range of the original data.

29. Answers will vary. Sample answer: Although it is likely that there is a cause-and-effect relationship between a registered nurse's years of experience and salary, the relationship between variables may also be influenced by other factors, such as work performance, level of education, or the number of years with an employer.

31a. $\hat{y} = -0.159x + 5.827$

b. $r \approx -0.852$

c.

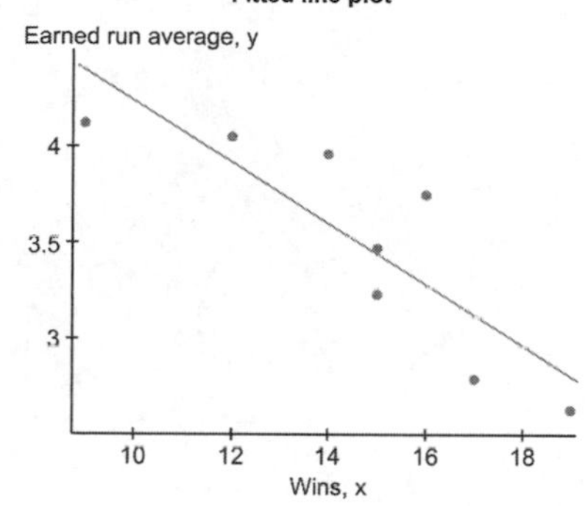

33. (a)

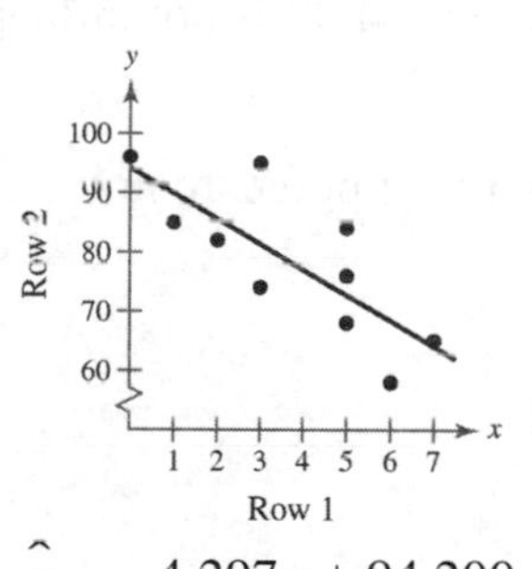

$\hat{y} = -4.297x + 94.200$

(b)

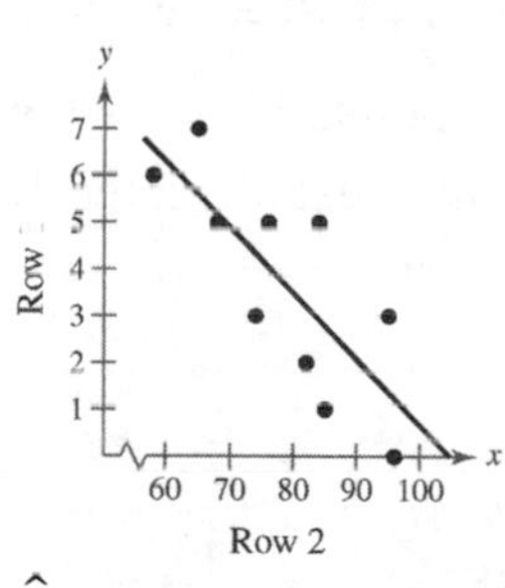

$\hat{y} = -0.1413x + 14.763$

(c) The slope of the line keeps the same sign, but the values of m and b change.

35. (a) $m \approx 0.139$

$b \approx 21.024$

$\hat{y} = 0.139x + 21.024$

(b)

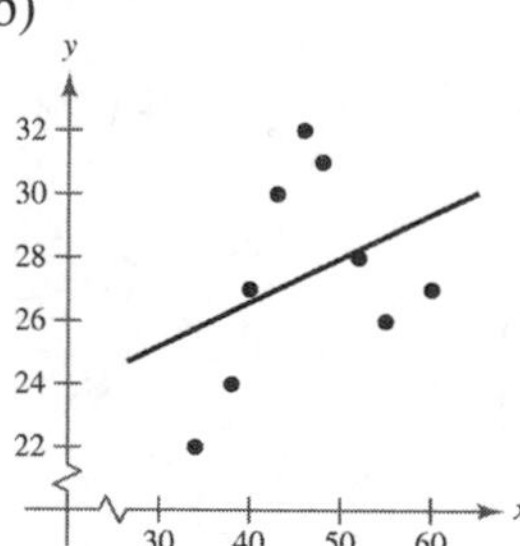

(c)

x	y	$\hat{y} = 0.139x + 21.024$	$y - \hat{y}$
38	24	26.306	−2.306
34	22	22.750	−3.750
40	27	26.584	0.416
46	32	27.418	4.582
43	30	27.001	2.999
48	31	27.696	3.304
60	27	29.364	−2.364
55	26	28.669	−2.669
52	28	28.252	−0.252

(d) The residual plot shows a pattern because the residuals do not fluctuate about 0. This implies that the regression line is not a good representation of the relationship between the variables.

37. (a)

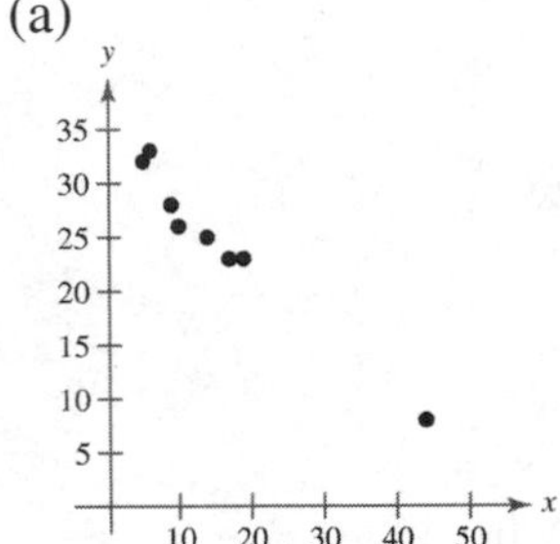

(b) The point (44, 8) may be an outlier.

(c) Excluding the point (44, 8) $\Rightarrow \hat{y} = -0.711x + 35.263$. The point (44, 8) is not influential because using all 8 points $\Rightarrow \hat{y} = -0.607x + 34.160$.
The slopes and y-intercepts with the point included and without the point are not significantly different.

39. $m \approx 654.536$

$b \approx -1214.857$

$\hat{y} = 654.536x - 1214.857$

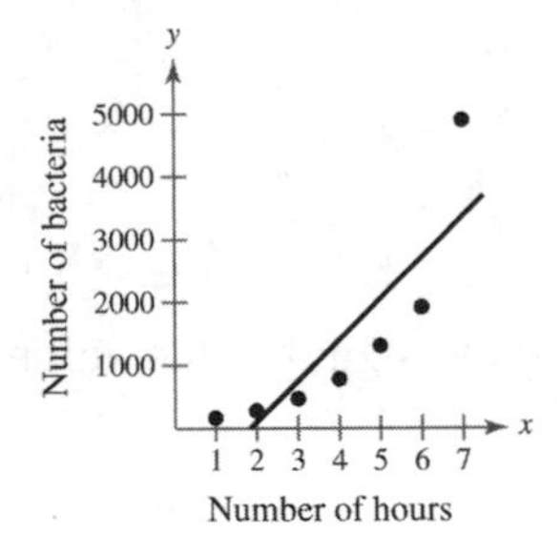

41. Using a technology tool $\Rightarrow y = 93.028(1.712)^x$.

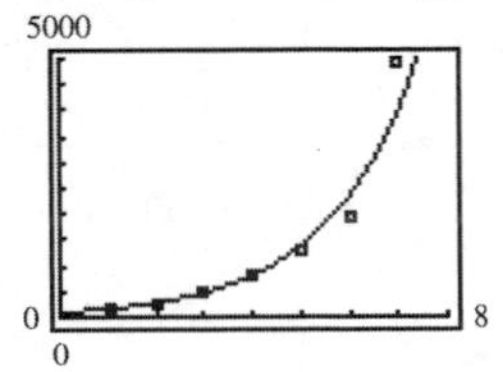

43. $m = -78.929$

$b = 576.179$

$\hat{y} = -78.929x + 576.179$

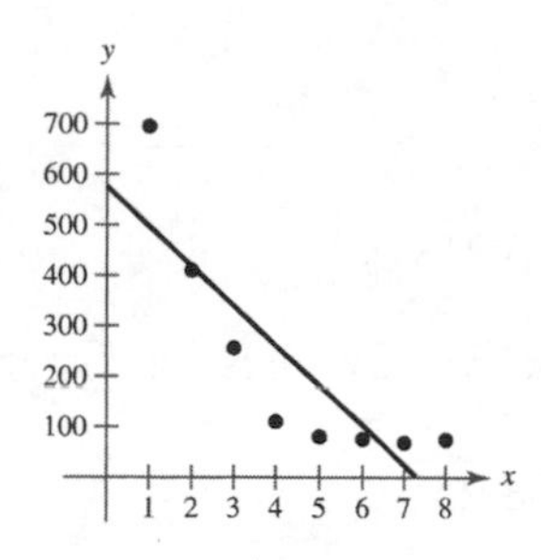

45. Using a technology tool $\Rightarrow y = 782.300x^{-1.251}$.

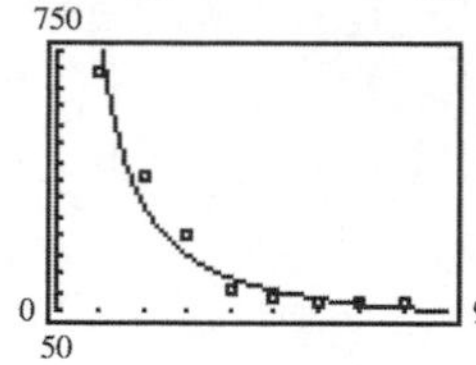

47. $y = a + b\ln x = 25.035 + 19.599\ln x$

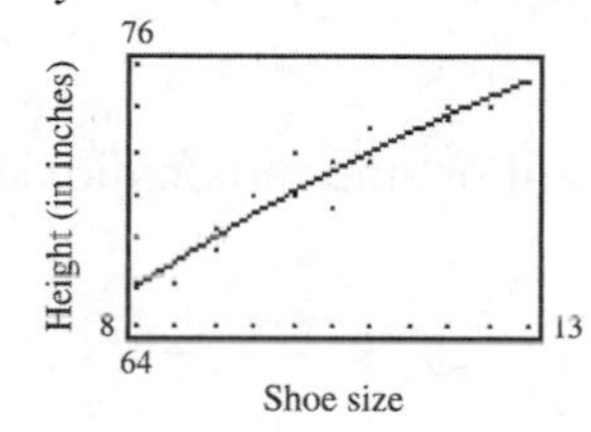

49. The logarithmic equation is a better model for the data. The graph of the logarithmic equation fits the data better than the regression line.

9.3 MEASURES OF REGRESSION AND PREDICTION INTERVALS

9.3 Try It Yourself Solutions

1a. $r \approx 0.979$ **b.** $r^2 \approx (0.979)^2 \approx 0.958$

c. About 95.8% of the variation in the times is explained. About 4.2% of the variation is unexplained.

2a.

x_i	y_i	$\hat{y}_i$	$y_i - \hat{y}_i$	$(y_i - \hat{y}_i)^2$
15	26	28.386	–2.386	5.693
20	32	35.411	–3.411	11.635
20	38	35.411	2.589	6.703
30	56	49.461	6.539	42.759
40	54	63.511	–9.511	90.459
45	78	70.536	7.464	55.711
50	80	77.561	2.439	5.949
60	88	91.611	–3.611	13.039
				$\Sigma = 231.948$

b. $n = 8$

c. $s_e = \sqrt{\dfrac{\Sigma\left(y_i - \hat{y}_i\right)^2}{n-2}} = \sqrt{\dfrac{231.948}{6}} \approx 6.218$

d. The standard error of estimate of the weekly sales for a specific radio ad time is about \$621.80.

3a. $n = 10$, d.f. $= 8$, $t_c = 2.306$, $s_e \approx 138.255$

b. $\hat{y} = 196.152(4.0) + 102.289 = 886.897$

c. $E = t_c s_e \sqrt{1 + \dfrac{1}{n} + \dfrac{n\left(x - \bar{x}\right)^2}{n\left(\Sigma x^2\right) - \left(\Sigma x\right)^2}}$

$$\approx (2.306)(138.255)\sqrt{1 + \frac{1}{10} + \frac{10(4.0 - 2.31)^2}{10(67.35) - (23.1)^2}}$$

$$\approx 364.088$$

d. $\hat{y} \pm E \rightarrow (522.809,\ 1250.985)$

e. You can be 95% confident that when the gross domestic product is \$4 trillion, the carbon dioxide emissions will be between 522.809 and 1250.985 million metric tons.

9.3 EXERCISE SOLUTIONS

1. Total variation $= \Sigma\left(y_i - \bar{y}\right)^2$; the sum of the squares of the differences between the y-values of each ordered pair and the mean of the y-values of the ordered pairs.

3. Unexplained variation $= \Sigma\left(y_i - \hat{y}_i\right)^2$; the sum of the squares of the differences between the observed y-values and the predicted y-values.

5. Two variables that have perfect positive or perfect negative linear correlation have a correlation coefficient of 1 or -1, respectively. In either case, the coefficient of determination is 1, which means that 100% of the variation in the response variable is explained by the variation in the explanatory variable.

7. $r^2 = (0.465)^2 \approx 0.216$; About 21.6% of the variation is explained. About 78.4% of the variation is unexplained.

9. $r^2 = (-0.957)^2 \approx 0.916$; About 91.6% of the variation is explained. About 8.4% of the variation is unexplained.

11. (a) $r^2 = \dfrac{\Sigma(\hat{y}_i - \bar{y})^2}{\Sigma(y_i - \bar{y})^2} \approx 0.798$

About 79.8% of the variation in proceeds can be explained by the variation in the number of issues, and about 20.2% of the variation is unexplained.

(b) $s_e = \sqrt{\dfrac{\Sigma(y_i - \hat{y}_i)^2}{n-2}} = \sqrt{\dfrac{650{,}383{,}054}{10}} \approx 8064.633$

The standard error of estimate of the proceeds for a specific number of issues is about \$8,064,633,000.

13. (a) $r^2 = \dfrac{\Sigma(\hat{y}_i - \bar{y})^2}{\Sigma(y_i - \bar{y})^2} \approx 0.981$

About 98.1% of the variation in sales can be explained by the variation in the total square footage, and about 1.9% of the variation is unexplained.

(b) $s_e = \sqrt{\dfrac{\Sigma(y_i - \hat{y}_i)^2}{n-2}} = \sqrt{\dfrac{8413.958}{9}} \approx 30.576$

The standard error of estimate of the sales for a specific total square footage is about 30,576,000,000.

15. (a) $r^2 = \dfrac{\Sigma(\hat{y}_i - \bar{y})^2}{\Sigma(y_i - \bar{y})^2} \approx 0.963$

About 96.3% of the variation in wages for federal government employees can be explained by the variation in wages for state government employees, and about 3.7% of the variation is unexplained.

(b) $s_e = \sqrt{\dfrac{\Sigma(y_i - \hat{y}_i)^2}{n-2}} = \sqrt{\dfrac{1614.4324}{4}} \approx 20.090$

The standard error of estimate of the average weekly wages for federal government employees for a specific average weekly wage for state government employees is about 20.09.

17. (a) $r^2 = \dfrac{\Sigma(\hat{y}_i - \bar{y})^2}{\Sigma(y_i - \bar{y})^2} \approx 0.790$

About 79.0% of the variation in the gross collection of corporate income taxes can be explained by the variation in the gross collections of individual income taxes, and about 21.0% of the variation is unexplained.

(b) $s_e = \sqrt{\dfrac{\Sigma(y_i - \hat{y}_i)^2}{n-2}} = \sqrt{\dfrac{8982.865}{5}} \approx 42.386$

The standard error of estimate of the gross collections of corporate income taxes for a specific gross collection of individual income taxes is about \$42,386,000,000.

19. $n = 12$, d.f. $= 10$, $t_c = 2.228$, $s_e \approx 8064.633$

$\hat{y} = 104.982x + 14{,}128.671 = 104.982(450) + 14{,}128.671 = 61{,}370.571$

$$E = t_c s_e \sqrt{1 + \frac{1}{n} + \frac{n(x - \bar{x})^2}{n(\sum x^2) - (\sum x)^2}}$$

$$\approx (2.228)(8064.633)\sqrt{1 + \frac{1}{12} + \frac{12(450 - 2142/12)^2}{12(615{,}732) - (2142)^2}}$$

$$\approx 21{,}253.747$$

$\hat{y} + E \to (40{,}116.824,\ 82{,}624.318) \to (40{,}116{,}824{,}000,\ 82{,}624{,}318{,}000)$

You can be 95% confident that the proceeds will be between 40,116,824,000 and 82,624,318,000 when the number of initial offerings is 450.

21. $n = 11$, d.f. $= 9$, $t_c = 1.833$, $s_e \approx 30.576$

$\hat{y} = 549.448x - 1881.694 = -549.448(5.75) - 1881.694 \approx 1277.632$

$$E = t_c s_e \sqrt{1 + \frac{1}{n} + \frac{n(x - \bar{x})^2}{n(\sum x^2) - (\sum x)^2}}$$

$$\approx (1.833)(30.576)\sqrt{1 + \frac{1}{11} + \frac{11(5.75 - 61.2/11)^2}{11(341.9) - (61.2)^2}}$$

$$\approx 59.197$$

$\hat{y} + E \to (1218.435,\ 1336.829) \to (1{,}218{,}435{,}000{,}000,\ 1{,}336{,}829{,}000{,}000)$

You can be 90% confident that the shopping center sales will be between \$1,218,435,000,000 and \$1,336,829,000,000 when the total square footage is 5.75 billion.

23. $n = 6$, d.f. $= 4$, $t_c = 4.604$, $s_e \approx 20.090$

$\hat{y} = 1.900x - 411.976 = 1.900(800) - 411.976 = 1108.024$

$$E = t_c s_e \sqrt{1 + \frac{1}{n} + \frac{n(x - \bar{x})^2}{n(\sum x^2) - (\sum x)^2}}$$

$$\approx (4.604)(20.090)\sqrt{1 + \frac{1}{6} + \frac{6(800 - 4854/6)^2}{6(3{,}938{,}466) - (4854)^2}}$$

$$\approx 100.204$$

$\hat{y} + E \to (1007.82,\ 1208.228)$

You can be 99% confident that the average weekly wages of federal government employees will be between 1007.82 and 1208.23 when the average weekly wage of state government employees is \$800.

25. $n = 7$, d.f. $= 5$, $t_c = 2.571$, $s_e \approx 42.386$

$\hat{y} = 0.415x - 186.626 = 0.415(1250) - 186.626 \approx 332.124$

$$E = t_c s_e \sqrt{1 + \frac{1}{n} + \frac{n(x - \bar{x})^2}{n(\sum x^2) - (\sum x)^2}}$$

$$\approx (2.571)(42.386)\sqrt{1 + \frac{1}{7} + \frac{7(1250 - 8151/7)^2}{7(9{,}686{,}505) - (8151)^2}}$$

$$\approx 118.395$$

$\hat{y} + E \to (213.729,\ 450.519)$

You can be 95% confident that the corporate income taxes collected by the U.S. Internal Revenue Service for a given year will be between \$213.729 million and \$450.519 million when the U.S. Internal Revenue Service collects \$1.250 million in individual income taxes that year.

27.

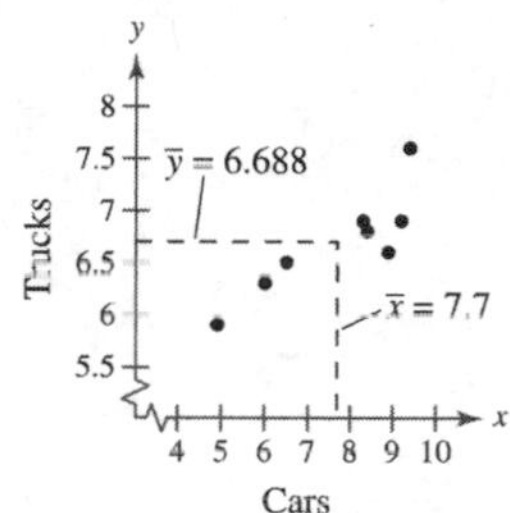

29.

x_i	y_i	$\hat{y}_i$	$\hat{y}_i - \bar{y}$	$y_i - \hat{y}_i$	$y_i - \bar{y}$
9.4	7.6	7.1252	0.4372	0.4748	0.912
9.2	6.9	7.0736	0.3856	−0.1736	0.212
8.9	6.6	6.9962	0.3082	−0.3962	−0.088
8.4	6.8	6.8672	0.1792	−0.0672	0.112
8.3	6.9	6.8414	0.1534	0.0586	0.212
6.5	6.5	6.377	−0.311	0.123	−0.188
6	6.3	6.248	−0.44	0.052	−0.388
4.9	5.9	5.9642	−0.7238	−0.0642	−0.788

31. $r^2 \approx 0.746$; About 74.6% of the variation in the median age of trucks can be explained by the variation in the median age of cars, and about 25.4% of the variation is unexplained.

33. $\hat{y} = 0.258x + 4.700 = 0.258(7.0) + 4.700 = 6.506$

$$E = t_c s_e \sqrt{1 + \frac{1}{n} + \frac{n(x - \bar{x})^2}{n(\sum x^2) - (\sum x)^2}}$$

$$\approx (2.447)(0.272)\sqrt{1 + \frac{1}{8} + \frac{8(7.0 - 61.6/8)^2}{8(493.92) - (61.6)^2}}$$

$$\approx 0.723$$

$\hat{y} + E \to (5.783,\ 7.229)$

35a. $r^2 \approx 0.671$

b. $s_e \approx 1.780$

c. $9.537 < y < 19.010$

37. critical value $= \pm 3.707$

$m \approx -0.205$

$s_e \approx 0.554$

$$t = \frac{m}{s_e}\sqrt{\sum x^2 - \frac{(\sum x)^2}{n}} \approx \frac{-0.205}{0.554}\sqrt{838.55 - \frac{(79.5)^2}{8}} \approx -2.578$$

Fail to reject $H_0: M = 0$.

39. $$E = t_c s_e \sqrt{\frac{1}{n} + \frac{\bar{x}^2}{\sum x^2 - [(\sum x)^2 / n]}} \approx (2.306)(138.255)\sqrt{\frac{1}{10} + \frac{\left(\frac{23.1}{10}\right)^2}{(67.35) - \frac{(23.1)^2}{10}}} \approx 221.216$$

$b \pm E \Rightarrow 102.289 \pm 221.216 = (-118.927,\ 323.505)$

$$E = \frac{t_c s_e}{\sqrt{\sum x^2 - \frac{(\sum x)^2}{n}}} \approx \frac{2.306(138.255)}{\sqrt{67.35 - \frac{(23.1)^2}{10}}} \approx 85.241$$

$m \pm E \Rightarrow 196.152 \pm 85.241 \Rightarrow (110.911,\ 281.393)$

9.4 MULTIPLE REGRESSION

9.4 Try It Yourself Solutions

1a. Enter the data.

b. $\hat{y} = 46.385 + 0.540x_1 - 4.897x_2$

2ab. (1) $\hat{y} = 46.385 + 0.540(89) - 4.897(1)$

(2) $\hat{y} = 46.385 + 0.540(78) - 4.897(3)$

(3) $\hat{y} = 46.385 + 0.540(83) - 4.897(2)$

c. (1) $\hat{y} = 89.548$ (2) $\hat{y} = 73.814$ (3) $\hat{y} = 81.411$

d. (1) 90 (2) 74 (3) 81

9.4 EXERCISE SOLUTIONS

1. $\hat{y} = 61{,}298 + 57.56x_1 - 78.45x_2$

(a) $\hat{y} = 61{,}298 + 57.56(1100) - 78.45(1090) = 39{,}103.5$ pounds per acre

(b) $\hat{y} = 61{,}298 + 57.56(1060) - 78.45(1050) = 39{,}939.1$ pounds per acre

(c) $\hat{y} = 61{,}298 + 57.56(1300) - 78.45(1250) = 38{,}063.5$ pounds per acre

(d) $\hat{y} = 61{,}298 + 57.56(1140) - 78.45(1120) = 39{,}052.4$ pounds per acre

3. $\hat{y} = -52.2 - 0.3x_1 + 4.5x_2$

(a) $\hat{y} = -52.2 - 0.3(70) + 4.5(8.6) = 7.5$ cubic feet

(b) $\hat{y} = -52.2 - 0.3(65) + 4.5(11.0) = 16.8$ cubic feet

(c) $\hat{y} = -52.2 - 0.3(83) + 4.5(17.6) = 51.9$ cubic feet

(d) $\hat{y} = -52.2 - 0.3(87) + 4.5(19.6) = 62.1$ cubic feet

5. $\hat{y} = -2518.364 + 126.822x_1 + 66.360x_2$

(a) $s_e = 28.489$; The standard error of estimate of the predicted sales given a specific total square footage and number of shopping centers is \$28.489 billion.

(b) $r^2 = 0.985$; The multiple regression model explains 98.5% of the variation in y.

7. $\hat{y} = -2518.364 + 126.822x_1 + 66.360x_2$

The equation is the same.

9. $n = 11,\ k = 2,\ r^2 = 0.985$

$$r_{adj}^2 = 1 - \left[\frac{(1-r^2)(n-1)}{n-k-1}\right] \approx 0.981$$

About 98.1% of the variation in y can be explained by the relationships between variables.

$r_{adj}^2 < r^2$

CHAPTER 9 REVIEW EXERCISE SOLUTIONS

1.

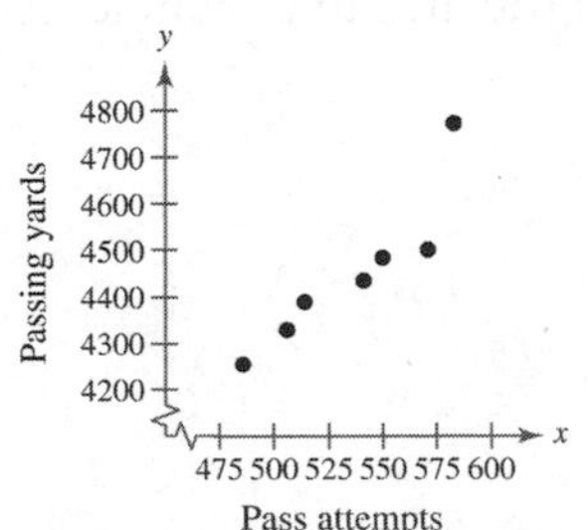

$r \approx 0.912$; strong positive linear correlation; the number of passing yards increases as the number of pass attempts increases.

3.

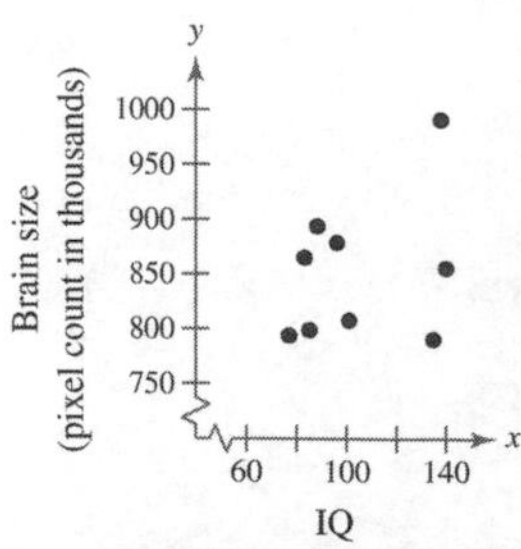

$r \approx 0.338$; weak positive linear correlation; brain size increases as IQ increases.

5. $H_0: \rho = 0;\ H_a: \rho \neq 0$

$\alpha = 0.01$, d.f. $= n - 2 = 24$

$t_0 = 2.797$

$$t = \frac{r}{\sqrt{\frac{1-r^2}{n-2}}} = \frac{0.24}{\sqrt{\frac{1-(0.24)^2}{26-2}}} \approx 1.211$$

Fail to reject H_0. There is not enough evidence at the 1% level of significance to conclude that a significant linear correlation exists.

7. $H_0: \rho = 0;\ H_a: \rho \neq 0$

$\alpha = 0.05$, d.f. $= n - 2 = 5$

$t_0 = \pm 2.571$

$$t = \frac{r}{\sqrt{\frac{1-r^2}{n-2}}} = \frac{0.912}{\sqrt{\frac{1-(0.912)^2}{7-2}}} \approx 4.972$$

Reject H_0. There is enough evidence at the 5% level of significance to conclude that a significant linear correlation exists between passing attempts and passing yards.

9. $H_0: \rho = 0;\ H_a: \rho \neq 0$

$\alpha = 0.01$, d.f. $= n - 2 = 7$

$t_0 = \pm 3.499$

$$t = \frac{r}{\sqrt{\frac{1-r^2}{n-2}}} \approx \frac{0.338}{\sqrt{\frac{1-(0.338)^2}{9-2}}} \approx 0.950$$

Fail to reject H_0. There is not enough evidence at the 1% level of significance to conclude that a significant linear correlation exists between brain size and IQ.

11. $\hat{y} = 0.038x - 3.529$

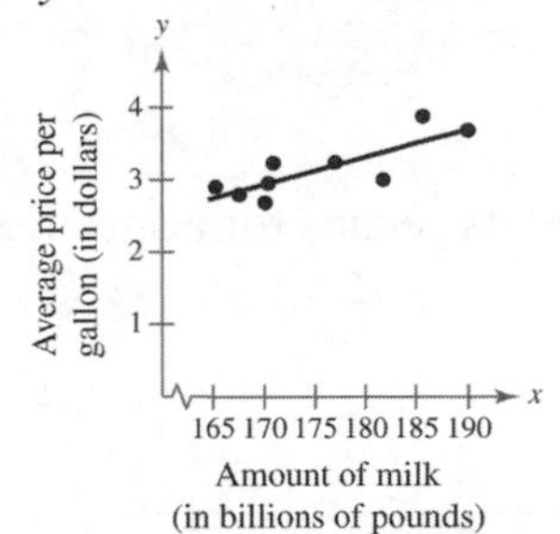

$r \approx 0.821$ (Strong positive linear correlation)

13. $\hat{y} = -0.086x + 10.450$

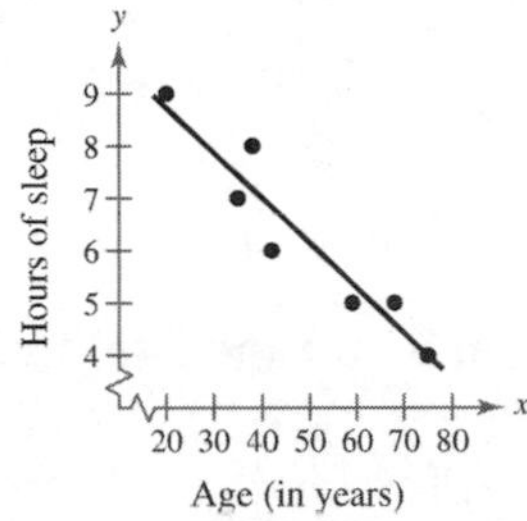

$r \approx -0.949$ (Strong negative linear correlation)

15. (a) It is not meaningful to predict the value of y for $x = 160$ because $x = 160$ is outside the range of the original data.

(b) $\hat{y} = 0.038(175) - 3.529 \approx \$3.12$

(c) $\hat{y} = 0.038(180) - 3.529 \approx \$3.31$

(d) It is not meaningful to predict the value of y for $x = 200$ because $x = 200$ is outside the range of the original data.

17. (a) It is not meaningful to predict the value of y for $x = 18$ because $x = 18$ is outside the range of the original data.

(b) $\hat{y} = 10.450 - 0.086x = 10.450 - 0.086(25) = 8.3$ hours

(c) It is not meaningful to predict the value of y for $x = 85$ because $x = 85$ is outside the range of the original data.

(d) $\hat{y} = 10.450 - 0.086x = 10.450 - 0.086(50) = 6.15$ hours

19. $r^2 = (-0.450)^2 \approx 0.203$

About 20.3% of the variation in y is explained.
About 79.7% of the variation in y is unexplained.

21. $r^2 = (0.642)^2 \approx 0.412$

About 41.2% of the variation in y is explained.
About 58.8% of the variation in y is unexplained.

23. (a) $r^2 \approx 0.679$

About 67.9% of the variation in y is explained.

About 32.1% of the variation in y is unexplained.

(b) $s_e \approx 1.138$

The standard error of estimate of the fuel efficiency of compact sports sedans for a specific price of compact sports sedan is about 1.138 miles per gallon.

25. $\hat{y} = 0.038(185) - 3.529 = 3.501$

$$E = t_c s_e \sqrt{1 + \frac{1}{n} + \frac{n(x-\bar{x})^2}{n(\sum x^2) - (\sum x)^2}} \approx (1.895)(0.246)\sqrt{1 + \frac{1}{9} + \frac{9(185 - 1578.8/9)^2}{9(277{,}555.56) - (1578.8)^2}}$$

$$\approx 0.524$$

$$\hat{y} - E < y < \hat{y} + E$$

$$3.501 - 0.524 < y < 3.501 + 0.524$$

$$2.997 < y < 4.025$$

You can be 90% confident that the price per gallon of milk will be between \$3.00 and \$4.03 when 185 billion pounds of milk is produced.

27. $\hat{y} = -0.086(45) + 10.450 = 6.580$

$$E = t_c s_e \sqrt{1 + \frac{1}{n} + \frac{n(x-\bar{x})^2}{n(\sum x^2) - (\sum x)^2}} \approx (2.571)(0.622)\sqrt{1 + \frac{1}{7} + \frac{7(45 - 337/7)^2}{7(18{,}563) - (337)^2}}$$

$$\approx 1.7127$$

$$\hat{y} - E < y < \hat{y} + E$$

$$6.580 - 1.7127 < y < 6.580 + 1.7127$$

$$4.867 < y < 8.293$$

You can be 95% confident that the hours slept will be between 4.867 and 8.293 hours for a person who is 45 years old.

29. $\hat{y} = -0.414(39.9) + 37.147 \approx 20.628$

$$E = t_c s_e \sqrt{1 + \frac{1}{n} + \frac{n(x-\bar{x})^2}{n(\sum x^2) - (\sum x)^2}} \approx (3.499)(1.138)\sqrt{1 + \frac{1}{9} + \frac{9(39.9 - 319.7/9)^2}{9(11{,}468.29) - (319.7)^2}}$$

$$\approx 4.509$$

$$\hat{y} - E < y < \hat{y} + E$$

$$20.628 - 4.509 < y < 20.628 + 4.509$$

$$16.119 < y < 25.137$$

You can be 99% confident that the fuel efficiency of the compact sports sedan that costs \$39,900 will be between 16.119 and 25.137 miles per gallon.

31. $\hat{y} = 3.674 + 1.287x_1 + (-7.531)x_2$

33. (a) 21.705 (b) 25.21 (c) 30.1 (d) 25.86

CHAPTER 9 QUIZ SOLUTIONS

1.

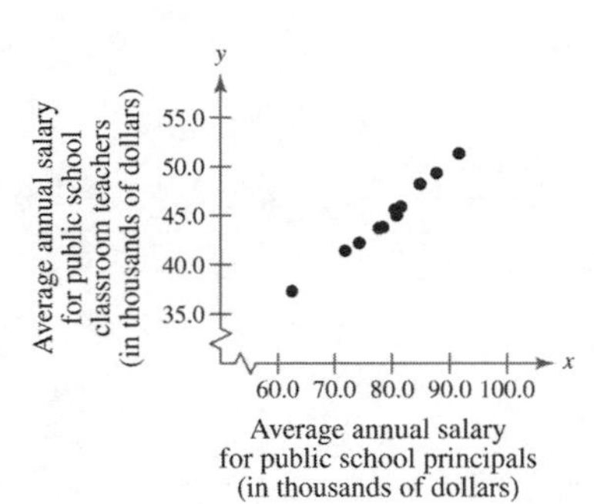

The data appear to have a positive linear correlation. As x increases, y tends to increase.

2. $r \approx 0.993$; Strong positive linear correlation; public school classroom teachers' salaries increase as public school principals' salaries increase.

3. $H_0: \rho = 0$; $H_a: \rho \neq 0$

$\alpha = 0.05$, d.f. $= n - 2 = 9$

$t_0 = \pm 2.262$

$$t = \frac{r}{\sqrt{\frac{1-r^2}{n-2}}} \approx \frac{0.993}{\sqrt{\frac{1-(0.993)^2}{11-2}}} \approx 25.22$$

Reject H_0. There is enough evidence at the 5% level of significance to conclude that a significant correlation exists.

4.

x	y	xy	x^2
62.5	37.3	2331.25	3906.25
71.9	41.4	2976.66	5169.61
74.4	42.2	3139.68	5535.36
77.8	43.7	3399.86	6052.84
78.4	43.8	3433.92	6146.56
80.8	45.0	3636.00	6528.64
80.5	45.6	3670.80	6480.25
81.5	45.9	3740.85	6642.25
84.8	48.2	4087.36	7191.04
87.7	49.3	4323.61	7691.29
91.6	51.3	4699.08	8390.56
$\sum x = 871.9$	$\sum y = 493.7$	$\sum xy = 39.439.08$	$\sum x^2 = 69{,}734.65$

$$m = \frac{n\sum xy - (\sum x)(\sum y)}{n\sum x^2 - (\sum x)^2} = \frac{11(39{,}439.08) - (871.9)(493.7)}{11(69{,}734.65) - (871.9)^2} = \frac{3372.74}{6871.54} \approx 0.490827$$

$$b = \bar{y} - m\bar{x} = \frac{\sum y}{n} - m\left(\frac{\sum x}{n}\right) \approx \frac{493.7}{11} - (0.490827)\left(\frac{871.9}{11}\right) \approx 5.9771$$

$\hat{y} = mx + b = 0.491x + 5.977$

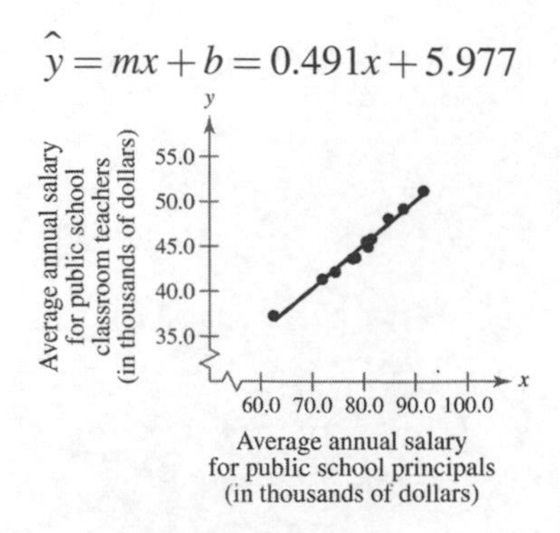

5. $\hat{y} = 0.491(90.5) + 5.977 = 50.4125 \Rightarrow \$50,412.50$

6. $r^2 \approx (0.993)^2 \approx 0.986$

About 98.6% of the variation in the average annual salaries of public school classroom teachers can be explained by the average annual salaries of public school principals, and about 1.4% of the variation is unexplained.

7. $s_e \approx 0.490$; The standard error of the estimate of the average annual salary of public school classroom teachers for a specific average annual salary of public school principals is about $490.

8. $\hat{y} = 0.491(85.75) + 5.977 \approx 48.080$

$$E = t_c s_e \sqrt{1 + \frac{1}{n} + \frac{n(x - \bar{x})^2}{n(\sum x^2) - (\sum x)^2}} \approx (2.262)(0.490)\sqrt{1 + \frac{1}{11} + \frac{11(85.75 - 871.9/11)^2}{11(69,734.65) - (871.9)^2}}$$

$$\approx 1.193$$

$$\hat{y} - E < y < \hat{y} + E$$

$$48.080 - 1.193 < y < 48.080 + 1.193$$

$$46.887 < y < 49.273$$

You can be 95% confident that the average annual salary of public school classroom teacher will be between $46,887 and $49,273 when the average annual salary of public school principals is $85,750.

9. (a) $\hat{y} = -47 + 5.91(22.7) - 1.99(14.0) = \$59.30$

(b) $\hat{y} = -47 + 5.91(17.9) - 1.99(14.2) = \$30.53$

(c) $\hat{y} = -47 + 5.91(20.9) - 1.99(15.5) = \$45.67$

(d) $\hat{y} = -47 + 5.91(19.1) - 1.99(15.1) = \$35.83$

CHAPTER 10

Chi-Square Tests and the *F*-Distribution

10.1 GOODNESS OF FIT TEST

10.1 Try It Yourself Solutions

1.

Tax Preparation Method	% of people	Expected frequency
Accountant	25%	500(0.25) = 125
By hand	20%	500(0.20) = 100
Computer software	35%	500(0.35) = 175
Friend/family	5%	500(0.05) = 25
Tax preparation service	15%	500(0.15) = 75

2a. The expected frequencies are 64, 80, 32, 56, 60, 48, 40, and 20, all of which are at least 5.

b. Claimed distribution:

Ages	Distribution
0–9	16%
10–19	20%
20–29	8%
30–39	14%
40–49	15%
50–59	12%
60–69	10%
70+	5%

H_0: The distribution of ages is as shown in table above.

H_a: The distribution of ages differs from the claimed distribution. (claim)

c. $\alpha = 0.05$ **d.** d.f. $= n - 1 = 7$

e. $\chi_0^2 = 14.067$; ion region: $\chi^2 > 14.067$

f.

Ages	Distribution	Observed	Expected	$\frac{(O-E)^2}{E}$
0–9	16%	76	64	2.250
10–19	20%	84	80	0.200
20–29	8%	30	32	0.125
30–39	14%	60	56	0.286
40–49	15%	54	60	0.600
50–59	12%	40	48	1.333
60–69	10%	42	40	0.100
70+	5%	14	20	1.800

$\chi^2 \approx 6.694$

g. Fail to reject H_0.

h. There is not enough evidence at the 5% level of significance to support the sociologist's claim that the distribution of ages differs from the age distribution 10 years ago.

3a. The expected frequency for each category is 30 which is at least 5.

b. Claimed distribution:

Color	Distribution
Brown	$16.\overline{6}\%$
Yellow	$16.\overline{6}\%$
Red	$16.\overline{6}\%$
Blue	$16.\overline{6}\%$
Orange	$16.\overline{6}\%$
Green	$16.\overline{6}\%$

H_0: The distribution of colors is uniform, as shown in the table above. (claim)

H_a: The distribution of colors is not uniform.

c. $\alpha = 0.05$

d. d.f. $= n - 1 = 5$

e. $\chi_0^2 = 11.071$; Rejection region: $\chi^2 > 11.071$

f.

Color	Distribution	Observed	Expected	$\frac{(O-E)^2}{E}$
Brown	16.6%	22	30	$2.1\overline{33}$
Yellow	16.6%	27	30	0.300
Red	16.6%	22	30	$2.1\overline{33}$
Blue	16.6%	41	30	$4.0\overline{33}$
Orange	16.6%	41	30	$4.0\overline{33}$
Green	16.6%	27	30	0.300
				12.933

$\chi^2 \approx 12.933$

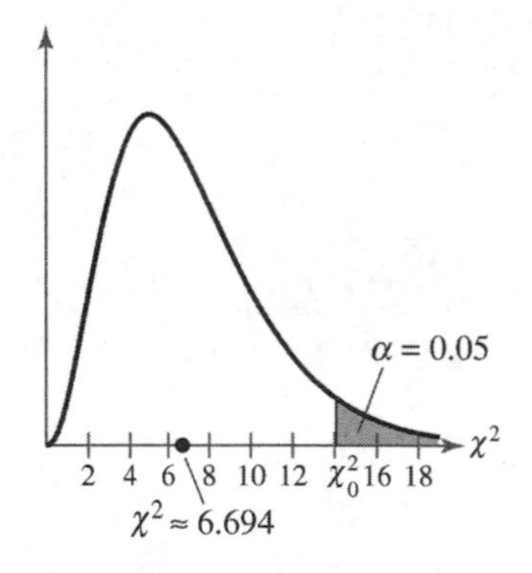

g. Reject H_0.

h. There is enough evidence at the 5% level of significance to dispute the claim that the distribution of different colored candies in bags of peanut M&M's is uniform.

10.1 EXERCISE SOLUTIONS

1. A multinomial experiment is a probability experiment consisting of a fixed number of independent trials in which there are more than two possible outcomes for each trial. The probability of each outcome is fixed, and each outcome is classified into categories.

3. $E_1 = np_1 = (150)(0.3) = 45$

5. $E_1 = np_1 = (230)(0.25) = 57.5$

7a. Claimed Distribution:

Age	Distribution
2–17	26.7%
18–24	19.8%
25–39	19.7%
40–49	14%
50+	19.8%

H_0: The distribution of the ages of moviegoers is 26.7% ages 2–17, 19.8% ages 18–24, 19.7% ages 25–39, 14% ages 40–49, and 19.8% ages 50+ (claim)

H_a: The distribution of the ages differs from the claimed or expected distribution.

b. $\chi_0^2 = 7.779$, Rejection region: $\chi^2 > 7.779$

c.

Age	Distribution	Observed	Expected	$\frac{(O-E)^2}{E}$
2–17	26.7%	240	267	2.730
18–24	19.8%	214	198	1.293
25–39	19.7%	183	197	0.995
40–49	14%	156	140	1.829
50+	19.8%	207	198	0.409
				7.256

$\chi^2 \approx 7.256$

d. Fail to reject H_0.

e. There is not enough evidence at the 10% level of significance to conclude that the distribution of the ages of movie goers has changed.

9a. Claimed distribution:

Day	Distribution
Sunday	7%
Monday	4%
Tuesday	6%
Wednesday	13%
Thursday	10%
Friday	36%
Saturday	24%

H_0 : The distribution of the days people order food for delivery is 7% Sunday, 4% Monday, 6% Tuesday, 13% Wednesday, 10% Thursday, 36% Friday, and 24% Saturday.

H_a : The distribution of days differs from the claimed or expected distribution.

b. $\chi_0^2 = 16.812$, Rejection region: $\chi^2 > 16.812$

c.

	Distribution	Observed	Expected	$\frac{(O-E)^2}{E}$
Sunday	7%	43	35	1.829
Monday	4%	16	20	0.800
Tuesday	6%	25	30	0.833
Wednesday	13%	49	65	3.938
Thursday	10%	46	50	0.320
Friday	36%	168	180	0.800
Saturday	24%	153	120	9.075
				17.595

$\chi^2 \approx 17.595$

d. Reject H_0.

e. There is enough evidence at the 1% level of significance to conclude that the distribution of delivery days has changed.

11a. Claimed distribution:

Season	Distribution
Spring	25%
Summer	25%
Fall	25%
Winter	25%

H_0 : The number of homicide crimes in California by season is uniform.

H_a : The number of homicide crimes in California by season is not uniform. (claim)

b. $\chi_0^2 = 7.815$, Rejection region: $\chi^2 > 7.815$

c.

Season	Distribution	Observed	Expected	$\frac{(O-E)^2}{E}$
Spring	25%	312	300	0.480
Summer	25%	299	300	$0.00\overline{3}$
Fall	25%	297	300	0.030
Winter	25%	292	300	$0.21\overline{3}$
				0.727

$\chi^2 \approx 0.727$

d. Fail to reject H_0.

e. There is not enough evidence at the 5% level of significance to reject the claim that the distribution of the number of homicide crimes in California by season is uniform.

13a. Claimed distribution:

Month	Distribution
Strongly agree	55%
Somewhat agree	30%
Neither agree or disagree	5%
Somewhat disagree	6%
Strongly disagree	4%

H_0: The distribution of the opinions of U.S. parents on whether a college education is worth the expense is 55% strongly agree, 70% somewhat agree, 5% neither agree or disagree, 6% somewhat disagree, 4% strongly disagree.

H_a: The distribution of opinions differs from the claimed or expected distribution. (claim)

b. $\chi_0^2 = 9.488$, Rejection region: $\chi^2 > 9.488$

c.

Month	Distribution	Observed	Expected	$\frac{(O-E)^2}{E}$
Strongly agree	55%	86	110	5.236
Somewhat agree	30%	62	60	0.067
Neither agree or disagree	5%	34	10	57.600
Somewhat disagree	6%	14	12	0.333
Strongly disagree	4%	4	8	2.000
				65.236

$\chi^2 \approx 65.236$

d. Reject H_0.

e. There is enough evidence at the 5% level of significance to conclude that the distribution of opinions of U.S. parents on whether a college education is worth the expense differs from the claimed or expected distribution.

15a. Claimed distribution:

Response	Distribution
Larger	33.3%
Same size	$33.\overline{3}\%$
Smaller	$33.\overline{3}\%$

H_0: The distribution of prospective home buyers by the size they want their next house to be is uniform. (claim)

H_a: The distribution of prospective home buyers by the size they want their next house to be is not uniform. (claim)

b. $\chi_0^2 = 5.991$, Rejection region: $\chi^2 > 5.991$

c.

Response	Distribution	Observed	Expected	$\frac{(O-E)^2}{E}$
Larger	33.3%	285	$266.\overline{66}$	1.260
Same size	$33.\overline{3}\%$	224	$266.\overline{66}$	6.827
Smaller	$33.\overline{3}\%$	291	$266.\overline{66}$	2.220
				10.308

$\chi^2 \approx 10.308$

d. Reject H_0.

e. There is enough evidence at the 5% level of significance to reject the claim that the distribution of prospective home buyers by the size they want their next house to be is uniform.

17. Chi-Square goodness-of-fit results:
Observed: Recent survey
Expected: Previous survey

N	DF	Chi-Square	P-Value
400	9	18.637629	0.0285

Reject H_0. There is enough evidence at the 10% level of significance to conclude that there has been a change in the claimed or expected distribution of U.S. adults' favorite sports.

19a. Frequency distribution: $\mu = 69.435$; $\sigma \approx 8.337$

Lower Boundary	Upper Boundary	Lower *z*-score	Upper *z*-score	Area
49.5	58.5	−2.39	−1.31	0.0867
58.5	67.5	−1.31	−0.23	0.3139
67.5	76.5	−0.23	0.85	0.3933
76.5	85.5	0.85	1.93	0.1709
85.5	94.5	1.93	3.01	0.0255

Class Boundaries	Distribution	Frequency	Expected	$\frac{(O-E)^2}{E}$
49.5–58.5	8.67%	19	17	0.2353
58.5–67.5	31.39%	61	63	0.0635
67.5–76.5	39.33%	82	79	0.1139
76.5–85.5	17.09%	34	34	0
85.5–94.5	2.55%	4	5	0.2000
		200		0.613

H_0: Test scores have a normal distribution. (claim)

H_a: Test scores do not have a normal distribution.

b. $\chi_0^2 = 13.277$; Rejection region . $\chi^2 > 13.277$

c. $\chi^2 = 0.613$

d. Fail to reject H_0.

e. There is not enough evidence at the 1% level of significance to reject the claim that the test scores are normally distributed.

10.2 INDEPENDENCE

10.2 Try It Yourself Solutions

1a.

	Hotel	Leg Room	Rental Size	Other	Total
Business	36	108	14	22	180
Leisure	38	54	14	14	120
Total	74	162	28	36	300

b. $n = 300$

c.

	Hotel	Leg Room	Rental Size	Other
Business	44.4	97.2	16.8	21.6
Leisure	29.6	64.8	11.2	14.4

2a. The claim is that "the travel concerns depend on the purpose of travel."

H_0 : Travel concern is independent of travel purpose.

H_a : Travel concern is dependent on travel purpose. (claim)

b. $\alpha = 0.01$ **c.** $(r-1)(c-1) = 3$

d. $\chi_0^2 = 11.345$; Rejection region: $\chi^2 > 11.345$

e.

O	E	$O-E$	$(O-E)^2$	$\frac{(O-E)^2}{E}$
36	44.4	−8.4	70.56	1.5892
108	97.2	10.8	116.64	1.2000
14	16.8	2.8	7.84	0.4667
22	21.6	0.4	0.16	0.0074
38	29.6	8.4	70.56	2.3838
54	64.8	−10.8	116.64	1.8000
14	11.2	2.8	7.84	0.7000
14	14.4	−0.4	0.16	0.0111
				8.1582

$\chi^2 \approx 8.158$

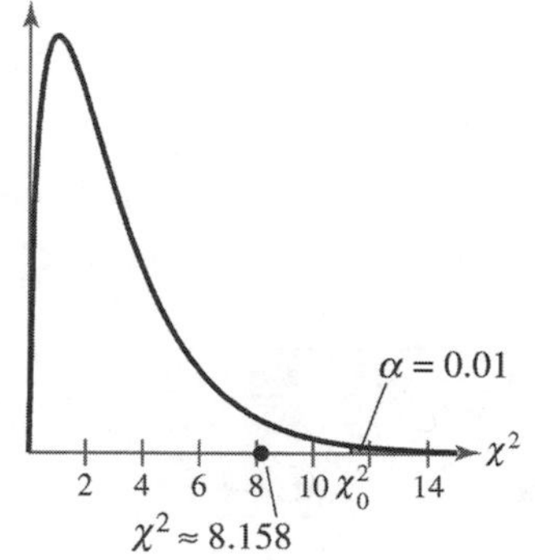

f. Fail to reject H_0.

g. There is not enough evidence to conclude that travel concern is dependent on travel purpose.

3a. H_0 : Whether or not a tax cut would influence an adult to purchase a hybrid vehicle is independent of age.

H_a : Whether or not a tax cut would influence an adult to purchase a hybrid vehicle is dependent on age. (claim)

b. Enter the data.

c. $\chi_0^2 = 9.210$; Rejection region: $\chi^2 > 9.210$

d. $\chi^2 \approx 15.306$

e. Reject H_0.

f. There is enough evidence at the 1% level of significance to conclude that whether or not a tax cut would influence an adult to purchase a hybrid vehicle is dependent on age.

10.2 EXERCISE SOLUTIONS

1. Find the sum of the row and the sum of the column in which the cell is located. Find the product of these sums. Divide the product by the sample size.

3. Answer will vary. *Sample answer:* For both the chi-square test for independence and the chi-square goodness-of-fit test, you are testing a claim about data that are in categories. However, the chi-square goodness-of-fit test has only one data value per category, while the chi-square test for independence has multiple data values per category.

Both tests compare observed and expected frequencies. However, the chi-square goodness-of-fit test simply compares the distributions, whereas the chi-square test for independence compares them and then draws a conclusion about the dependence or independence of the variables.

5. False. If the two variables of a chi-square test for independence are dependent, then you can expect a large difference between the observed frequencies and the expected frequencies.

7a.

	Athlete has		
Result	**Stretched**	**Not stretched**	**Total**
Inquiry	18	22	40
No Inquiry	211	189	400
	229	211	440

b.

	Athlete has	
Result	**Stretched**	**Not stretched**
Inquiry	20.82	19.18
No Inquiry	208.18	191.82

9a.

	Preference			
Bank employee	**New procedure**	**Old Procedure**	**No preference**	**Total**
Teller	92	351	50	493
Customer service	76	42	8	126
Total	168	393	58	619

b.

	Preference		
Bank employee	**New procedure**	**Old Procedure**	**No preference**
Teller	133.80	313.00	46.19
Customer service	34.20	80.00	11.81

11a.

	Type of car				
Gender	**Compact**	**Full-size**	**SUV**	**Truck/Van**	**Total**
Male	28	39	21	22	110
Female	24	32	20	14	90
	52	71	41	36	200

b.

	Type of car			
Gender	**Compact**	**Full-size**	**SUV**	**Truck/Van**
Male	28.6	39.05	22.55	19.8
Female	23.4	31.95	18.45	16.2

13a. The claim is "achieving a basic skill level is related to the location of the school."

H_0 : Skill level in a subject is independent of location. (claim)

H_a : Skill level in a subject is dependent on location.

b. d.f. $= (r-1)(c-1) = 2$

$\chi_0^2 = 9.210$; Rejection region: $\chi^2 > 9.210$

c.

O	E	$O-E$	$(O-E)^2$	$\frac{(O-E)^2}{E}$
43	41.129	1.871	3.500641	0.0851
42	41.905	0.095	0.009025	0.0002
38	39.965	−1.965	3.861225	0.0966
63	64.871	−1.871	3.500641	0.0540
66	66.095	−0.095	0.009025	0.0001
65	63.035	1.965	3.861225	0.0613
				0.2973

$\chi^2 \approx 0.297$

d. Fail to reject H_0. There is not enough evidence at the 1% level of significance to reject the claim that skill level in a subject is independent of location.

15a. The claim is "the number of times former smokers tried to quit before they were habit-free is related to gender."

H_0 : The number of times former smokers tried to quit is independent of gender.

H_a : The number of times former smokers tried to quit is dependent of gender. (claim)

b. d.f. $= (r-1)(c-1) = 2$

$\chi_0^2 = 5.991$; Rejection region: $\chi^2 = 5.991$

c.

O	E	$O-E$	$(O-E)^2$	$\frac{(O-E)^2}{E}$
271	270.930	0.070	0.004900	0
257	257.286	−0.286	0.081796	0.0003
149	148.784	0.216	0.046656	0.0003
146	146.070	−0.070	0.004900	0
139	138.714	0.286	0.081796	0.0006
80	80.216	−0.216	0.046656	0.0006
				0.0018

$\chi^2 \approx 0.002$

d. Fail to reject H_0. There is not enough evidence at the 5% level of significance to conclude that the number of times former smokers tried to quit is dependent on gender.

17a. The claim is "the treatment. is related to the result"

H_0: Results are independent of the type of treatment.

H_a: Results are dependent on the type of treatment. (claim)

b. d.f. $= (r-1)(c-1) = 1$

$\chi_0^2 = 2.706$; Rejection region: $\chi^2 > 2.706$

c.

O	E	$O-E$	$(O-E)^2$	$\frac{(O-E)^2}{E}$
39	31.660	7.34	53.8756	1.7017
25	32.340	−7.34	53.8756	1.6659
54	61.340	−7.34	53.8756	0.8783
70	62.660	7.34	53.8756	0.8598
				5.1057

$\chi^2 \approx 5.106$

d. Reject H_0. There is enough evidence at the 10% level of significance to conclude that results are dependent on the type of treatment.

Sample answer: Because the results are significant at the 10% level of significance, you may consider recommending the drug. However, you may want to perform more tests before recommending the drug.

19a. The claim is "the reason and the type of worker are dependent."

H_0: Reasons are independent of the type of worker.

H_a: Reasons are dependent on the type of worker. (claim)

b. d.f. $= (r-1)(c-1) = 2$

$\chi_0^2 = 9.210$; Rejection region: $\chi^2 = 9.210$

c.

O	E	$O-E$	$(O-E)^2$	$\frac{(O-E)^2}{E}$
30	39.421	−9.421	88.755241	2.2515
36	31.230	4.770	22.752900	0.7286
41	36.349	4.651	21.631801	0.5951
47	37.579	9.421	88.755241	2.3618
25	29.770	−4.770	22.752900	0.7643
30	34.651	−4.651	21.631801	0.6243
				7.3256

$\chi^2 \approx 7.326$

d. Fail to reject H_0. There is not enough evidence at the 1% level of significance to conclude that the reason(s) for continuing education are dependent on the type or worker. Based on these results, marketing strategies should not differ between technical and non-technical audiences in regard to reason(s) for continuing education.

21a. The claim is "the type of crash depends on the type of vehicle."

H_0 : The type of crash is independent of the type of vehicle.

H_a : The type of crash is dependent on the type of vehicle. (claim)

b. d.f. $= (r-1)(c-1) = 2$

$\chi_0^2 = 5.991$; Rejection region: $\chi^2 > 5.991$

c.

O	E	$O-E$	$(O-E)^2$	$\frac{(O-E)^2}{E}$
1,237	1425.637	−188.637	35,583.917770	24.9600
547	452.600	94.400	8,911.360000	19.6893
479	384.763	94.237	8,880.612169	23.0807
1,453	1264.363	188.637	35,583.917770	28.1438
307	401.400	−94.400	8,911.360000	22.2007
247	341.237	−94.237	8,800.612169	26.0248
				144.0990

$\chi^2 \approx 144.099$

d. Reject H_0. There is enough evidence at the 5% level of significance to conclude that the type of crash is dependent on the type of vehicle.

23a–b. Contingency table results:

Rows: Expected income
Columns: None

Cell format
Count Expected count

	More Likely	Less Likely	Did not make a difference
Less than $35,000	37 33.33	10 6.836	22 27.99
35,000 to $50,000	28 25.17	12 5.164	15 21.14
$50,000 to $100,000	55 62.75	9 12.87	65 52.7
Greater than $100,000	36 34.75	1 7.127	29 29.18
Total	156	32	131

	Did not consider it	Total
Less than $35,000	25 25.85	94
35,000 to $50,000	16 19.52	71
$50,000 to $100,000	48 48.68	177
Greater than $100,000	32 26.95	98
Total	121	440

Statistic	DF	Value	P-value
Chi-square	9	26.22966	0.0019

c. Reject H_0. There is enough evidence at the 1% level of significance to conclude that the decision to borrow money is dependent on the child's expected income after graduation.

25. The claim is "the proportions of motor vehicle crash deaths involving males or females are the same for each age group."

H_0 : The proportions are equal. (claim)

H_a : At least one of the proportions is different from the others.

d.f. = $(r-1)(c-1) = 7$

$\chi_0^2 = 14.067$; Rejection region: $\chi^2 > 14.067$

O	E	$O-E$	$(O-E)^2$	$\frac{(O-E)^2}{E}$
123	121.632	1.368	1.871424	0.01539
97	89.965	7.035	49.491225	0.55012
82	79.169	2.831	8.014561	0.10123
82	82.048	−0.048	0.002304	0.00003
56	56.138	−0.138	0.109044	0.00034
31	35.266	−4.266	18.198756	0.51604
26	31.668	−5.668	32.126224	1.01447
14	15.114	−1.114	1.240996	0.08211
46	47.368	−1.368	1.871424	0.03951
28	35.035	−7.035	49.491225	1.41262
28	30.831	−2.831	8.014561	0.25995
32	31.952	0.048	0.002304	0.00007
22	21.862	0.138	0.019044	0.00087
18	13.734	4.266	18.198756	1.32509
18	12.332	5.668	32.126224	2.60511
7	5.886	1.114	1.240996	0.21084
				8.13379

$\chi^2 \approx 8.134$

Fail to reject H_0. There is not enough evidence at the 5% level of significance to reject the claim that the proportions of motor vehicle crash deaths involving males or females are the same for each age group.

27. Right-tailed

29.

	Educational attainment			
Status	**Not a high school graduate**	**High school graduate**	**Some college, no degree**	**Associate's, bachelor's, or advanced degree**
Employed	0.055	0.183	0.114	0.290
Unemployed	0.006	0.011	0.005	0.007
Not in the labor force	0.073	0.118	0.053	0.085

31. Several of the expected frequencies are less than 5.

33. 45.2%

35.

	Educational attainment			
Status	**Not a high school graduate**	**High school graduate**	**Some college, no degree**	**Associate's, bachelor's, or advanced degree**
Employed	0.411	0.587	0.660	0.759
Unemployed	0.046	0.036	0.030	0.019
Not in the labor force	0.544	0.377	0.311	0.223

37. 4.6%

39. Answers will vary. *Sample answer:* As educational attainment increases, employment increases.

10.3 COMPARING TWO VARIANCES

10.3 Try It Yourself Solutions

1a. $\alpha = 0.05$ **b.** $F_0 = 2.45$

2a. $\alpha = 0.01$ **b.** $F_0 = 18.31$

3a. The claim is "the variance of the time required for nutrients to enter the bloodstream is less with the specially treated intravenous solution than the variance of the time without the solution."
$H_0: \sigma_1^2 \leq \sigma_2^2$; $H_a: \sigma_1^2 > \sigma_2^2$ (claim)

b. $\alpha = 0.01$

c. $\text{d.f.}_N = n_1 - 1 = 24$
$\text{d.f.}_D = n_2 - 1 = 19$

d. $F_0 = 2.92$; Rejection region: $F > 2.92$

e. $F = \frac{s_1^2}{s_2^2} = \frac{180}{56} \approx 3.21$

f. Reject H_0.

g. There is enough evidence at the 1% level of significance to support the researcher's claim that a specially treated intravenous solution decreases the variance of the time required for nutrients to enter the bloodstream.

4a. The claim is "the pH levels of the soil in two geographic regions have equal standard deviations."

b. $\alpha = .01$

c. $\text{d.f.}_N = n_1 - 1 = 15$
$\text{d.f.}_D = n_2 - 1 = 21$

d. $F_0 = 3.43$; Rejection region: $F > 3.43$

e. $F = \frac{s_1^2}{s_2^2} = \frac{(0.95)^2}{(0.78)^2} \approx 1.48$

f. Fail to reject H_0.

g. There is not enough evidence at the 1% level of significance to reject the biologist's claim that the pH levels of the soil in the two geographic regions have equal standard deviations.

10.3 EXERCISE SOLUTIONS

1. Specify the level of significance α. Determine the degrees of freedom for the numerator and denominator. Use Table 7 in Appendix B to find the critical value *F*.

3. (1) The samples must be randomly selected, (2) the samples must be independent, and (3) each population must have a normal distribution.

5. $F_0 = 2.54$ **7.** $F_0 = 2.06$ **9.** $F_0 = 9.16$ **11.** $F_0 = 1.80$

13. $H_0: \sigma_1^2 \le \sigma_2^2$; $H_a: \sigma_1^2 > \sigma_2^2$ (claim)

$\text{d.f.}_N = 4$

$\text{d.f.}_D = 5$

$F_0 = 3.52$; Rejection region: $F > 3.52$

$F = \frac{s_1^2}{s_2^2} = \frac{773}{765} \approx 1.010$

Fail to reject H_0. There is not enough evidence at the 10% level of significance to support the claim.

15. $H_0: \sigma_1^2 \le \sigma_2^2$ (claim); $H_a: \sigma_1^2 > \sigma_2^2$

$\text{d.f.}_N = 10$

$\text{d.f.}_D = 9$

$F_0 = 5.26$; Rejection region: $F > 5.26$

$F = \frac{s_1^2}{s_2^2} = \frac{842}{836} \approx 1.007$

Fail to reject H_0. There is not enough evidence at the 1% level of significance to reject the claim.

17. $H_0: \sigma_1^2 = \sigma_2^2$ (claim); $H_a: \sigma_1^2 \ne \sigma_2^2$

$\text{d.f.}_N = 12$

$\text{d.f.}_D = 19$

$F_0 = 3.76$; Rejection region: $F > 3.76$

$F = \frac{s_1^2}{s_2^2} = \frac{9.8}{2.5} \approx 3.920$

Reject H_0. There is enough evidence at the 1% level of significance to reject the claim.

19. Population 1: Company B
Population 2: Company A

a. The claim is "the variance of the life of Company A's appliances is less than the variance of the life of Company B's appliances."

$H_0: \sigma_1^2 \le \sigma_2^2$; $H_a: \sigma_1^2 > \sigma_2^2$ (claim)

b. $\text{d.f.}_N = 24$

$\text{d.f.}_D = 19$

$F_0 = 2.11$; Rejection region: $F > 2.11$

c. $F = \dfrac{s_1^2}{s_2^2} = \dfrac{2.8}{2.6} \approx 1.08$

d. Fail to reject H_0.

e. There is not enough evidence at the 5% level of significance to support Company A's claim that the variance of life of its appliances is less than the variance of life of Company B appliances.

21. Population 1: Company B
Population 2: Company A

a. The claim is "the variance of the prices differ between the two companies."

$H_0: \sigma_1^2 = \sigma_2^2$; $H_a: \sigma_1^2 \ne \sigma_2^2$ (claim)

b. $\text{d.f.}_N = 4$

$\text{d.f.}_D = 6$

$F_0 = 6.23$; Rejection region: $F > 6.23$

c. $F = \dfrac{s_1^2}{s_2^2} = \dfrac{30,445}{14,490.48} \approx 2.10$

d. Fail to reject H_0.

e. There is not enough evidence at the 5% level of significance to conclude that the variances of the prices differ between the two companies.

23. Population 1: District 1
Population 2: District 2

a. The claim is "the standard deviations of science achievement test scores for eighth grade students are the same in Districts 1 and 2."

$H_0: \sigma_1^2 = \sigma_2^2$(claim); $H_a: \sigma_1^2 \ne \sigma_2^2$

b. $\text{d.f.}_N = 11$

$\text{d.f.}_D = 13$

$F_0 = 2.635$; Rejection region: $F > 2.635$

c. $F = \dfrac{s_1^2}{s_2^2} = \dfrac{(36.8)^2}{(32.5)^2} \approx 1.282$

d. Fail to reject H_0.

e. There is not enough evidence at the 10% level of significance to reject the administrator's claim that the standard deviations of science assessment test scores for eighth grade students are the same in Districts 1 and 2.

25. Population 1: New York
Population 2: California

a. The claim is "the standard deviation of the annual salaries for actuaries is greater in New York than in California."

$H_0 : \sigma_1^2 \le \sigma_2^2;\ H_a : \sigma_1^2 > \sigma_2^2$ (claim)

b. $\text{d.f.}_N = 15$

$\text{d.f.}_D = 16$

$F_0 = 2.35$; Rejection region: $F > 2.35$

c. $F = \dfrac{s_1^2}{s_2^2} = \dfrac{(14{,}900)^2}{(9{,}600)^2} \approx 2.41$

d. Reject H_0.

e. There is enough evidence at the 5% level of significance to conclude the standard deviation of the annual salaries for actuaries is greater in New York than in California.

27. Hypothesis test results:

σ_1^2 : variance of population 1

σ_2^2 : variance of population 2

σ_1^2 / σ_2^2 : variance ratio

$H_0 : \sigma_1^2 / \sigma_2^2 = 1$

$H_A : \sigma_1^2 / \sigma_2^2 \ne 1$

Ratio	n1	n2	Sample Ratio	F-Stat	P-value
σ_1^2 / σ_2^2	15	18	0.5281571	0.5281571	0.2333

$P \approx 0.2333 > 0.10$, so fail to reject H_0. There is not enough evidence at the 10% level of significance to reject the claim.

29. Hypothesis test results:

σ_1^2 : variance of population 1

σ_2^2 : variance of population 2

σ_1^2 / σ_2^2 : variance ratio

$H_0 : \sigma_1^2 / \sigma_2^2 = 1$

$H_A : \sigma_1^2 / \sigma_2^2 > 1$

Ratio	n1	n2	Sample Ratio	F-Stat	P-value
σ_1^2 / σ_2^2	22	29	2.153926	2.153926	0.0293

$P \approx 0.0293 < 0.05$, so reject H_0. There is enough evidence at the 5% level of significance to reject the claim.

31. Right-tailed: $F_R = 14.73$

Left-tailed:

(1) $\text{d.f.}_N = 3$ and $\text{d.f.}_D = 6$

(2) $F = 6.60$

(3) Critical value is $\frac{1}{F} = \frac{1}{6.60} \approx 0.15$.

33. $\frac{s_1^2}{s_2^2} F_L < \frac{\sigma_1^2}{\sigma_2^2} < \frac{s_1^2}{s_2^2} F_R \rightarrow \left(\frac{10.89}{9.61}\right)(0.331) < \frac{\sigma_1^2}{\sigma_2^2} < \left(\frac{10.89}{9.61}\right)(3.33) \rightarrow 0.375 < \frac{\sigma_1^2}{\sigma_2^2} < 3.774$

10.4 ANALYSIS OF VARIANCE

10.4 Try It Yourself Solutions

1a. The claim is "there is a difference in the mean a monthly sales among the sales regions."

$H_0: \mu_1 = \mu_2 = \mu_3 = \mu_4$

H_a: At least one mean is different from the others. (claim)

b. $\alpha = 0.05$

c. $\text{d.f.}_N = 3$

$\text{d.f.}_D = 14$

d. $F_0 = 3.34$; Rejection region: $F > 3.34$

e.

Variation	Sum of Squares	Degrees of Freedom	Mean Squares	*F*
Between	549.8	3	183.3	4.22
Within	608.0	14	43.4	

$F \approx 4.22$

f. Reject H_0.

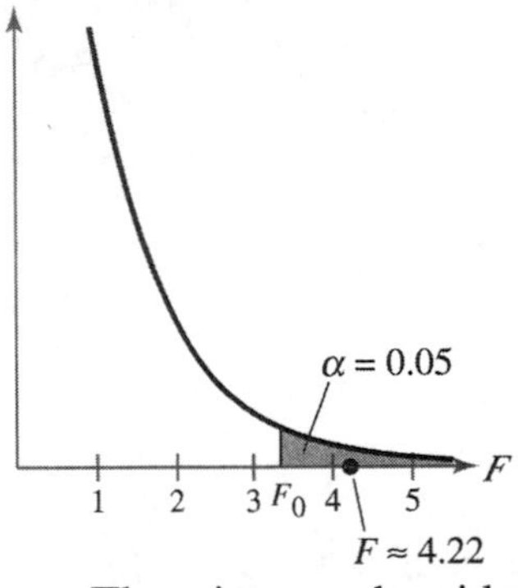

g. There is enough evidence at the 5% level of significance to conclude that there is a difference in the mean monthly sales among the sales regions.

2a. Enter the data.

b. $H_0: \mu_1 = \mu_2 = \mu_3 = \mu_4$

H_a: At least one mean is different from the others. (claim)

Variation	Sum of Squares	Degrees of Freedom	Mean Squares	F
Between	0.584	3	0.195	1.34
Within	4.360	30	0.145	

$F = 1.34 \rightarrow P\text{-value} = 0.280$

c. $0.280 > 0.05$

d. Fail to reject H_0. There is not enough evidence to conclude that at least one mean is different from the others.

10.4 EXERCISE SOLUTIONS

1. $H_0: \mu_1 = \mu_2 = \ldots = \mu_4$

H_a: At least one mean is different from the others.

3. The MS_B measures the differences related to the treatment given to each sample.

The MS_W measures the differences related to entries within the same sample.

5a. The claim is that "the mean costs per ounce are different."

$H_0: \mu_1 = \mu_2 = \mu_3$

H_a: At least one mean is different from the others. (claim)

b. $\text{d.f.}_N = k - 1 = 2$

$\text{d.f.}_D = N - k = 26$

$F_0 = 3.37$; Rejection region: $F > 3.37$

c.

Variation	Sum of Squares	Degrees of Freedom	Mean Squares	F
Between	0.518	2	0.259	1.02
Within	6.629	26	0.255	

$F \approx 1.02$

d. Fail to reject H_0.

e. There is not enough evidence at the 5% level of significance to conclude that the mean costs per ounce are different.

7a. The claim is "at least one mean salary is different."

$H_0: \mu_1 = \mu_2 = \mu_3$

H_a: At least one mean is different from the others. (claim)

b. $\text{d.f.}_N = k - 1 = 2$

$\text{d.f.}_D = N - k = 27$

$F_0 = 5.49$; Rejection region: $F > 5.49$

c.

Variation	Sum of Squares	Degrees of Freedom	Mean Squares	F
Between	3125.521	2	1562.761	21.99
Within	1918.441	27	71.053	

$F \approx 21.99$

d. Reject H_0.

e. There is enough evidence at the 1% level of significance to conclude that at least one mean is different.

9a. The claim is "at least one mean cost is different."

$H_0 : \mu_1 = \mu_2 = \mu_3 = \mu_4 = \mu_5$

H_a : At least one mean is different from the others. (claim)

b. $\text{d.f.}_N = k - 1 = 4$

$\text{d.f.}_D = N - k = 21$

$F_0 = 4.37$; Rejection region: $F > 4.37$

c.

Variation	Sum of Squares	Degrees of Freedom	Mean Squares	F
Between	2841.1	4	710.275	12.61
Within	1182.9	21	56.329	

$F \approx 12.61$

d. Reject H_0.

e. There is enough evidence at the 1% level of significance to conclude that at least one mean cost per miles is different.

11a. The claim is "the mean number of days patients spend at the hospital is the same for all four regions."

$H_0 : \mu_1 = \mu_2 = \mu_3 = \mu_4$ (claim)

H_a : At least one mean is different from the others. (claim)

b. $\text{d.f.}_N = k - 1 = 3$

$\text{d.f.}_D = N - k = 29$

$F_0 = 4.54$; Rejection region: $F > 4.54$

c.

Variation	Sum of Squares	Degrees of Freedom	Mean Squares	F
Between	5.608	3	1.869	0.56
Within	97.302	29	3.355	

$F \approx 0.56$

d. Fail to reject H_0.

e. There is not enough evidence at the 1% level of significance for the company to reject the claim that the mean number of days patients spend at the hospital is the same for all four regions.

13a. The claim is "the mean energy consumption in at least one region is different from the others."

$H_0: \mu_1 = \mu_2 = \mu_3 = \mu_4$

H_a: At least one mean is different from the others. (claim)

b. $\text{d.f.}_N = k - 1 = 3$

$\text{d.f.}_D = N - k = 35$

$F_0 = 2.247$; Rejection region: $F > 2.247$

c.

Variation	Sum of Squares	Degrees of Freedom	Mean Squares	*F*
Between	15,340.657	3	5113.552	3.107
Within	57,609.672	35	1645.991	

$F \approx 3.107$

d. Reject H_0.

e. There is enough evidence at the 10% level of significance to conclude that the mean energy consumption of at least one region is different from the others.

15. Analysis of Variance results:
Data stored in separate columns.

Column means

Column	N	Means	Std. Error
Grade 9	8	84.375	9.531784
Grade 10	8	79.25	9.090321
Grade 11	8	76.625	7.648383
Grade 12	8	70.75	6.9224014

ANOVA table

Source	df	SS	MS	F-stat	P-value
Treatments	3	771.25	257.08334	0.45923114	0.7129
Error	28	15674.675	559.8125		
Total	31	16446			

Fail to reject H_0. There is not enough evidence at the 1% level of significance to reject the claim that the mean numbers of female students who played on a sports team are equal for all grades.

17. H_0: Advertising medium has to effect on mean ratings.

H_a: Advertising medium has an effect on mean ratings.

H_0: Length of ad has no effect on mean ratings.

H_a: Length of ad has an effect on mean ratings.

H_0: There is no interaction effect between advertising medium and length of ad on mean ratings.

H_a: There is an interaction effect between advertising medium and length of ad on mean ratings.

Source	d.f.	SS	MS	F	P
Ad medium	1	1.25	1.25	0.57	0.459
Length of ad	1	0.45	0.45	0.21	0.655
Interaction	1	0.45	0.45	0.21	0.655
Error	16	34.80	2.17		
Total	19	36.95			

None of the null hypotheses can be rejected at the 10% level of significance.

19. H_0 : Age has no effect on mean GPA.
H_a : Age has an effect on mean GPA.

H_0 : Gender has no effect on mean GPA.
H_a :Gender has an effect on mean GPA.

H_0 : There is no interaction effect between age and gender on mean GPA.
H_a : There is no interaction effect between age and gender on mean GPA.

Source	d.f.	SS	MS	F	P
Age	3	0.41	0.14	0.12	0.948
Gender	1	00.18	0.18	0.16	0.697
Interaction	3	0.29	0.10	0.08	0.968
Error	16	18.66	1.17		
Total	23	19.55			

None of the null hypotheses can be rejected at the 10% level of significance.

21.

	Mean	Size
Pop 1	66.36	10
Pop2	48.52	10
Pop3	42.27	10

$SS_W = 1918.441$

$\sum (n_i - 1) = N - k = 27$

$F_0 = 5.49 \rightarrow \text{CV}_{\text{Scheffe}'} = 5.49(3-1) = 10.98$

$$\frac{(\bar{x}_1 - \bar{x}_2)^2}{\frac{SS_W}{\sum (n_i - 1)}\left[\frac{1}{n_1} + \frac{1}{n_2}\right]} \approx 23.396 \rightarrow \text{ Significant difference}$$

$$\frac{(\bar{x}_1 - \bar{x}_3)^2}{\frac{SS_W}{\sum (n_i - 1)}\left[\frac{1}{n_1} + \frac{1}{n_3}\right]} \approx 40.837 \rightarrow \text{Significant difference}$$

$$\frac{(\overline{x}_2 - \overline{x}_3)^2}{\frac{SS_W}{\sum(n_i - 1)}\left[\frac{1}{n_2} + \frac{1}{n_3}\right]} \approx 2.749 \rightarrow \text{No difference}$$

23.

	Mean	Size
Pop 1	$65.9\overline{3}$	6
Pop2	66.1	8
Pop3	$64.8\overline{63}$	11
Pop 4	$66.9\overline{8}$	9

$SS_W = 89.448$

$\sum(n_i - 1) = N - k = 30$

$F_0 = 2.28 \rightarrow \text{CV}_{\text{Scheffe'}} = 2.28(4 - 1) = 6.84$

$$\frac{(\overline{x}_1 - \overline{x}_2)^2}{\frac{SS_W}{\sum(n_i - 1)}\left[\frac{1}{n_1} + \frac{1}{n_2}\right]} \approx 0.032 \rightarrow \text{No difference}$$

$$\frac{(\overline{x}_1 - \overline{x}_3)^2}{\frac{SS_W}{\sum(n_i - 1)}\left[\frac{1}{n_1} + \frac{1}{n_3}\right]} \approx 1.490 \rightarrow \text{No difference}$$

$$\frac{(\overline{x}_1 - \overline{x}_4)^2}{\frac{SS_W}{\sum(n_i - 1)}\left[\frac{1}{n_1} + \frac{1}{n_4}\right]} \approx 1.345 \rightarrow \text{No difference}$$

$$\frac{(\overline{x}_2 - \overline{x}_3)^2}{\frac{SS_W}{\sum(n_i - 1)}\left[\frac{1}{n_2} + \frac{1}{n_3}\right]} \approx 2.374 \rightarrow \text{No difference}$$

$$\frac{(\overline{x}_2 - \overline{x}_4)^2}{\frac{SS_w}{\sum(n_i - 1)}\left[\frac{1}{n_2} + \frac{1}{n_4}\right]} \approx 1.122 \rightarrow \text{No difference}$$

$$\frac{(\overline{x}_3 - \overline{x}_4)^2}{\frac{SS_w}{\sum(n_i - 1)}\left[\frac{1}{n_3} + \frac{1}{n_4}\right]} \approx 7.499 \rightarrow \text{Significant difference}$$

CHAPTER 10 REVIEW EXERCISE SOLUTIONS

1a.

Response	Distribution
Less than \$10	29%
\$10 to \$20	16%
More than \$21	9%
Don't give one/other	46%

H_0 : The distribution of the allowance amounts is 29% less than \$10, 16% \$10 to \$20, 9% more than \$21, and 46% don't give one/other.

H_a : The distribution of amounts differ from the claimed or expected distribution. (claim)

b. $\chi_0^2 = 6.251$; Rejection region: $\chi^2 > 6.251$

c.

Response	Distribution	Observed	Expected	$\frac{(O-E)^2}{E}$
Less than \$10	29%	353	319.87	3.431
\$10 to \$20	16%	167	176.48	0.509
More than \$21	9%	94	99.27	0.280
Don't give one/other	46%	489	507.38	0.666
				4.886

$\chi^2 \approx 4.886$

d. Fail to reject H_0.

e. There is not enough evidence at the 10% level of significance to conclude that there has been a change in the claimed or expected distribution.

3a.

Response	Distribution
Short-game shots	65%
Approach and swing	22%
Driver shots	9%
Putting	4%

H_0 : The distribution of responses from golf students about what they need the most help with is 22% approach and swing, 9% driver shots, 4% putting and 65% short-game shots. (claim)

H_a : The distribution of responses differs from the claimed or expected distribution.

b. $\chi_0^2 = 7.815$; Rejection region: $\chi^2 > 7.815$

c.

Response	Distribution	Observed	Expected	$\frac{(O-E)^2}{E}$
Short-game shots	65%	276	282.75	0.161
Approach and swing	22%	99	95.70	0.114
Driver shots	9%	42	39.15	0.207
Putting	4%	18	17.40	0.021
				0.503

$\chi^2 \approx 0.503$

d. Fail to reject H_0.

e. There is not enough evidence at the 5% level of significance to conclude that the distribution of golf students' responses is the same as the claimed or expected distribution.

5a. Expected frequencies:

	Years of full-time teaching experience				
Gender	**Less than 3 years**	**3–9 years**	**10–20 years**	**20 years or more**	**Total**
Male	63	356.4	319.8	310.8	1050
Female	147	831.6	746.2	725.2	2450
Total	210	1188	1066	1036	3500

b. H_0: Years of full-time teaching experience is independent of gender.

H_a: Years of full-time teaching experience is dependent on gender.

$\chi_0^2 = 11.345$

O	E	$O-E$	$(O-E)^2$	$\frac{(O-E)^2}{E}$
58	63.0	–5.0	25.00	0.3968
377	356.4	20.6	424.36	1.1907
280	319.8	–39.8	1584.04	4.9532
335	310.8	24.2	585.64	1.8843
152	147	5.0	25.00	0.1701
811	831.6	–20.6	424.36	0.5103
786	746.2	39.8	1584.04	2.1228
701	725.2	–24.2	585.64	0.8076
				12.0358

$\chi^2 \approx 12.036$

Reject H_0.

c. There is enough evidence at the 1% level of significance to conclude that public school teachers' genders and years of full-time teaching experience are related.

7a. Expected frequencies:

	Vertebrae Group					
Status	**Mammals**	**Birds**	**Reptiles**	**Amphibians**	**Fish**	**Total**
Endangered	54.86	40.38	22.10	16.00	26.67	160.01
Threatened	17.14	12.62	6.90	5.00	8.33	49.99
Total	72	53	29	21	35	210

b. H_0: Species' status (endangered or threatened) is independent of vertebrate group.

H_a: Species' status (endangered or threatened) is dependent on vertebrate group.

$\chi_0^2 = 13.277$

O	E	$O-E$	$(O-E)^2$	$\frac{(O-E)^2}{E}$
62	54.857	7.143	51.022449	0.9301
48	40.381	7.619	58.049161	1.4375
17	22.095	5.095	25.959025	1.1749
12	16.000	−4.000	16.000000	1.0000
21	26.667	−5.667	32.114889	1.2043
10	17.143	−7.143	51.022449	2.9763
5	12.619	−7.619	58.049161	4.6001
12	6.905	5.095	25.959025	3.7595
9	5.000	4.000	16.000000	3.2000
14	8.333	5.667	32.114889	3.8539
				24.1366

$\chi^2 \approx 24.154$

Reject H_0.

c. There is enough evidence at the 1% level of significance to conclude that species' status (endangered or threatened) is dependent on vertebrate group.

9. $F_0 \approx 2.286$ **11.** $F_0 = 2.39$ **13.** $F_0 \approx 2.06$ **15.** $F_0 = 2.08$

17. $H_0: \sigma_1^2 \le \sigma_2^2$(claim); $H_a: \sigma_1^2 > \sigma_2^2$

d.f.$_N = 15$

d.f.$_D = 20$

$F_0 = 3.09$; Rejection region: $F > 3.09$

$$F = \frac{s_1^2}{s_2^2} = \frac{653}{270} \approx 2.419$$

Fail to reject H_0. There is not enough evidence at the 1% level of significance to reject the claim.

19. Population 1: Garfield County

Populations 2: Kay County

The claim is "the variation in wheat production is greater in Garfield County than in Kay County."

$H_0: \sigma_1^2 \le \sigma_2^2$; $H_a: \sigma_1^2 > \sigma_2^2$ (claim)

d.f.$_N = 20$

d.f.$_D = 15$

$F_0 = 1.92$; Rejection region: $F > 1.92$

$$F = \frac{s_1^2}{s_2^2} = \frac{(0.76)^2}{(0.58)^2} \approx 1.717$$

Fail to reject H_0. There is not enough evidence at the 10% level of significance to support the claim that the variation in wheat production is greater in Garfield County than in Kay County.

21. Population 1: Male $\rightarrow s_1^2 = 20{,}002.56$

Population 2: Female $\rightarrow s_2^2 = 17{,}136.11$

$H_0: \sigma_1^2 = \sigma_2^2;\ H_a: \sigma_1^2 \neq \sigma_2^2$ (claim)

$\text{d.f.}_N = 12$

$\text{d.f.}_D = 8$

$F_0 = 7.015$; Rejection region: $F > 7.015$

$$F = \frac{s_1^2}{s_2^2} = \frac{20{,}002.56}{17{,}136.11} \approx 1.167$$

Fail to reject H_0. There is not enough evidence at the 1% level of significance to support the claim that the test score variance for females is different than that for males.

23. The claim is "the mean energy consumption in at least one region is different from the others."

$H_0: \mu_1 = \mu_2 = \mu_3 = \mu_4$

H_a: At least one mean is different from the others. (claim)

$\text{d.f.}_N = k - 1 = 3$

$\text{d.f.}_D = N - k = 28$

$F_0 = 2.29$; Rejection region: $F > 2.29$

Variation	Sum of Squares	Degrees of Freedom	Mean Squares	*F*
Between	512.457	3	170.819	8.51
Within	562.162	28	20.077	

$F \approx 8.508$

Reject H_0. There is enough evidence at the 10% level of significance to conclude that at least one of the mean costs is different from the others.

CHAPTER 10 QUIZ SOLUTIONS

1. Population 1: San Francisco $\rightarrow s_1^2 = 88.664$

Population 2: Baltimore $\rightarrow s_2^2 = 41.792$

a. $H_0: \sigma_1^2 = \sigma_2^2;\ H_a: \sigma_1^2 \neq \sigma_2^2$ (claim)

b. $\alpha = 0.01$

c. $\text{d.f.}_N = 12$, $\text{d.f.}_D = 14 \Rightarrow F_0 = 4.43$

d. Rejection region: $F > 4.43$

e. $F = \dfrac{s_1^2}{s_2^2} = \dfrac{88.644}{41.792} \approx 2.12$

f. Fail to reject H_o.

g. There is not enough evidence at the 1% level of significance to conclude the variances in annual wages for San Francisco, CA and Baltimore, MD are different.

2a. $H_0: \mu_1 = \mu_2 = \mu_3 = \mu_4$

H_a: At least one mean is different from the others. (claim)

b. $\alpha = 0.10$

c. $\text{d.f.}_N = 2$, $\text{d.f.}_D = 40 \Rightarrow F_0 = 2.44$

d. Rejection region: $F > 2.44$

e.

Variation	Sum of Squares	Degrees of Freedom	Mean Squares	F
Between	3423.12	2	1711.56	27.48
Within	2491.79	40	62.29	

$F \approx 27.48$

f. Reject H_o.

g. There is enough evidence at the 10% level of significance to reject the claim that the mean annual wages are equal for all three cities.

3a. H_0: The distribution of educational achievement for people in the United States ages 35–44 is 13.4% not a high school graduate, 31.2% high school graduate, 17.2% some college, no degree; 8.8% associate's degree, 19.1% bachelor's degree, and 10.3% advanced degree.

H_a: The distribution of educational achievement for people in the United States ages 35–44 differs from the claimed distribution. (claim)

b. $\alpha = 0.05$

c. $\chi_0^2 = 11.071$

d. Rejection region: $\chi^2 > 11.071$

e.

	Distribution	Observed	Expected	$\frac{(O-E)^2}{E}$
Not a H.S. graduate	13.4%	36	43.014	1.144
H.S. graduate	31.2%	92	100.152	0.664
Some college no degree	17.2%	55	55.212	0.001
Associate's degree	8.8%	32	28.248	0.498
Bachelor's degree	19.1%	70	61.311	1.231
Advanced degree	10.3%	36	33.063	0.261
		321		3.799

$\chi^2 \approx 3.799$

f. Fail to reject H_0.

g. There is not enough evidence at the 5% level of significance to conclude that the distribution for people in the United States ages 35–44 differs from the distribution for ages 25 and older.

4a. H_0: The distribution of educational achievement for people in the United States ages 65–74 is 13.4% not a high school graduate, 31.2% high school graduate, 17.2% some college, no degree; 8.8% associate's degree, 19.1% bachelor's degree, and 19.1% advanced degree.

H_a: The distribution of educational achievement for people in the United States ages 65–74 differs from the claimed distribution. (claim)

b. $\alpha = 0.01$

c. $\chi_0^2 = 15.086$

d. Rejection region: $\chi^2 > 15.086$

e.

	Distribution	Observed	Expected	$\frac{(O-E)^2}{E}$
Not a H.S. graduate	13.4%	86	60.434	10.8154
H.S. graduate	31.2%	161	140.712	2.9251
Some college no degree	17.2%	72	77.572	0.4002
Associate's degree	8.8%	27	39.688	4.0563
Bachelor's degree	19.1%	60	86.141	7.9329
Advanced degree	10.3%	45	46.453	0.0454
		451		26.175

$\chi^2 \approx 26.175$

f. Reject H_0.

g. There is enough evidence at the 1% level of significance to conclude that the distribution for people in the United States ages 65–74 differs from the distribution for people ages 25 and older.

Nonparametric Tests

11.1 THE SIGN TEST

11.1 Try It Yourself Solutions

1a. The claim is "the median number of days a home is on the market in its city is greater than 120."
H_0: median ≤ 120; H_a: median > 120 (claim)

b. $\alpha = 0.025$

c. $n = 24$

d. The critical value is 6.

e. $x = 6$

f. Reject H_0.

g. There is enough evidence at the 2.5% level of significance to support the agency's claim that the median number of days a home is on the market in its city is greater than 120.

2a. The claim is "the median age of automobiles in operation in the United States is 9.4 years."
H_0: median $= 9.4$; H_a: median > 9.4 (claim)

b. $\alpha = 0.10$

c. $n = 92$

d. The critical value is $z_0 = -1.645$.

e. $x = 41$

$$z = \frac{(x+0.5) = 0.5(n)}{\frac{\sqrt{n}}{2}} = \frac{(41+0.5) - 0.5(92)}{\frac{\sqrt{92}}{2}} \approx \frac{-4.5}{4.80} \approx -0.94$$

f. Fail to reject H_0.

g. There is not enough evidence at the 10% level of significance to reject the organization's claim that the median age of automobiles in operation in the United States is 9.4 years.

3a. The claim is "a new vaccine will decrease the number of colds in adults."
H_0: The number of colds will not decrease.
H_a: The number of colds will decrease. (claim)

b. $\alpha = 0.05$

c. $n = 11$

d. The critical value is 2.

e. $x = 2$

f. Reject H_0.

g. There is enough evidence at the 5% level of significance to support the researcher's claim that the new vaccine will decrease the number of colds in adults.

11.1 EXERCISE SOLUTIONS

1. A nonparametric test is a hypothesis test that does not require any specific conditions concerning the shapes of populations or the values of any population parameters.

A nonparametric test is usually easier to perform than its corresponding parametric test, but the nonparametric test is usually less efficient.

3. When n is less than or equal to 25, the test statistic is equal to x (the smaller number of + or – signs). When n is greater than 25, the test statistic is equal to

$$z = \frac{(x+0.5)-0.5n}{\frac{\sqrt{n}}{2}}.$$

5. Identify the claim and state H_0 and H_a. Identify the level of significance and sample size. Find the critical value using Table 8 (if $n \le 25$) or Table 4 ($n > 25$). Calculate the test statistic. Make a decision and interpret in the context of the problem.

7. (a) The claim is "the median amount of new credit card charges for the previous month was more than $300."
H_0: median $\le$ \$300; H_a: median $>$ \$300 (claim)
(b) The critical value is 1.
(c) $x = 5$
(d) Fail to reject H_0.
(e) There is not enough evidence at the 1% level of significance for the accountant to conclude that the median amount of new credit charges for the previous month was more than $300.

9. (a) The claim is "the median sales price of new privately owned one-family homes sold in the past year is $198,000 or less."
H_0: median $\le$ \$198,000; H_a: median $>$ \$198,000 (claim)
(b) The critical value is 1.
(c) $x = 4$
(d) Fail to reject H_0.
(e) There is not enough evidence at the 5% level of significance to reject the claim that the median sales price of new privately owned one-family homes sold in the past year is $198,000 or less.

11. (a) The claim is "the median amount of credit card debt for families holding such debt is at least $3000."
H_0: median $\ge$ \$3000 (claim); H_a: median $<$ \$3000
(b) The critical value is $z_0 = -2.05$.
(c) $x = 44$

$$z = \frac{(x+0.5)-0.5(n)}{\frac{\sqrt{n}}{2}} = \frac{(44+0.5)-0.5(104)}{\frac{\sqrt{104}}{2}} \approx \frac{-7.5}{5.099} \approx -1.47$$

(d) Fail to reject H_0.

(e) There is not enough evidence at the 2% level of significance to reject the institution's claim that the median amount of credit card debt for families holding such debts is at least \$3000.

13. (a) The claim is "the median age of Twitter® users is greater than 30 years old."
H_0 : median ≤ 30; H_a : median > 30 (claim)
(b) The critical value is 4.
(c) $x = 10$
(d) Fail to reject H_0.
(e) There is not enough evidence at the 1% level of significance to support the research group's claim that the median age of Twitter users is greater than 30 years old.

15. (a) The claim is "the median number of rooms in renter-occupied units is four."
H_0 : median = 4; (claim) H_a : median $\neq 4$
(b) The critical value is $z_0 = -1.96$.
(c) $x = 31$

$$z = \frac{(x+0.5)-0.5(n)}{\frac{\sqrt{n}}{2}} = \frac{(31+0.5)-0.5(80)}{\frac{\sqrt{80}}{2}} \approx \frac{-8.5}{4.47} \approx -1.90$$

(d) Fail to reject H_0.
(e) There is not enough evidence at the 5% level of significance to reject the organization's claim that the median number of rooms in renter-occupied units is 4.

17. (a) The claim is "the median hourly wage of computer systems analysts is \$37.06."
H_0 : median = \$37.06; (claim) H_a : median $\neq$ \$37.06
(b) The critical value is $z_0 = -2.575$.
(c) $x = 18$

$$z = \frac{(x+0.5)-0.5(n)}{\frac{\sqrt{n}}{2}} = \frac{(18+0.5)-0.5(43)}{\frac{\sqrt{43}}{2}} \approx \frac{-3}{3.28} \approx -0.91$$

(d) Fail to reject H_0.
(e) There is not enough evidence at the 1% level of significance to reject the labor organization's claim that the median hourly earnings of computer systems analysts is \$37.06.

19. (a) The claim is "the lower back pain intensity scores decreased after acupuncture."
H_0 : The lower back pain intensity scores have not decreased.
H_a : The lower back pain intensity scores have decreased. (claim)
(b) The critical value is 1.
(c) $x = 0$
(d) Reject H_0.
(e) There is enough evidence at the 5% level of significance to conclude that the lower back pain intensity scores were lower after acupuncture.

21. (a) The claim is "the student's critical reading SAT scores improved."

H_0: The SAT scores have not improved.

H_a: The SAT scores have improved. (claim)

(b) The critical value is 2.

(c) $x = 4$

(d) Fail to reject H_0.

(e) There is not enough evidence at the 5% level of significance to conclude that the critical reading SAT scores improved.

23. (a) The claim is "the proportion of adults who feel older than their real age is equal to the proportion of adults who feel younger than their real age."

H_0: The proportion of adults who feel older than their real age is equal to the proportion of adults who feel younger than their real age. (claim)

H_a: The proportion of adults who feel older than their real age is different from the proportion of adults who feel younger than their real age.

The critical value is 3.

$x = 3$

Reject H_0.

(b) There is enough evidence at the 5% level of significance to reject the claim that the proportion of adults who feel older than their real age is equal to the proportion of adults who feel younger than their real age.

25. Hypothesis test results:

Parameter: median of variable

H_0: Parameter = 22.55

H_A: Parameter ≠ 22.55

Variable	a	n for test
Hourly wages (in dollars)	14	13

Sample Median	Below	Equal	Above	P-value
26.075	2	1	11	0.0225

$P \approx 0.0225 < 0.05$, so reject H_0. There is enough evidence at the 5% level of significance to reject the labor organization's claim that the median hourly wage of tool and die makers is $22.25.

27. (a) The claim is "the median weekly earnings of female workers are less than or equal to $638."

H_0: median ≤ $638; (claim) H_a: median > $638

(b) The critical value is $z_0 = 2.33$.

(c) $x = 29$

$$z = \frac{(x+0.5)-0.5(n)}{\frac{\sqrt{n}}{2}} = \frac{(29+0.5)-0.5(47)}{\frac{\sqrt{47}}{2}} \approx \frac{5}{3.428} \approx 1.46$$

(d) Fail to reject H_0.

(e) There is not enough evidence at the 1% level of significance to reject the organization's claim that the median weekly earnings of female workers is less than or equal to $638.

29. (a) The claim is "the median age of brides at the time of their first marriage is less than or equal to 26 years."

H_0: median ≤ 26; H_a: median > 26 (claim)

(b) The critical value is $z_0 = 1.645$.

(c) $x = 35$

$$z = \frac{(x+0.5)-0.5(n)}{\frac{\sqrt{n}}{2}} = \frac{(35+0.5)-0.5(59)}{\frac{\sqrt{59}}{2}} \approx \frac{5}{3.841} \approx 1.302$$

(d) Fail to reject H_0.

(e) There is not enough evidence at the 5% level of significance to reject the counselor's claim that the median age of brides at the time of their first marriage is less than or equal to 26 years.

11.2 THE WILCOXON TESTS

11.2 Try It Yourself Solutions

1a. The claim is "a spray-on water repellant is effective."

H_0: There is no difference in the amounts of water repelled.

H_a: There is a difference in the amounts of water repelled.

b. $\alpha = 0.01$

c. $n = 11$

d. The critical value is 5.

e.

No repellent	Repellent applied	Difference	Absolute value	Rank	Signed rank
8	15	−7	7	11	−11
7	12	−5	5	9	−9
7	11	−4	4	7.5	−7.5
4	6	−2	2	3.5	−3.5
6	6	0	0	—	—
10	8	2	2	3.5	3.5
9	8	1	1	1.5	1.5
5	6	−1	1	1.5	−1.5
9	12	−3	3	5.5	−5.5
11	8	3	3	5.5	5.5
8	14	−6	6	10	−10
4	8	−4	4	7.5	−7.5

Sum of negative ranks = −55.5

Sum of positive ranks = 10.5

$w_s = 10.5$

f. Fail to reject H_0.

g. There is not enough evidence at the 1% level of significance for the quality control inspector to conclude that the spray-on water repellent is effective.

2a. The claim is "there is no difference in the claims paid by paid by the companies."

H_0: There is no difference in the claims paid by paid by the companies.

H_a: There is a difference in the claims paid by paid by the companies. (claim)

b. $\alpha = 0.05$

c. The critical values are $z_0 = \pm 1.96$.

d. $n_1 = 12$ and $n_2 = 12$

e.

Ordered data	Sample	Rank	Ordered data	Sample	Rank
1.7	B	1	5.3	B	13
1.8	B	2	5.6	B	14
2.2	B	3	5.8	A	15
2.5	A	4	6.0	A	16
3.0	A	5.5	6.2	A	17
3.0	B	5.5	6.3	A	18
3.4	B	7	6.5	A	19
3.9	A	8	7.3	B	20
4.1	B	9	7.4	A	21
4.4	B	10	9.9	A	22
4.5	A	11	10.6	A	23
4.7	B	12	10.8	B	24

R = sum ranks of company B = 120.5

f. $\mu_R = \dfrac{n_1(n_1 + n_2 + 1)}{2} = \dfrac{12(12+12+1)}{2} = 150$

$\sigma_R = \sqrt{\dfrac{n_1 n_2 (n_1 + n_2 + 1)}{12}} = \sqrt{\dfrac{(12)(12)(12+12+1)}{12}} \approx 17.321$

$z = \dfrac{R - \mu_R}{\sigma_R} \approx \dfrac{120.5 - 150}{17.321} \approx -1.703$

g. Fail to reject H_0.

h. There is not enough evidence at the 5% level of significance to conclude that there is a difference in the claims paid by the companies.

11.2 EXERCISE SOLUTIONS

1. If the samples are dependent, use the Wilcoxon signed-rank test. If the samples are independent, use the Wilcoxon rank sum test.

3. (a) The claim is "there was no reduction in diastolic blood pressure."

H_0: There is no reduction in diastolic blood pressure. (claim)

H_a: There is a reduction in diastolic blood pressure.

(b) Wilcoxon signed-rank test

(c) The critical value is 10.

(d)

Before treatment	After treatment	Difference	Absolute difference	Rank	Signed rank
108	99	9	9	8	8
109	115	–6	6	4.5	–4.5
120	105	15	15	12	12
129	116	13	13	10.5	10.5
112	115	–3	3	2	–2
111	117	–6	6	4.5	–4.5
117	108	9	9	8	8
135	122	13	13	10.5	10.5
124	120	4	4	3	3
118	126	–8	8	6	–6
130	128	2	2	1	1
115	106	9	9	8	8

Sum of negative ranks = –17
Sum of positive ranks = 61
$w_s = 17$

(e) Fail to reject H_0.

(f) There is not enough evidence at the 1% level of significance to reject the claim that there was no reduction in diastolic blood pressure.

5. (a) The claim is "the cost of prescription drugs is lower in Canada than in the United States."
H_0: The cost of prescription drugs is not lower in Canada than in the United States.
H_a: The cost of prescription drugs is lower in Canada than in the United States. (claim)

(b) Wilcoxon signed-rank test

(c) The critical value is 4.

(d)

Cost in US.	Cost in Canada	Difference	Absolute difference	Rank	Signed rank
1.26	1.04	0.22	0.22	1	1
1.76	0.82	0.94	0.94	3	3
4.19	2.22	1.97	1.97	7	7
3.36	2.22	1.14	1.14	4	4
1.80	1.31	0.49	0.49	2	2
9.91	11.47	–1.56	1.56	6	–6
3.95	2.63	1.32	1.32	5	5

Sum of negative ranks = –6
Sum of positive ranks = 22
$w_s = 6$

(e) Fail to reject H_0.

(f) There is not enough evidence at the 5% level of significance for the researcher to conclude that the cost of prescription drugs is lower in Canada than in the United States.

7. (a) The claim is "there is a difference in the salaries earned by teachers in Wisconsin and Michigan."

H_0: There is not a difference in the salaries.

H_a: There is a difference in the salaries. (claim)

(b) Wilcoxon rank sum test

(c) The critical values are $z_0 = \pm 1.96$.

(d)

Ordered data	Sample	Rank
46	WI	1
49	MI	2
50	WI	3.5
50	WI	3.5
51	WI	6.5
51	WI	6.5
51	WI	6.5
51	MI	6.5
52	WI	9
53	WI	10.5
53	MI	10.5
54	MI	12
55	WI	13.5
55	MI	13.5
57	MI	15.5
57	MI	15.5
58	MI	17
59	WI	18
61	MI	19
63	MI	20.5
63	MI	20.5
64	WI	22
72	MI	23

R = sum ranks of Wisconsin = 100.5

$$\mu_R = \frac{n_1(n_1+n_2+1)}{2} = \frac{11(11+12+1)}{2} = 132$$

$$\sigma_R = \sqrt{\frac{n_1 n_2(n_1+n_2+1)}{12}} = \sqrt{\frac{(11)(12)(11+12+1)}{12}} \approx 16.248$$

$$z = \frac{R-\mu_R}{\sigma_R} \approx \frac{100.5-132}{16.248} \approx -1.94$$

(e) Fail to reject H_0.

(f) There is not enough evidence at the 5% level of significance to support the representative's claim that there is a difference in the salaries earned by teachers in Wisconsin and Michigan.

9. The claim is "a certain fuel additive improves a car's gas mileage."

H_0: The fuel additive does not improve gas mileage.

H_a: The fuel additive does improve gas mileage. (claim)

The critical value is $z_0 = 1.28$.

Before	After	Difference	Absolute value	Rank	Signed rank
36.4	36.7	−0.3	0.3	4.5	−4.5
36.4	36.9	−0.5	0.5	11	−11
36.6	37.0	−0.4	0.4	7	−7
36.6	37.5	−0.9	0.9	17	−17
36.8	38.0	−1.2	1.2	19.5	−19.5
36.9	38.1	−1.2	1.2	19.5	−19.5
37.0	38.4	−1.4	1.4	25	−25
37.1	38.7	−1.6	1.6	30.5	−30.5
37.2	38.8	−1.6	1.6	30.5	−30.5
37.2	38.9	−1.7	1.7	32	−32
36.7	36.3	0.4	0.4	7	7
37.5	38.9	−1.4	1.4	25	−25
37.6	39.0	−1.4	1.4	25	−25
37.8	39.1	−1.3	1.3	21.5	−21.5
37.9	39.4	−1.5	1.5	28.5	−28.5
37.9	39.4	−1.5	1.5	28.5	−28.5
38.1	39.5	−1.4	1.4	25	−25
38.4	39.8	−1.4	1.4	25	−25
40.2	40.0	0.2	0.2	2.5	2.5
40.5	40.0	0.5	0.5	11	11
40.9	40.1	0.8	0.8	16	16
35.0	36.3	−1.3	1.3	21.5	−21.5
32.7	32.8	−0.1	0.1	1	−1
33.6	34.2	−0.6	0.6	14.5	−14.5
34.2	34.7	−0.5	0.5	11	−11
35.1	34.9	0.2	0.2	2.5	2.5
35.2	34.9	0.3	0.3	4.5	4.5
35.3	35.3	0	0	—	—
35.5	35.9	−0.4	0.4	7	−7
35.9	36.4	−0.5	0.5	11	−11
36.0	36.6	−0.6	0.6	14.5	−14.5
36.1	36.6	−0.5	0.5	11	−11
37.2	38.3	−1.1	1.1	18	−18

Sum of negative ranks = −484.5
Sum of positive ranks = 43.5
$w_s = 43.5$

$$z = \frac{w_s - \frac{n(n+1)}{4}}{\sqrt{\frac{n(n+1)(2n+1)}{24}}} = \frac{43.5 - \frac{32(32+1)}{4}}{\sqrt{\frac{32(32+1)[(2)32+1]}{24}}} = \frac{-220.5}{\sqrt{2860}} = -4.123$$

Note: $n = 32$ because one of the differences is zero and should be discarded.
Reject H_0. There is enough evidence at the 10% level of significance for the engineer to conclude that the gas mileage is improved.

11.3 THE KRUSKAL-WALLIS TEST

11.3 Try It Yourself Solutions

1a. The claim is "the distribution of the veterinarians' salaries in these three states are different."

H_0: There is no difference in the salaries in the three states.

H_a: There is a difference in the salaries in the three states. (claim)

b. $\alpha = 0.05$

c. d.f. $= k - 1 = 2$

d. $\chi_0^2 = 5.991$; Rejection region: $\chi^2 > 5.991$

e.

Ordered data	State	Rank
88.28	CA	1
88.80	PA	2
92.50	NY	3
93.10	NY	4
94.40	PA	5
95.15	CA	6
96.25	PA	7
97.25	PA	8
97.44	CA	9.5
97.50	NY	9.5
97.50	NY	11
97.89	CA	12
98.85	NY	13
99.20	PA	14

Ordered data	State	Rank
99.70	CA	15
99.75	NY	16
99.95	CA	17
99.99	PA	18
100.55	PA	19
100.75	CA	20
101.20	CA	21
101.55	NY	22
101.97	NY	23
102.35	NY	24
103.20	CA	25
103.70	PA	26
110.45	PA	27
113.90	CA	28

$R_1 = 157.5$

$R_2 = 119$

$R_3 = 129$

f. $$H = \frac{12}{N(N+1)}\left(\frac{R_1^2}{n_1} + \frac{R_2^2}{n_2} + \frac{R_3^2}{n_3}\right) - 3(N+1)$$

$$= \frac{12}{28(29)}\left(\frac{(157.5)^2}{10} + \frac{(129)^2}{9} + \frac{(119.5)^2}{9}\right) - 3(29)$$

$$\approx 0.433$$

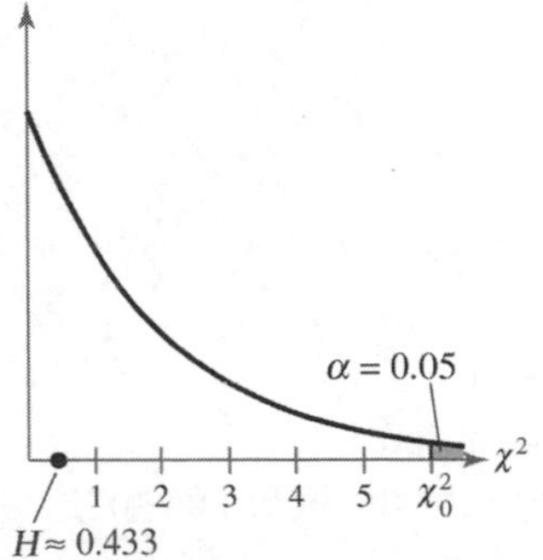

g. Fail to reject H_0.

h. There is not enough evidence at the 5% level of significance to conclude that the distributions of the veterinarians' salaries in these three states are different.

11.3 EXERCISE SOLUTIONS

1. The conditions for using a Kruskal-Wallis test are that each sample must be randomly selected and the size of each sample must be at least 5.

3. (a) The claim is "the distributions of the annual premiums in these three states are different."

H_0: There is no difference in the premiums.

H_a: There is a difference in the premiums. (claim)

(b) The critical value is 5.991.

(c)

Ordered data	Sample	Rank
535	VA	1
546	VA	2
618	VA	3
625	VA	4
725	CT	5
730	VA	6
757	MA	7
806	CT	8
815	VA	9
889	MA	10
890	CT	11
912	VA	12
930	CT	13
947	CT	14
980	MA	15
1025	MA	16
1040	CT	17
1105	MA	18
1110	MA	19
1165	CT	20
1295	MA	21

$R_1 = 88,\ R_2 = 106,\ R_3 = 37$

$$H = \frac{12}{N(N+1)}\left(\frac{R_1^2}{n_1} + \frac{R_2^2}{n_2} + \frac{R_3^2}{n_3}\right) - 3(N+1)$$

$$= \frac{12}{21(21+1)}\left(\frac{(88)^2}{7} + \frac{(106)^2}{7} + \frac{(37)^2}{7}\right) - 3(21+1)$$

$$\approx 9.506$$

(d) Reject H_0.

(e) There is enough evidence at the 5% level of significance to conclude that the distributions of the annual premiums in Connecticut, Massachusetts, and Virginia are different.

5. The claim is "the distributions of the annual salaries in these four states are different."

H_0: There is no difference in the salaries.

H_a: There is a difference in the salaries. (claim)

(b) The critical value is 6.251.

(c)

Ordered data	Sample	Rank
25.5	KT	1
27.1	WV	2
27.8	SC	3
28.9	WV	4
30.4	WV	5
30.9	KT	6
32.5	KT	7.5
32.5	NC	7.5
32.7	SC	9
33.6	NC	10
34.1	SC	11
34.2	KT	12
35.4	SC	13
36.6	NC	14
37.4	WV	15
38.2	WV	16
38.9	NC	17
40.5	NC	18
40.9	SC	19
41.5	SC	20
42.6	WV	21
43.1	KT	22
51.3	NC	23
54.7	KT	24

$R_1 = 72.5,\ R_2 = 89.5,\ R_3 = 75,\ R_4 = 63$

$$H = \frac{12}{N(N+1)}\left(\frac{R_1^2}{n_1} + \frac{R_2^2}{n_2} + \frac{R_3^2}{n_3} + \frac{R_4^2}{n_4}\right) - 3(N+1)$$

$$= \frac{12}{24(24+1)}\left(\frac{(72.5)^2}{6} + \frac{(89.5)^2}{6} + \frac{(75)^2}{6} + \frac{(63)^2}{6}\right) - 3(24+1)$$

$$\approx 1.202$$

$H = 1.202$

(d) Fail to reject H_0.

(e) There is not enough evidence at the 10% level of significance to conclude that the distributions of the annual salaries in the four states are different.

7. **Kruskal-Wallis results:**
Data stored in separate columns.
Chi Square = 8.0965185 (adjusted for ties)
DF = 2
P-value = 0.0175

Column	N	Median	Ave. Rank
A	6	5	6.75
B	6	8.5	14.5
C	6	5	7.25

$P \approx 0.0175 > 0.01$, so fail to reject H_0. There is not enough evidence at the 1% level of significance to conclude that the distributions of the number of job offers at Colleges A, B, and C are different.

9. (a) The claim is "the mean number of days patients spend in a hospital differs according to the region of the United States in which the patient lives."
H_0: There is no difference in the number of days spent in the hospital.
H_a: There is a difference in the number of days spent in the hospital. (claim)
The critical value is 11.345.

Ordered data	Sample	Rank
1	NE	2.5
1	MW	2.5
1	S	2.5
1	S	2.5
2	W	5
3	NE	8.5
3	NE	8.5
3	MW	8.5
3	MW	8.5
3	W	8.5
3	W	8.5
4	MW	13.5
4	MW	13.5
4	MW	13.5
4	W	13.5
5	NE	19
5	MW	19
5	S	19
5	S	19
5	S	19
5	W	19
5	W	19
6	NE	26
6	NE	26
6	NE	26
6	MW	26
6	W	26
6	W	26
6	W	26
7	MW	30.5
7	S	30.5
8	NE	33.5
8	NE	33.5
8	S	33.5
8	S	33.5
9	MW	36
11	NE	37

$R_1 = 220.5,\ R_2 = 171.5,\ R_3 = 159.5,\ R_4 = 151.5$

$$H = \frac{12}{N(N+1)}\left(\frac{R_1^2}{n_1}+\frac{R_2^2}{n_2}+\frac{R_3^2}{n_3}+\frac{R_4^2}{n_4}\right) - 3(N+1)$$

$$= \frac{12}{37(37+1)}\left(\frac{(220.5)^2}{10}+\frac{(171.5)^2}{10}+\frac{(159.5)^2}{8}+\frac{(151.5)^2}{9}\right) - 3(37+1)$$

$$\approx 1.507$$

$H \approx 1.507$;

Fail to reject H_0. There is not enough evidence at the 1% level of significance to support the underwriter's claim that there is a difference in the mean number of days spent in the hospital.

(b)

Variation	Sum of squares	Degrees of freedom	Mean squares	F
Between	9.17	3	3.06	0.52
Within	194.72	33	5.90	

Because $\alpha = 0.01$ and the test is two-tailed, use the $\frac{1}{2}\alpha = 0.005$ table. The critical value is about 5.24. Because $F \approx 0.52$ is less than the critical value, the decision is to fail to reject H_0. There is not enough evidence at the 1% level of significance to support the underwriter's claim that there is a difference in the mean number of days spent in the hospital.

11.4 RANK CORRELATION

11.4 Try It Yourself Solutions

1a. The claim is "there is a correlation between the number of males and females who receive doctoral degrees."

H_0: $\rho_s = 0$; H_a: $\rho_s \neq 0$ (claim)

b. $\alpha = 0.01$

c. The critical value is 0.929.

d.

Male	Rank	Female	Rank	d	d^2
25	3.5	20	1.5	2	4
24	1.5	20	1.5	0	0
24	1.5	22	3.0	−1.5	2.25
25	3.5	23	4.0	−0.5	0.25
27	5.0	26	5.0	0	0
29	6.0	27	6.0	0	0
30	7.0	30	7.0	0	0
					$\sum d^2 = 6.5$

$\sum d^2 = 6.5$

e. $r_c \approx 1 - \dfrac{6\sum d^2}{n(n^2-1)} = 1 - \dfrac{6(6.5)}{7(7^2-1)} \approx 0.884$

f. Fail to reject H_0.

g. There is not enough evidence at the 1% level of significance to conclude that a significant correlation exists between the number of males and females who received doctoral degrees.

11.4 EXERCISE SOLUTIONS

1. The Spearman rank correlation coefficient can be used to describe the relationship between linear and nonlinear data. Also, it can be used for data at the ordinal level and it is easier to calculate by hand than the Pearson correlation coefficient.

3. The ranks of corresponding data pairs are identical when r_s is equal to 1.
The ranks of corresponding data pairs are in reverse order when r_s is equal to −1.
The ranks of corresponding data pairs have no relationship when r_s is equal to 0.

5. (a) The claim is "there is a correlation between debt and income in the farming business."
$H_0\colon \rho_s = 0;\ H_a\colon \rho_s \neq 0$ (claim)
(b) The critical value is 0.929.
(c)

Debt	Rank	Income	Rank	d	d²
19,955	7	28,926	7	0	0
10.480	4	8,630	2	2	4
14.434	6	12,942	5	1	1
9,982	2	8,807	3	−1	1
10,085	3	11,028	4	−1	1
4,235	1	7,008	1	0	0
13,286	5	15,268	6	−1	1
					$\sum d^2 = 8$

$\sum d^2 = 8$

$$r_c \approx 1 - \frac{6\sum d^2}{n(n^2-1)} \approx 0.857$$

(d) Fail to reject H_0.
(e) There is not enough evidence at the 1% level of significance to support the claim that there is a correlation between debt and income in the farming business.

7. (a) The claim is "there is a correlation between wheat and oat prices."
$H_0\colon \rho_s = 0;\ H_a\colon \rho_s \neq 0$ (claim)
(b) The critical value is 0.833.
(c)

Oat	Rank	Wheat	Rank	d	d²
1.10	1	2.62	1	0	0
1.59	4	2.78	2	2	4
1.81	6	3.56	6	0	0
1.48	2.5	3.40	3.5	−1	1
1.48	2.5	3.40	3.5	−1	1
1.63	5	3.42	5	0	0
1.87	7	4.26	7	0	0
2.63	8	6.48	8	0	0
3.10	9	6.80	9	0	0
					$\sum d^2 = 6$

$\sum d^2 = 6$

$$r_c \approx 1 - \frac{6\sum d^2}{n(n^2-1)} \approx 0.950$$

(d) Reject H_0.

(e) There is enough evidence at the 1% level of significance to conclude that there is a correlation between the oat and wheat prices.

9. The claim is "there is a correlation between science achievement scores and GNI."
H_0: $\rho_s = 0$; H_a: $\rho_s \neq 0$ (claim)
The critical value is 0.700.

Science average	Rank	GNI	Rank	d	d^2
534	9	1307	3	6	36
495	5	2467	6	−1	1
516	7	3207	7	0	0
475	2	1988	5	−3	9
531	8	4829	8	0	0
410	1	989	2	−1	1
488	3	1314	4	−1	1
503	6	438	1	5	25
489	4	13,886	9	−5	25
					$\sum d^2 = 98$

$\sum d^2 = 98$

$$r_c \approx 1 - \frac{6\sum d^2}{n(n^2 - 1)} \approx 0.183$$

Fail to reject H_0. There is not enough evidence at the 5% level of significance to conclude that there is a correlation between science achievement scores and GNI.

11. The claim is "there is a correlation between science and mathematics achievement scores."
H_0: $\rho_s = 0$; H_a: $\rho_s \neq 0$ (claim)
The critical value is 0.700.

Science average	Rank	Science average	Rank	d	d^2
534	9	527	9	0	0
495	5	496	5	0	0
516	7	504	7	0	0
475	2	462	2	0	0
531	8	523	8	0	0
410	1	406	1	0	0
488	3	480	4	−1	1
503	6	502	6	0	0
489	4	474	3	1	1
					$\sum d^2 = 2$

$\sum d^2 = 2$

$$r_c \approx 1 - \frac{6\sum d^2}{n(n^2 - 1)} \approx 0.983$$

Reject H_0. There is enough evidence at the 5% level of significance to conclude that there is a correlation between science and mathematics achievement scores.

13. The claim is "there is a correlation between average hours worked and the number of on-the-job injuries."

$H_0: \rho_s = 0$; $H_a: \rho_s \neq 0$ (claim)

The critical values are $\frac{\pm z}{\sqrt{n-1}} = \frac{\pm 1.96}{\sqrt{33-1}} \approx \pm 0.346$.

Hours worked	Rank	Injuries	Rank	D	D^2
47.6	31	16	2	29	841
44.1	11	33	31.5	−20.5	420.25
45.6	24	25	17.5	6.5	42.25
45.5	21.5	33	31.5	−10	100
44.5	14	18	3.5	10.5	110.25
47.3	30	20	7	23	529
44.6	16	21	10	6	36
45.9	25	18	3.5	21.5	462.25
45.5	21.5	21	10	11.5	132.25
43.7	9.5	28	26	−16.5	272.25
44.8	17.5	15	1	16.5	272.25
42.5	3	26	21	−18	324
46.5	26	34	33	−7	49
42.3	2	32	30	−28	784
45.5	21.5	26	21	0.5	0.25
41.8	1	28	26	−25	625
43.1	5.5	22	12	−6.5	42.25
44.4	12	19	5	7	49
44.5	14	23	13.5	0.5	0.25
43.7	9.5	20	7	2.5	6.25
44.9	19	28	26	−7	49
47.8	32	24	15.5	16.5	272.25
46.6	27	26	21	6	36
45.5	21.5	29	29	−7.5	56.25
43.5	7.5	21	10	−2.5	6.25
42.8	4	28	26	−22	484
44.8	17.5	23	13.5	4	16
43.5	7.5	26	21	−13.5	182.25
47.0	29	24	15.5	13.5	182.25
44.5	14	20	7	7	49
50.1	33	28	26	7	49
46.7	28	26	21	7	49
43.1	5.5	25	17.5	−12	144
					$\sum d^2 = 6673$

$\sum d^2 = 6673$

$$r_c \approx 1 - \frac{6\sum d^2}{n(n^2-1)} \approx -0.115$$

Fail to reject H_0. There is not enough evidence at the 5% level of significance to conclude that there is a correlation between average hours worked and the number of on-the-job injuries.

11.5 THE RUNS TEST

11.5 Try It Yourself Solutions

1a. *PPP F P F PPPP FF P F PP FFF PPP F PPP*

b. 13 groups $\Rightarrow$ 13 runs

c. 3, 1, 1, 1, 4, 2, 1, 1, 2, 3, 3, 1, 3

2a. The claim is "the sequence of genders is not random."

H_0: The sequence of genders is random.

H_a: The sequence of genders is not random. (claim)

b. $\alpha = 0.05$

c. *M FFF MM FF M F MM FFF*

n_1 = number of F's = 9

n_2 = number of M's = 6

G = number of runs = 8

d. lower critical value = 4

upper critical value = 13

e. $G = 8$

f. Fail to reject H_0.

g. There is not enough evidence at the 5% level of significance to support the claim that the sequence of genders is not random.

3a. The claim is "the sequence of weather conditions is not random."

H_0: The sequence of weather conditions is random.

H_a: The sequence of weather conditions is not random. (claim)

b. $\alpha = 0.05$

c. n_1 = number of N's = 21

n_2 = number of S's = 10

G = number of runs = 17

d. critical values = ± 1.96

e. $$\mu_G = \frac{2n_1n_2}{n_1+n_2} + 1 = \frac{2(21)(10)}{21+10} + 1 = 14.55$$

$$\sigma_G = \sqrt{\frac{2n_1n_2(2n_1n_2 - n_1 - n_2)}{(n_1+n_2)^2(n_1+n_2-1)}} = \sqrt{\frac{2(21)(10)(2(21)(10) - 21 - 10)}{(21+10)^2(21+10-1)}} \approx 2.4$$

$$z = \frac{G - \mu_G}{\sigma_G} \approx \frac{17 - 14.55}{2.38} \approx 1.03$$

f. Fail to reject H_0.

g. There is not enough evidence at the 5%level of significance to support the claim that the sequence of weather conditions is not random.

11.5 EXERCISE SOLUTIONS

1. Answers will vary. *Sample answer:* It is called the runs test because it considers the number of runs of data in a sample to determine whether the sequence of data was randomly selected.

3. Number of runs = 8
Run lengths = 1, 1, 1, 1, 3, 3, 1, 1

5. Number of runs = 9
Run lengths = 1, 1, 1, 1, 1, 6, 3, 2, 4

7. n_1 = number of T's = 6
n_2 = number of F's = 6

9. n_1 = number of M's = 10
n_2 = number of F's = 10

11. n_1 = number of T's = 6
n_1 = number of F's = 6
too high: 11; too low: 3

13. n_1 = number of N's = 11
n_1 = number of S's = 7
too high: 14; too low: 5

15. (a) The claim is "the tosses were not random."
H_0: The coin tosses were random.
H_a: The coin tosses were not random. (claim)
(b) n_1 = number of H's = 7
n_2 = number of T's = 9
lower critical value = 4
upper critical value = 14
(c) $G = 9$ runs
(d) Fail to reject H_0.
(e) There is not enough evidence at the 5% level of significance to support the claim that the coin tosses were not random.

17. (a) The claim is "the sequence of World Series winning teams is not random."
H_0: The sequence of leagues of winning teams is random.
H_a: The sequence of leagues of winning teams is not random. (claim)
(b) n_1 = number of N's = 17
n_2 = number of A's = 23
$z_0 = \pm 1.96$
(c) $G = 26$ runs

$$\mu_G = \frac{2n_1n_2}{n_1 + n_2} + 1 = \frac{2(17)(23)}{17 + 23} + 1 = 20.55$$

$$\sigma_G = \sqrt{\frac{2n_1n_2(2n_1n_2 - n_1 - n_2)}{(n_1 + n_2)^2(n_1 + n_2 - 1)}} = \sqrt{\frac{2(17)(23)(2(17)(23) - 17 - 23)}{(17 + 23)^2(78 + 23 - 1)}} \approx 3.05$$

$$z = \frac{G - \mu_G}{\sigma_G} \approx \frac{26 - 20.55}{3.05} \approx 1.79$$

(d) Fail to reject H_0.
(e) There is not enough evidence at the 5% level of significance to conclude that the sequence of leagues of World Series winning teams is not random.

19. (a) The claim is "the microchips are random by gender."

H_0: The microchips are random by gender.

H_a: The microchips are not random by gender. (claim)

(b) n_1 = number of M's = 9

n_2 = number of F's = 20

lower critical value = 8; upper critical value = 18

(c) $G = 12$ runs

(d) Fail to reject H_0.

(e) There is not enough evidence at the 5% level of significance to reject the claim that the microchips are random by gender.

21. The claim is "the daily high temperatures do not occur randomly."

H_0: Daily high temperatures occur randomly.

H_a: Daily high temperatures do not occur randomly. (claim)

median = 87

n_1 = number of F's = 15

n_2 = number of A's = 13

lower critical value = 9

upper critical value = 21

$G = 11$ runs

Fail to reject H_0.

There is not enough evidence at the 5% level of significance to support the claim that the daily high temperatures do not occur randomly.

23. Answers will vary.

CHAPTER 11 REVIEW EXERCISE SOLUTIONS

1. (a) The claim is "the median number of customers per day is not more than 650."

H_0: median ≤ 650 (claim); H_a: median > 650

(b) The critical value is 2.

(c) $x = 7$

(d) Fail to reject H_0.

(e) There is not enough evidence at the 1% level of significance to reject the bank manager's claim that the median number of customers per day is no more than 650.

3. (a) The claim is "median sentence length for all federal prisoners is 2 years."

H_0: median = 2 (claim); H_a: median $\neq 2$

(b) The critical value is $z_0 = -1.645$.

$$z = \frac{(x+0.5)-0.5(n)}{\frac{\sqrt{n}}{2}} = \frac{(65+0.5)-0.5(174)}{\frac{\sqrt{174}}{2}} \approx \frac{-21.5}{6.595} \approx -3.26$$

(c) $x = 65$ (d) Reject H_0.

(e) There is enough evidence at the 10% level of significance to reject the agency's claim that the median sentence length for all federal prisoner's is 2 years.

5. (a) The claim is "there was no reduction in diastolic blood pressure."

H_0: There is no reduction in diastolic blood pressure. (claim)

H_a: There is a reduction in diastolic blood pressure.

(b) The critical value is 2.

(c) $x = 3$

(d) Fail to reject H_0.

(e) There is not enough evidence at the 5% level of significance to reject the claim that there was no reduction in diastolic blood pressure.

7. (a) Independent; Wilcoxon rank sum test

(b) The claim is "there is a difference in the total times required to earn a doctorate degree by female and male graduate students."

H_0: There is no difference in the total times to earn a doctorate degree by female and male graduate students.

H_a: There is a difference in the total times to earn a doctorate degree by female and male graduate students.

(c) The critical values are $z_0 \pm 2.575$.

(d)

Ordered data	Sample	Rank
7	F	1.5
7	F	1.5
8	M	4.5
8	M	4.5
8	M	4.5
8	M	4.5
9	F	8.5
9	F	8.5
9	M	8.5
9	M	8.5
10	F	13
10	F	13
10	M	13
10	M	13
10	M	13
11	F	17.5
11	M	17.5
11	M	17.5
11	M	17.5
12	F	20.5
12	F	20.5
13	F	22.5
13	F	22.5
14	F	24

$$\mu_R = \frac{n_1(n_1 + n_2 + 1)}{2} = \frac{12(12 + 12 + 1)}{2} = 150$$

$$\sigma_R = \sqrt{\frac{n_1 n_2 (n_1 + n_2 + 1)}{12}} \sqrt{\frac{(12)(12)(12 + 12 + 1)}{12}} \approx 17.321$$

$$z = \frac{R - \mu_R}{\sigma_R} \approx \frac{126.5 - 150}{17.321} \approx -1.357$$

(e) Fail to reject H_0.

(f) There is not enough evidence at the 1% level of significance to support the claim that there is a difference in the total times to earn a doctorate degree by female and male graduate students.

9. (a) The claim is "the distributions of the ages of the doctorate recipients in these three fields of study are different."

H_0: There is no difference in ages of doctorate recipients among the fields of study.

H_a: There is a difference in ages of doctorate recipients among the fields of study. (claim)

(b) The critical value is 9.210.

(c)

Ordered data	Sample	Rank
29	L	1.5
29	P	1.5
30	L	5.5
30	P	5.5
30	P	5.5
30	P	5.5
30	P	5.5
30	S	5.5
31	L	12.5
31	L	12.5
31	L	12.5
31	P	12.5
31	P	12.5
31	P	12.5
31	S	12.5
31	S	12.5
32	L	20
32	L	20
32	L	20
32	P	20
32	P	20
32	S	20
32	S	20
33	P	25
33	S	25
33	S	25
34	L	28
34	L	28
34	S	28
35	L	31
35	S	31
35	S	31
36	S	33

$R_1 = 191.5,\ R_2 = 126,\ R_3 = 243.5$

$$H = \frac{12}{N(N+1)}\left(\frac{R_1^2}{n_1} + \frac{R_2^2}{n_2} + \frac{R_3^2}{n_3}\right) - 3(N+1)$$

$$= \frac{12}{33(33+1)}\left(\frac{(191.5)^2}{11} + \frac{(126)^2}{11} + \frac{(243.5)^2}{11}\right) - 3(33+1)$$

$$\approx 6.741$$

(d) Fail to reject H_0.

(e) There is not enough evidence at the 1% level of significance to conclude that the distributions of ages of the doctorate recipients in these three fields are different.

11. (a) The claim is "there is a correlation between overall score and price."
H_0: $\rho_s = 0$; H_a: $\rho_s \neq 0$ (claim)

(b) The critical value is 0.786.

(c)

Score	Rank	Price	Rank	d	d^2
93	7	500	6.5	0.5	0.25
91	6	300	5	1	1
90	5	500	6.5	−1.5	2.25
87	4	150	2	2	4
85	2	250	4	−1	1
74	3	200	3	−1	1
69	1	130	1	0	0
					$\sum d^2 = 9.5$

$\sum d^2 = 9.5$

$$r_s = 1 - \frac{6\sum d^2}{n(n^2-1)} = 1 - \frac{6(9.5)}{7(7^2-1)} = 0.8301$$

(d) Reject H_0.

(e) There is enough evidence at the 5% level of significance to conclude that there is a correlation between overall score and price.

13. (a) The claim is "the stops were not random by gender."
H_0: The traffic stops were random by gender.
H_a: The traffic stops were not random by gender. (claim)

(b) n_1 = number of F's = 12
n_2 = number of M's = 13
lower critical value = 8
upper critical value = 19

(c) $G = 14$ runs

(d) Fail to reject H_0.

(e) There is not enough evidence at the 5% level of significance support the claim that the traffic stops were not random by gender.

CHAPTER 11 QUIZ SOLUTIONS

1. (a) The claim is "there is a difference in the hourly earnings of union and nonunion workers in state and local governments."

H_0: There is no difference in the hourly earnings.

H_a: There is a difference in the hourly earnings. (claim)

(b) Wilcoxon rank sum test

(c) The critical values are $z_0 \pm 1.645$.

(d)

Ordered data	Sample	Rank
19.10	N	1
19.85	N	2
20.05	N	3
20.70	N	4
20.90	N	5
21.15	N	6
21.75	N	7
23.40	N	8
24.20	U	9
24.80	U	10.5
24.80	N	10.5
25.05	U	12
25.30	U	13
25.60	U	14.5
25.60	N	14.5
26.50	U	16
27.20	U	17
29.75	U	18
30.33	U	19
32.97	U	20

R = sum ranks of nonunion workers = 61

$$\mu_R = \frac{n_1(n_1+n_2+1)}{2} = \frac{10(10+10+1)}{2} = 105$$

$$\sigma_R = \sqrt{\frac{n_1 n_2(n_1+n_2+1)}{12}}\sqrt{\frac{(10)(10)(10+10+1)}{12}} \approx 13.229$$

$$z = \frac{R-\mu_R}{\sigma_R} \approx \frac{61-105}{13.229} \approx -3.33$$

(e) Reject H_0.

(f) There is enough evidence at the 10% level of significance to support the claim that there is a difference in the hourly earnings of union and nonunion workers in state and local governments.

2. (a) The claim is "the median number of annual volunteer hours is 52."
H_0: median is 52 (claim); H_a: median $\neq 52$

(b) Sign test

(c) The critical values are ± 1.96.

(d) $x = 23$

$$z = \frac{(x+0.5) - 0.5n}{\frac{\sqrt{n}}{2}}$$

$$= \frac{(23+0.5) - 0.5(70)}{\frac{\sqrt{70}}{2}} = -2.75$$

(e) Reject H_0.

(f) There is enough evidence at the 5% level of significance to reject the organization's claim that the median number of annual volunteer hours is 52.

3. (a) The claim is "the distributions of sales prices in these four regions are different."
H_0: There is no difference in sales prices among the regions.
H_a: There is a difference in sales prices among the regions. (claim)

(b) Kruskal-Wallis test

(c) The critical value is 11.345.

(d)

Ordered data	Sample	Rank
149.8	S	1
150.9	S	2
161	S	3
164.6	S	4
169.5	S	5
170.5	MW	6
172.6	S	7
175.5	S	8
175.9	MW	9
185.3	MW	10
187.1	MW	11
188.9	MW	12
189.9	W	13
190.5	S	14
191.9	MW	15
200.9	MW	16
201.9	W	17
205.1	MW	18
206.3	W	19
218.5	W	20
220	W	21
225.7	W	22
230	W	23
237.9	NE	24
238.6	NE	25
245.5	NE	26
250	NE	27
252.5	NE	28
255.7	W	29
259.4	NE	30
265.9	NE	31
270.2	NE	32

$R_1 = 223,\ R_2 = 97,\ R_3 = 44,\ R_4 = 164$

$$H = \frac{12}{N(N+1)}\left(\frac{R_1^2}{n_1} + \frac{R_2^2}{n_2} + \frac{R_3^2}{n_3} + \frac{R_4^2}{n_4}\right) - 3(N+1)$$

$$= \frac{12}{32(32+1)}\left(\frac{(223)^2}{8} + \frac{(97)^2}{8} + \frac{(44)^2}{8} + \frac{(164)}{8}\right) - 3(32+1)$$

$$\approx 25.957$$

(e) Reject H_0.

(f) There is enough evidence at the level of significance to conclude that the distributions of the sales prices in these regions are different.

4. (a) The claim is "days with rain are not random."

H_0: The days with rain are random.

H_a: The rains with rain are not random. (claim)

(b) Runs test

(c) n_1 = number of N's = 15

n_2 = number of R's = 15

lower critical value = 10

upper critical value = 22

(d) $G = 16$ runs

(e) Fail to reject H_0.

(f) There is not enough evidence at the 5% level of significance for the meteorologist to conclude that days with rain are not random.

5. (a) $H_0: \rho_s = 0$; $H_a: \rho_s \neq 0$ (claim)

(b) Spearman rank correlation coefficient.

(c) The critical value is 0.829.

Larceny	Rank	Motor vehicle	Rank	d	d^2
1403	1	161	1	0	0
1506	2	608	2	0	0
2937	4	659	3	1	1
3449	6	897	4	1	1
2728	3	774	5	−1	1
3042	5	945	6	−1	1
					$\sum d^2 = 4$

$\sum d^2 = 4$

$$r_c = 1 - \frac{6\sum d^2}{n(n^2-1)} = 1 - \frac{6(4)}{6(6^2-1)} \approx 0.886$$

(e) Reject H_0.

(f) There is enough evidence at the 10% level of significance to conclude that there is a correlation between the number of larceny-thefts and the number of motor vehicle thefts.

CUMULATIVE REVIEW FOR CHAPTERS 9–11

1. (a)

$r \approx 0.815$

There is a strong positive linear correlation.

(b) H_0: $\rho = 0$

H_a: $\rho \neq 0$ (claim)

$t_0 = 2.110$

$$t = \frac{r}{\sqrt{\frac{1-r^2}{n-2}}} \approx \frac{0.815}{\sqrt{\frac{1-(0.815)^2}{19-2}}} \approx 5.799$$

Reject H_0. There is enough evidence at the 5% level of significance to conclude that there is a significant linear correlation between the men's and women's winning 100-meter times.

(c) $\hat{y} = 1.264x - 1.581$

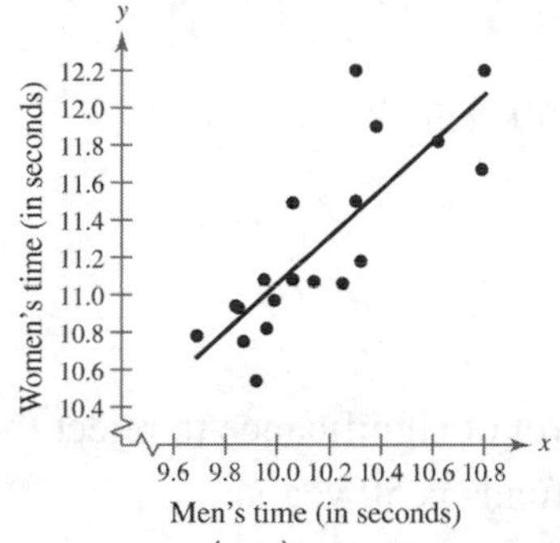

(d) $\hat{y} = 1.264(9.9) - 1.581 = 10.93$ seconds

2. The claim is "there is a difference in the weekly earnings of workers who are union members and workers who are not union members."

H_0: Median (Union) = Median (Nonunion)

H_a: Median (Union) $\neq$ Median (Nonunion) (claim)

The critical values are $z_0 = \pm 1.96$.

Ordered data	Sample	Rank
557	N	1
638	N	2
655	N	3
691	N	4
692	U	5
758	N	6
800	U	7
803	N	8
814	N	9
855	U	10
862	N	11
884	U	12
904	U	13
930	U	14
991	U	15
994	U	16
1040	U	17

R = sum ranks of nonunion workers = 44

$$\mu_R = \frac{n_1(n_1 + n_2 + 1)}{2} = \frac{8(8+9+1)}{2} = 72$$

$$\sigma_R = \sqrt{\frac{n_1 n_2 (n_1 + n_2 + 1)}{12}} \sqrt{\frac{8(9)(8+9+1)}{12}} \approx 10.392$$

$$z = \frac{R - \mu_R}{\sigma_R} \approx \frac{44 - 72}{10.392} \approx -2.69$$

Reject H_0. There is enough evidence at the 5% level of significance to support the agency's claim that there is a difference in the weekly earnings of workers who are union members and workers who are not union members.

3. The claim is "the median age of people with mutual funds is 50 years."

H_0: Median = 50 (claim)

H_a: Median $\neq 50$

The critical value is 3.

$x = 6$

Fail to reject H_0. There is not enough evidence at the 1% level of significance to reject the company's claim that the median age of people with mutual funds is 50 years.

4. The claim is "the mean expenditures are equal for all four regions."

H_0: $\mu_1 = \mu_2 = \mu_3 = \mu_4$ (claim)

H_a: At least one μ is different.

$F_0 = 2.29$

Variation	Sum of squares	Degrees of freedom	Mean squares	F
Between	1,028,888.6	3	342,962.88	4.119
Within	2,331,130.2	28	83,254.65	

$F = 4.119$

Reject H_0. There is enough evidence at the 10% level of significance to reject the claim that the expenditures are equal for all four regions.

5. (a) $\hat{y} = 11{,}182 + 174.53(91) - 104.41(88) = 17{,}876.15$ pounds per ounce.

(b) $\hat{y} = 11{,}182 + 174.53(110) - 104.41(98) = 20{,}148.12$ pounds per ounce.

6. The claim is "the standard deviations reading test scores for eigth grade students are the same in Colorado and Utah."

H_0: $\sigma_1^2 = \sigma_2^2$ (claim)

H_a: $\sigma_1^2 \neq \sigma_2^2$

$F_0 = 2.46$

$$F = \frac{s_1^2}{s_2^2} = \frac{34.6^2}{33.2^2} = 1.086$$

Fail to reject H_0. There is not enough evidence at the 10% level of significance to reject the claim that the standard deviations of reading test scores for eighth grade students are the same in Colorado and Utah.

7. The claim is "the distributions of the annual household incomes in these regions are different."
H_0: The medians are all equal.
H_a: The medians are not all equal. (claim)
The critical value is 11.345.

Ordered data	Sample	Rank
41.5	S	1
44.4	S	2
45.2	MW	3
45.6	S	4
46.4	S	5
47.0	S	6
47.1	NE	7
48.5	MW	8
49.2	S	9.5
49.2	S	9.5
49.3	MW	11
50.0	NE	12
50.7	MW	13
51.4	W	14
51.6	NE	15
51.8	MW	16
52.0	MW	17
52.4	W	18
52.5	NE	19
53.5	W	20
54.0	W	21
54.3	NE	22
54.4	MW	23
54.7	W	24
54.8	NE	25
55.7	NE	26
55.9	W	27
56.8	W	28

$R_1 = 126,\ R_2 = 91,\ R_3 = 37,\ R_4 = 152$

$$H = \frac{12}{N(N+1)}\left(\frac{R_1^2}{n_1} + \frac{R_2^2}{n_2} + \frac{R_3^2}{n_3} + \frac{R_4^2}{n_4}\right) - 3(N+1)$$

$$= \frac{12}{28(28+1)}\left(\frac{(126)^2}{7} + \frac{(91)^2}{7} + \frac{(37)^2}{7} + \frac{(152)}{7}\right) - 3(28+1)$$

$$\approx 15.671$$

Reject H_0. There is enough evidence at the 1% level of significance to support the claim to conclude that the distributions of annual household incomes in these regions are different.

8. Claimed distributions:

Response	Distribution
None	5%
Little	16%
Half	31%
Most	33%
All	15%

H_0: The distribution is as claimed. (claim)

H_a: The distribution is not as claimed.

Response	Distribution	Observed	Expected	$\frac{(O-E)^2}{E}$
None	5%	31	45	4.356
Little	16%	164	144	2.778
Half	31%	277	279	0.014
Most	33%	305	297	0.215
All	15%	123	135	1.067
		900		$x^2 \approx 8.430$

$\chi_0^2 = 9,488$

$\chi \approx 8.430$

Fail to reject H_0. There is not enough evidence at the 5% level of significance to conclude that the distribution of how much parents intend to contribute to their children's college costs differs from the claimed or expected distribution.

9. (a) $r^2 \approx 0.733$

Metacarpal bone length explains about 73.3% of the variability in height. About 26.7% of the variation is unexplained.

(b) $s_e \approx 4.255$

The standard error of estimate of the height for a specific metacarpal bone length is about 4.255 centimeters.

(c) $\hat{y} = 94.428 + 1.700(50) = 179.428$

$$E = t_c S_e \sqrt{1 + \frac{1}{n} + \frac{n(x_0 - \bar{x})^2}{n\sum x^2 - \left(\sum x\right)^2}} \approx 2.365(4.255)\sqrt{1 + \frac{1}{9} + \frac{9(50 - 45.444)^2}{9(18,707) - (409)^2}} \approx 11.402$$

$\hat{y} \pm E \Rightarrow (168.026,\ 190.83)$

You can be 95% confident that the height will be between 168.026 centimeters and 190.83 centimeters when the metacarpal bone length is 50 centimeters.

10. The claim is "there is a correlation between overall score and the price."

$H_0: \rho_s = 0$; $H_a: \rho_s \neq 0$ (claim)

The critical value is 0.643.

Score	Rank	Price	Rank	d	d^2
74	3.5	77	3.5	0	0
82	7	96	6	1	1
78	5	77	3.5	1.5	2.25
84	8	116	8	0	0
80	6	98	7	−1	1
64	1	67	1	0	0
70	2	70	2	0	0
74	3.5	81	5	−1.5	2.25
					$\sum d^2 = 6.5$

$$r_c = 1 - \frac{6\sum d^2}{n(n^2 - 1)} \approx 0.923$$

Reject H_0.

There is enough evidence at the 10% level of significance to conclude that there is a correlation between the overall score and the price.